Klaus Varrentrapp

A Practical Framework for Adaptive Metaheuristics

Klaus Varrentrapp

A Practical Framework
for Adaptive
Metaheuristics

VDM Verlag Dr. Müller

Imprint

Bibliographic information by the German National Library: The German National Library lists this publication at the German National Bibliography; detailed bibliographic information is available on the Internet at http://dnb.d-nb.de.

Cover image: www.purestockx.com

Publisher:
VDM Verlag Dr. Müller Aktiengesellschaft & Co. KG, Dudweiler Landstr. 125 a, 66123 Saarbrücken, Germany,
Phone +49 681 9100-698, Fax +49 681 9100-988,
Email: info@vdm-verlag.de

Produced in USA and UK by:
Lightning Source Inc., La Vergne, Tennessee, USA
Lightning Source UK Ltd., Milton Keynes, UK
BookSurge LLC, 5341 Dorchester Road, Suite 16, North Charleston, SC 29418, USA

ISBN: 978-3-639-03456-1

Extended Abstract

Local search methods are useful tools for tackling hard problems such as many combinatorial optimization problems (COP). Local search methods work by probing (in contrast to enumerating) the search space. So-called trajectory-based local search methods start from some initial solution and iteratively replace the current solution by a neighboring one which differs only in some local changes. The possible changes to the current solution are typically defined by local search operators and potentially are randomized. The long-term goal is to find a global or a very good local optimum.

Experience from mathematics has shown that exploiting regularities in problem solving is beneficial. Consequently, identifying and exploiting regularities in the context of local search methods is deemed to be desirable, too. Due to the complexity of the COPs tackled, regularities might better be detected and learned automatically. This can be achieved by means of machine learning techniques to extend existing local search methods. Learning requires feedback, but in the context of local search methods, instructive feedback is not available, since global or very good local optima are not known in advance. Instead, evaluative feedback can be derived from the cost function of COPs evaluating single solutions, for example.

Reinforcement learning (RL) is a machine learning technique that only needs evaluative feedback. It is based on the notion of an agent that moves from state to state by applying actions and receiving a scalar evaluative feedback for each move called reward. The rewards of an agent's interaction in the form of sequences of actions and moves from one state to the next are accumulated in a sum of discounted rewards called return. Maximizing the return is the long-term goal of an agent. In order to achieve this goal, a particular RL method called Q-learning established a so-called action-value function. Action-value functions evaluate in each state the applicable actions in terms of how useful their application in this state would be in maximizing the return. Due to the size of the state spaces tackled, action-value functions are represented by function approximators that work on a selection of predictive real-valued state characteristics called features. Solving a RL problem can be done by learning a useful action-value function during an agent's interaction and next deriving a search strategy called policy by applying in each state the best-valued action.

The present thesis attempts to develop learning local search methods in a general and practical manner. One possibility to enhance local search methods with learning capabilities is by using RL methods. RL techniques can be applied to Markov decision processes (MDP). The direct application of existing RL techniques for extending existing local search methods is enabled by the concept of a local search agent (LSA). The advancement of a trajectory-based local search method can be regarded as the interaction of a virtual agent whose states basically consist of solutions and whose actions are composed of arbitrary hierarchical compositions of local search operators, altogether yielding the same setting as an MDP. The resulting LSA using RL can then be called a learning LSA. The changes in cost for each move of a learning LSA can be used as reward. Based on these, returns can be computed such that maximizing the return reflects the goal of finding a global or a very good local optimum. The hierarchical structure of LSA actions allows to use so-called ILS-actions. ILS-actions coincide with the application of one iteration of the well-known

Iterated Local Search (ILS) metaheuristic. The advantage of this metaheuristic and this kind of action is that only solutions from the subset of local optima – which must contain any acceptable solution – are considered and thus introduces a search space abstraction which in turn can improve performance. A learning LSA that employs ILS-actions iteratively will visit local optima in a guided and adaptive manner. The resulting theoretical framework is called Guided Adaptive Iterated Local Search (GAILS).

In order to evaluate randomized GAILS algorithms, empirical experiments have to be conducted. Each GAILS algorithm thereby consists of three, mainly independent parts. The first part comprises the actions of a learning LSA which are specific to a problem type. The LSA actions being arbitrary hierarchical compositions of local search operators are implemented through basic local search operators. The second part represents the RL techniques used, which in turn transparently use actions and hence are problem type independent. The third part consists of the function approximators used by RL techniques to implement policies. The function approximators only require as input a vector of real-valued features and this way are independent from the first two parts. Empirical experiments can be supported by providing a framework that can decouple these three main parts in any GAILS algorithm program instantiation, thus allowing for an arbitrary reuse and combination enabling rapid prototyping. The GAILS implementation framework is such an application framework which is designed to rapidly implement learning LSAs reflecting the separation of a learning LSA into its three main parts. It provides generic interfaces between components of the three parts and this way provides for a separation of problem type specific states from search control. It also provides for a separation of search control from the state of the search control unit. Hierarchically built actions are mapped to object hierarchies.

Two GAILS algorithms according to Q-learning algorithm $Q(0)$ and $Q(\lambda)$ that are based on ILS-actions were developed, built and compared to corresponding standard implementations of the ILS metaheuristic. These so-called Q-ILS algorithms were tested for two problems using different function approximators. The results showed that learning useful policies and transfer of what was learned across multiple problem instances, even of different sizes, is possible and useful.

Contents

3.2.5   Estimation Methods   53

3.2.6   Temporal Difference Learning   55

3.2.7   Learning Policies Directly   58

3.2.8   Other Approaches and Variants   59

3.3   Function Approximation   61

3.3.1   Generalization and Normalization   61

3.3.2   Strategy Function Representation   62

3.3.3   General Function Approximation Issues   64

3.3.4   Function Approximator Types   66

4   The GAILS Method   69

4.1   Concept   69

4.1.1   Idea   70

4.1.2   Iterating Local Search   71

4.1.3   Action-Hierarchies   73

4.2   Learning Perspective   80

4.2.1   Learning Opportunities   80

4.2.2   Actions   82

4.2.3   Learning Scenarios   84

4.3   Goal Design   89

4.3.1   Agent Moves   90

4.3.2   Episodes   91

4.3.3   Reward Design   92

4.3.4   Move Costs 102

4.3.5   Summary and Conclusion 104

4.4   Related Work 106

4.4.1   Adaptive Metaheuristics 107

4.4.2   Genetic Algorithms and Reinforcement Learning 111

4.4.3   Reinforcement Learning Based Combinatorial Optimization 114

4.4.4   Other Adaptive Approaches to Combinatorial Optimization 117

4.4.5   Hierarchical Reinforcement Learning 118

4.4.6   Comparison 120

List of Figures

List of Tables

List of Algorithms

Chapter 1

Introduction

1.1 Motivation

One important class of problems that arise in Artifical Intelligence, Computer Science, and other fields are combinatorial optimization problems (COPs) which have great practical relevance. COPs of different types frequently arise in scientific, industrial and business environments where planning, scheduling, time-tabling, sequencing and other tasks have to be carried out [NW88, CCPS97]. One decisive trait of COPs is that solutions to an instance of a COP can be separated into discrete solution components and emerge as complete assignments of values to the solution components. The assignment or rather solution construction thereby is subject to constraints, first of all that the values assigned stem from given discrete and typically finite domains, potentially one per solution component. If all constraints are fulfilled, a feasible solution results. All other assignments merely are candidate solutions. Solutions to COP instances additionally are assigned a cost by means of a cost function. The goal is to find not only a feasible solution but a globally optimal feasible solution, also called global optimum. Solving a COP means finding a global optimum [Hoo98, Hoo99b, HS04]. As an example consider the Traveling Salesman Problem (TSP) [JM97, GP02]. A TSP instance consists of a number of cities with a travel distance given between any two cities. The goal is to find a round trip starting at a city which visits any other city exactly once and finally returns to the starting city. A round trip can be represented as an ordered list of cities which will be visited in the given order. The round trip is completed by a final travel from the last city to the first of the ordered list. The positions of the ordered list correspond to the solution components and get assigned a value in the form of a city (from the domain of all cities). The length of the ordered list is equal to the number of cities. The constraints for a feasible solution require that each city is assigned to exactly one position.

COPs are easy to formulate, but often are very hard to solve in practice. Although a global optimum can be found by enumerating all candidate solutions, checking their constraints and comparing costs, this approach typically will take exponential time in the number of solution components of an instance. Exponential time, in contrast to polynomial time, is considered to be intractable in practice. Unfortunately, for a large class of such problems, called *NP*-hard problems, no algorithm is known that solves them in polynomial time in the worst case and it is expected by many researchers that there does not exist one for any of these problems [GJ79]. Many methods have been proposed to remedy this dilemma. All of them trade off the guarantee to find a global optimum for better runtime. Some so-called approximate methods guarantee to find near-optimal or approximate feasible solutions within a given bound in polynomial time. Other methods work heuristically and do not give any guarantee for solution quality at all but work well in practice yielding acceptable solution quality. Prominent among the heuristic methods are so-called local search methods. They do not search comprehensively, but produce a chain of feasible or candidate solutions that can be viewed as samples. An important subgroup of local search methods are metaheuristics [LO96, OL96, OK96, CDG99b, VMOR99, HR01, GK02, BR03, RdS03].

Local search methods work by probing (in contrast to enumerating) the search space of solutions. So-called trajectory-based local search methods start from some initial solution and iteratively replace the current solution by a neighboring one which differs only in some local changes. The possible changes to the current solution are typically defined by some local search operators and potentially are randomized. The long-term goal is to find a global or a very good local optimum. The locality of changes gives rise to the notion of local search. The stepwise procedure is directed by the cost function: In each step, called local search step, the new, perhaps partial, solution is supposed to have better cost than the old one. In each local search step, the systematic changes are applied in any possible variation, each variation yielding a new potential solution called neighbor. A neighbor among the set of all neighbors with better cost finally becomes the new solution. This process of iterative improvement is called local search procedure and does not necessarily, in practice almost never, find a global optimum, but gets stuck in a locally optimal solution, called local optimum, where no neighbor with better cost exists. In order to continue the search, these simplest iterative improvement schemes, called local search procedures, have to escape local optima. Therefore, meta-strategies or -heuristics for how to escape local optima are required. Local search methods in the form of local search procedures that are enhanced with such meta-strategies are called metaheuristics.

Local search methods work well in practice. The abandonment of optimality guarantees often is acceptable, since the quality of feasible solutions found in practice by local search methods typically suffices. Nevertheless, the aim of many researchers is to improve performance even more. Because of the complexity COP search spaces exhibit and the fact that most local search methods are randomized, theoretical and also empirical analysis of them intended to reveal which ingredients and inner workings influence performance and in which way is extremely difficult. Current research practice therefore concentrates on inventing new search strategies and on constructing new local search methods that work better in practice as verified in an empirical manner and by statistical methods. The strategies invented thereby remain rather static during the search. They do not adapt themselves to, nor consider specialties or regularities of, the COPs or even the problem instance they are supposed to solve. If they do, these are detected and implemented in the invention process by the constructor a priori. Or, the newly invented local search methods are implicitly adapted by an a priori parameter tuning process. Although identifying regularities and their underlying laws by means of learning, and subsequently using this knowledge to improve behavior is a very successful general technique for enhancing problem solving capability (see humans as role model), it has not been truly integrated in automated form in local search methods and metaheuristics yet. The hypothesis is that automated learning eventually must be integrated in order to substantially improve local search methods. This hypothesis stems from the observation that improvements in performance of any exact and approximation methods, including local search methods, is due to an increased learning of the properties of COPs types and instances, typically in the form of mathematical propositions. Such knowledge can be used to exclude substantial parts of the search space from the search since it is known that the solutions contained in these parts are provably of inferior cost. Learning this way mostly is done a priori and by human researchers, though. Consequently, one obvious extension and direction of improvement of local search methods is to incorporate automated machine learning techniques to make local search methods learning and hence adaptive themselves and turn static strategies into flexible guidance.

When inventing new methods and ideas, they have to be evaluated. In the context of local search methods, which are highly randomized and complex and therefore cannot be evaluated analytically, evaluation has to be done empirically by conducting experiments. Different algorithm variations (or algorithms for short) resulting from a new method have to be tested by instantiating them in the form of executable programs, running the program, and subsequently analyze the results by statistical methods. The number of experiments and tested algorithms thereby cannot be high enough along the lines: The more evidence is collected, the more reliable are insights and hypothesis. Accordingly, research success is not only dependent on good ideas, but also on whether these ideas can be implemented and tested efficiently and comprehensively. The research procedure typically starts with first experiments that are supposed to reveal the inner workings of a new method or at least those ingredients and variable parameters that truly influence performance. At first, the goal is not to find the best and most highly tuned algorithm, but rather to test whether a proposed method works in principle, and why and under which circumstances it performs. Being able to rapidly prototype algorithmic ideas and thus being able to immediately evaluate the ideas then can immensely support the design and invention of methods and algorithms. Consequently, what is needed are means to quickly instantiate prototypical algorithms into executable programs. Rapid prototyping thereby requires reuse of code and then enables rapid implementation. By reusing code, external influences on performance such as coding skills can be eliminated and hence will not blur experimental results as well. Rapid prototyping can be achieved by so-called application frameworks or frameworks for short [FS87, Fay99, FSJ99, FSJ99]. Roughly, such frameworks try to separate individual parts of programs into independent and invariant pieces of code which can be reused and extended. Hence, it is appropriate to accompany the invention and investigation of a new approach such as integrating machine learning methods into local search methods with a framework that supports the implementation and investigation of respective algorithms.

1.2 Objectives and Contribution

The objective of this thesis is the development of a method that incorporates machine learning techniques into local search methods in order to enhance these with true learning capability and make them truly adaptive. The proposed method is to be investigated theoretically as well as empirically. To support the second enterprise, an application framework – here to be called implementation framework to emphasize the algorithm implementation aspect – is to be designed and implemented in order to quickly instantiate algorithms according to the proposed method. This is to enable comprehensive experimentation later. In some first experiments it has to be verified that the framework in fact can support rapid algorithm instantiation and hence experimentation and that the method proposed in fact does learn something useful and can improve performance. The ideas and core concepts of the method proposed and of the corresponding implementation framework are outlined briefly next, followed by an overview of the contributions of this thesis.

How to incorporate learning into local search methods? Supervised learning is not possible in the context of local search, since no instructive feedback is available. It is not known which steps will lead to a global optimum or at least some good local optima. Even the quality of local optima found during a search process in absolute terms cannot be estimated since the cost of a global optimum is not known in advance. Nevertheless, the cost of solutions can be used as evaluative feedback. The same problem of having only evaluative feedback available emerges in a class of learning problems called reinforcement learning [SB98]. Reinforcement learning considers problems where a so-called agent moves from state to state by applying actions from a set of applicable actions and tries to achieve some (long-term) goal such as visiting certain states, so-called goal states. For each move to a new state, evaluative feedback called reward is obtained. The goal is to learn a policy in the form of a mapping from states to actions that maximizes the expected accumulated weighted rewards. The accumulation in turn is set up such that maximizing the accumulated rewards reflects the goal an agent wants to achieve. The ingredients necessary to apply reinforcement learning techniques – moves, states, actions, and rewards – are available for local search methods also, for example in the form of local search steps, solutions, several procedures for how to do local search steps, and derivatives of the cost of solutions, respectively, for example. A global optimum then becomes a goal state and a local search method itself can be regarded as an agent, a so-called local search agent (LSA). An LSA can learn a policy which effectively means to learn to take the proper action, e.g. a local search step, for each move. In doing so, the learned policy will have to capture regularities of the search space and exploit them. An LSA effectively becomes a learning LSA.

The space of local optima is smaller than the original search space of candidate or feasible solutions, yet it must contain any global optimum also. Consequently, it seems very favorable to elevate the search process to a higher level of abstraction by making an LSA move in the space of local optima only by using appropriately designed actions. By abstracting to a search space that only consists of local optima, the most important decisions to be taken by local search methods during the search can be learned more directly. The method presented in this thesis is designed to support the direction of abstracting to the set of local optima. This is done by varying LSA actions: Depending on the local search method used to yield an LSA, actions of an LSA can differ. The abstraction to a search space consisting only of local optima is adopted from a metaheuristic called Iterated Local Search (ILS) [LMS01, LMS02, BR03], which is one of the simplest, yet most powerful metaheuristics. It employs a local search procedure, a so-called perturbation, and a so-called acceptance criterion. Starting in some initial local optimum found by the local search procedure, it repeatedly applies the perturbation which incurs major changes to the current local optimum effectively escaping it such that the subsequently applied local search procedure hopefully will yield a new local optimum. The acceptance criterion finally decides, whether to continue from the new local optimum found or whether to back up to an old one. In terms of such an ILS being an LSA, the actions employed, called ILS-actions, are a composition of perturbation, local search procedure, and acceptance criterion. They make a respective LSA move exactly in the space of local optima. In fact, any local search method must repeatedly or iteratively visit local optima in order to ensure optimal performance. Consequently, the process of iteratively visiting local optima is a must and therefore generic for any local search method. Hence, the iterative visits of local optima as done by the ILS metaheuristic is a role-model for almost any local search method and, in some sense, ILS itself can be considered to be a role-model for almost any local search method. The method proposed here is named *Guided Adaptive Iterated Local Search* (GAILS). The name extension is due to the fact that reinforcement learning techniques and the LSA concepts are used to learn a policy which adaptively guides an LSA in moving from local optimum to local optimum. Such a movement effectively involves iterative visits of local optima by means of local search.

As was outlined in Section 1.1 investigating a new method such as the GAILS method is greatly supported by an accompanying framework. In case of the GAILS method, and accompanying framework called GAILS framework is to be designed and implemented also. Now, what are the actual requirements for such a framework? For the first phase of investigation, it is useful to start simple. The emphasis is more on implementing and testing principles in many combinations rather than on high efficiency. Therefore, the GAILS framework concentrates in its first version to support the development of LSAs based on ILS and other so-called trajectory methods [BR03] such as Simulated Annealing [Dow93, AKv97], Tabu Search [GTdW93, GL97], and so on. Those components of any actual algorithm according to the GAILS method, called GAILS algorithms, that are independent from each other in principle should be made independent and arbitrarily exchangeable and combinable in instantiating actual executable programs also. These independent components fall into three classes or parts. The first part contains all components that are problem type specific. The second part contains all components that implement learning and search strategies in the form of conventional or reinforcement learning enhanced local search methods. They use problem specific components only as black-boxes and hence are independent of a specific COP type. The third part contains those components that eventually do learn in the form of function approximators which realize learned policies.

The contribution of this thesis on the one hand is the invention of the GAILS method and on the other hand the design and implementation of the GAILS framework. The GAILS method is a new and very general method to incorporate true learning capability and adaptiveness into local search methods, and metaheuristics in particular. Using reinforcement learning techniques in the context of local search is not new in principle. In contrast to previous approaches (cf. Section 4.4), the GAILS method is the first to learn a central steering component for searching completely anew and as a whole. Also, it uses abstraction in the form of supporting any kinds of actions for LSAs such as those inducing local optima based moves. Finally, the transfer of reinforcement learning techniques to local search methods is realized rigorously, yet in a simple way, as manifested by the concept of a learning LSA. The GAILS method is motivated, explained, and justified in depth. Its potentials and hazards are analyzed theoretically a priori by presenting learning opportunities and scenarios, and other design possibilities and their likely consequences with respect to performance and learning. The a priori analysis also includes potential benefits of the GAILS method such as computer supported human learning and visions how the GAILS method can accelerate research in combinatorial optimization. First experiments then are conducted showing that GAILS algorithms in fact do learn and in fact can show improved behavior to comparable non-learning variants.

The GAILS framework enables rapid prototyping of general learning enhanced local search methods, and of GAILS algorithms in particular. Although frameworks are not new in the context of local search methods, they all do not support learning (cf. Section 5.4). The concepts of a learning LSA and of actions for LSAs from the GAILS method in contrast yield a framework quite differently in design from existing frameworks for local search methods. The GAILS framework provides an infrastructure that practically accelerates local search method and metaheuristic construction, while simultaneously keeping things simple and usable. This usability is shown by instantiating several GAILS algorithms for several combinations of COP types to work on, several metaheuristic and learning variants thereof, and several function approximators. The resulting various GAILS algorithms then are used for first experiments investigating the GAILS method.

Altogether, a "A Practical Framework for Adaptive Metaheuristics" has been invented where the notion framework both denotes an implementation framework and the conceptual framework of a methodology.

1.3   Organization

This document is organized as follows:

- **Chapter 2** introduces the basics of combinatorial optimization, local search methods, meta-heuristics, ILS, machine learning and learning in general, and their mutual connections, finally yielding the concept of an LSA and that of actions for an LSA.

- **Chapter 3** provides all necessary information about the (reinforcement) learning approach that is employed by the GAILS method. The problem definition and other ingredients for reinforcement learning are presented as well as solution methods. Furthermore, function approximation in the context of reinforcement learning is discussed.

- **Chapter 4** presents the GAILS method itself. The idea, important concepts, potentials, and hazards are discussed providing an overview how to properly design algorithms according to the GAILS method. Related work, a summary, and final conclusions accompany the presentation and discussion of the GAILS method.

- **Chapter 5** discusses the GAILS implementation framework. The framework requirements are derived and a design is presented which aims at fulfilling the identified requirements. The design presentation goes as far as presenting some of the most important implementation issues. Related work is presented also.

- **Chapter 6** explains the first experiments conducted with GAILS algorithms and shows their results, and draws and justifies conclusions.

- **Chapter 7** finally wraps up the collected results and insights and presents the overall conclusions and contributions. At last, a look into the future lists possible improvements for the GAILS method and framework and the experiments presented here and gives further perspectives for the GAILS method.

General note: Variables and function identifier are used consistently within a section. Upon their first introduction their domain will be given in parentheses. The domain then is valid for the rest of the section. In a subsequent section another domain might be assigned which then is valid for the respective section.

Chapter 2

Local Search and Machine Learning

This chapter introduces and relates basic concepts and notions from the fields of combinatorial optimization, local search, and machine learning. In doing so, the general motivation for incorporating machine learning techniques into local search methods will be exhibited which then will lead to the invention of the Guided Adaptive Iterated Local Search (GAILS) method later. Besides the motivation, ways how to accomplish such an incorporation will be indicated and several useful concepts in this respect will be established.

Concepts from combinatorial optimization and local search are defined in sections 2.1, 2.2, and 2.3. One important class of local search methods, metaheuristics, and in particular the Iterated Local Search (ILS) metaheuristic will be introduced in sections 2.4 and 2.5, respectively. Next, relevant notions from learning and machine learning are presented in Section 2.6. The role of machine learning for search is discussed in Section 2.7 resulting in the motivation to incorporate machine learning techniques in search methods and local search methods in particular. Finally, the concept of an agent and more specific of an LSA are provided in Section 2.8 which are the starting points for the venture of incorporating machine learning techniques into local search methods eventually yielding the GAILS method.

2.1 Combinatorial Problems

Many problems are of combinatorial nature. A *solution* to these problems can be imagined as a decomposition of the whole information a solution contains into single discrete pieces or components. Each such discrete component is called *solution component* and is assigned a value from a domain of values, possibly a different domain for each component. Thus, it is generic in nature. The values of the domains are the concrete elements of a solution and are called *solution elements*. If the set of different values assignable to components is in fact discrete also, such a problem is called *combinatorial problem* (CP). Each complete solution component assignment is called *complete* or *candidate solution*. The number of solution components for a solution of a COP is called the *problem instance size* or *instance size* for short. Any incomplete solution component assignment is called *incomplete* or *partial solution*. For a given solution, the actual values assigned to the solution components are also called solution elements. The space of all candidate solutions is the space of all possible variations of solution element assignments for the solution components. The solution element assignments to the solution components are restricted by means of constraints. These constraints involve some or all solution components. The set of all valid or rather *feasible solutions* is the set of those complete assignments that fulfill all constraints. The aim in solving CPs is to find a feasible solution. More generally then, solving a combinatorial problem typically involves finding a grouping, ordering, or assignment of a discrete set of components or objects (the values) which satisfies certain constraints.

In the context of local search, single solutions can be regarded as states of the search process, too. Solutions are therefore called *search states*. The different kinds of solutions are transferred to states also yielding *candidate* and *feasible search states*. The space of all candidate solutions is also called *search space* or *search state space*. The number of all candidate states is the size of the search space. In the context of this document, candidate solutions or search state are often only denoted solutions or search states for short, respectively. If feasible solutions or search states are meant, this will be stated explicitly.

In order to process a problem with a computer a formal representation of the search state space in terms of a formal representation of single search states must be undertaken; any candidate search state must be encoded somehow. Consequently, any search state centrally consists of a *solution encoding*. The solution components of the search state can be thought of being represented by variables, since they are discrete as well as the respective domains of assignable values in the form of solution elements. A solution encoding then simply is a set of variables with values assigned. More formally, following the definition of [BR03], a *problem instance* $\mathcal{P}^{(n)}$ of *size* n ($n \in \mathbb{N}^+$) for a CP can be defined as:

- a set of variables X, $X := \{x_1, \ldots, x_n\}$,

- variable domains $D_1, \ldots, D_n$,

- constraints C among variables, and

- $\mathcal{P}^{(n)} := (X, \{D_1, \ldots, D_n\}, C)$.

The set of all possible assignments S is:

$$S := \{s = \{(x_1, y_1), \ldots, (x_n, y_n)\} \mid y_i \in D_i, i \in \{1, \ldots, n\}\}$$

This set is called the *set of candidate solutions* or *set of candidate search states* and represents the search space of potential solutions and hence search states just mentioned.

The (sub-) set of all feasible assignments S' is

$$S' := \{\{(x_1, y_1), \ldots, (x_n, y_n)\} \mid x_i \in X, y_i \in D_i, i \in \{1, \ldots, n\} \wedge \forall c \in C . s \text{ satisfies } c\} \subseteq S$$

S' is called the (sub-) *set of feasible* or *admissible solutions* and hence also (sub-) set of *feasible* or *admissible search states*. Variables x_i ($x_i \in X_i, i \in \{1, \ldots, n\}$) are the solution components, the elements of $D_1, \ldots, D_n$ are the solution elements. To any solution component x_i, a solution element y_i ($y_i \in D_i$) is assigned. For many CPs, search methods will work on feasible search states only, so the set of all candidate search state there will be equal to the set of feasible search states: $S = S'$.

The aim in solving a combinatorial problem instance is to find a feasible search state s ($s \in S'$) i.e. a feasible variable assignment. A set $\mathcal{P}$, $\mathcal{P} := \bigcup_{n \in \mathbb{N}^+} \mathcal{P}^{(n)}$, of all problem instances that coherently share the same principle constraints and domains is called *problem type* or *problem* for short. Typically, it will be spoken of *the* problem type, since only one problem type per context typically is sensible.

2.2 Combinatorial Optimization Problems

In many cases solving problems can be regarded as searching a space of candidate search states with the aim to find a search state that is admissible. Often, a *cost* or *objective function* assigns an additional cost to candidate search states. Then, in addition to finding a feasible search state, a preferably optimal search state with minimal or maximal cost is searched for. These combinatorial problems are called *combinatorial optimization problems* (COP) [NW88, CCPS97]. A problem instance $\mathcal{P}_{\mathcal{O}}^{(n)}$ of size n ($n \in \mathbb{N}^+$) for a COP is a problem instance $\mathcal{P}^{(n)}$ ($\mathcal{P}^{(n)} = (X, \{D_1, \ldots, D_n\}, C)$) of size n for a CP with an associated objective or cost function f ($f \colon S \to \mathbb{R}^+$) such that $\mathcal{P}_{\mathcal{O}}^{(n)} := (X, \{D_1, \ldots, D_n\}, C, f)$. The cost function f basically is a mapping $f \colon D_1 \times \ldots \times D_n \to \mathbb{R}^+$. A problem instance for a COP is shortly also called COP. Again, problem instances for COPs are grouped into problem types just like problem instances for CPs. The problem type in the context of COPs is also denoted by *COP type*.

The aim in solving a COP is to find a global optimal. A global optimum is a feasible search state that has minimal or maximal cost, depending on whether the objective is to maximize or minimize cost. Without loss of generality, it is assumed from now on that any COP has the objective to minimize cost, since any maximization problem can be transformed into an equivalent minimization problem by simply multiplying the cost function with -1. Any global optimum remains a global optimum afterwards and no new global optima are introduced anew. Similarly, this is true for local optima presented later in Section 2.3 also. Then, the aim in solving a COP is to find a minimal search state s^{**}. Note that in the context of search states, attributes such as good, bad, best, worst, average, better, worse, improving, worsening, minimal, maximal, suboptimal, quality and so on always refer to the cost of a search state.

Formally, the set of all global optima S^{**} is[1]

$$S^{**} := \{s \in S' \,|\, \forall \, s' \in S' \,.\, f(s) \leq f(s')\} \subseteq S' \subseteq S$$

For a minimal search state then holds: $s^{**} \in S^{**}$. Note that this definition of combinatorial optimization problems only considers minimizing optimization problems. Note also that non-feasible search states cannot be globally optimal. To prevent from having non-feasible search states with lower cost than feasible search states, the cost function can be altered such that it assigns penalties to non-feasible search states such that *any* non-feasible search state has (substantially) higher cost than *any* feasible search state. The set of global optima then can be written as

$$S^{**} = \{s \in S \,|\, \forall \, s' \in S \,.\, f(s) \leq f(s')\} \subseteq S'$$

COPs occur frequently in practice and comprise, for example,

- finding shortest or cheapest round trips (Traveling Salesman Problem (TSP), [JM97, GP02]) and routing (Vehicle Routing (VR), [CW64]),

- finding models of propositional formulae (Satisfiability Problem (SAT), [DGP97]),

- planning/scheduling [Pin95, Bru98, Nar01] (Flow/Job Shop Problem (FSP/JSP), [Fre82, GP96, Pra02]), sequencing [Bak74], ordering (Linear Ordering Problem (LOP), [SS03]), and assignment (Quadratic Assignment Problem (QAP), [PW94, BÇPP98, Çel98, SD01]),

- resource allocation (graph coloring (GC), [BCT94, JT94, CDS03]), and time-tabling (TT) [dW85, CLL96, Sch99],

---

[1]The set of local optima, as will be defined later, is denoted by S^*, so the set of global optima, which is a subset of the set of all local optima, S^*, is therefore denoted S^{**}.

- protein structure or folding prediction [KPL⁺98, BL98, SH03], and molecule docking prediction [RHHB97], and so on.

As an example, consider the TSP. Its problem formulation is given as a weighted graph representing locations and distances or rather traveling times between different locations. Solving a
TSP (instance) requires to find a round trip through the nodes that visits each node exactly once,
i.e. requires to find a Hamilton Circuit, such that the sum of the weights of the edges included
in the round trip is minimal. If the sum of weights is viewed as the cost of a round trip, the
objective is to minimize the cost of such a round trip. The TSP will be used as a role model for
COPs and often used to in examples throughout this document. Figure 2.1 shows an example of
a TSP problem instance with 532 cities on a map. An individual search state, i.e. a round trip
or tour is represented as a permutation of a default numbering of the cities defining the current
order in which the cities are to be visited. The variables then are the positions of the represented
tour and the values assigned are the city numbers. The edges of a tour then are the edges from
the weighted graph that was given as problem formulation that are really used for the round trip.
The constraints have to ensure the round trip property. Ensuring the permutation property is not
difficult, so most methods for solving the TSP work on the set of feasible search states only which
is the set of all correct permutations.

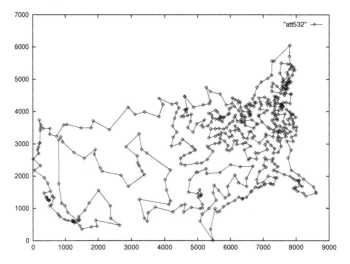

Figure 2.1:   TSP problem instance with 532 cities in the USA

Many COPs are very hard to solve in practice, for example since they are *NP*-complete or *NP*-
hard [GJ79]. Instead of insisting on a guarantee to find a global optimal search state, the goal
of combinatorial optimization is to solve COPs as good as possible and is called *goal in solving
COPs*. As a secondary goal, the solution process has to be quick. As a result, COPs really are
multi-objective optimization problems with two dimensions of optimality: search state quality as
measured by the cost function and runtime. A *goal search state* is a global optimum or a reasonably
good feasible search state. A reasonably good search state is a search state with a search state
quality acceptable to work with in practice, e.g. as determined by given quality bounds. Note that
there can be more than one goal search state, since there can be more than one global optimum
already. Also, the notion "reasonably good" was chosen deliberately to indicate that the needed

quality of a goal search state is dependent on the actual case in practice. It cannot be determined in advance and can also comprise several (feasible) search states. Not at all can this notion be defined for all COPs in advance for *all* practical cases.

The optimality criteria search state quality and runtime have to be traded off, if they are not tackled simultaneously in a pareto-optimal manner [Ste86, Ehr00]. This approach is not taken here. Instead, as is common practice, the two objectives actually are traded-off against each other. Here, the runtime optimality criterion is regarded to be of secondary nature only. This is simply because the learning and guiding that is attempted by the GAILS method presented in this document might add computational overhead. The additional computational overhead might at first decrease the chances of the approach to competitively solve COPs with respect to runtime compared to state of the art metaheuristics such as ILS. Nevertheless, there are several COPs that employ a cost function which is very computation-expensive to compute [EEE01, VBKG03, Kno04, KH05], for example when training neural networks by means of local search methods, where the computation of cost function values basically requires to train and evaluate a neural network sufficiently [AC04]. The computational overhead introduced by learning and applying a strategy function might be negligible compared to the effort to compute cost function values in these cases. In contrast, it might pay off especially to exploit search space regularities to speed-up the search and reduce the number of cost function evaluations. The resulting goal in solving COPs to begin with then will be to only find a goal search state. This goal stand-alone is of scientific relevance nonetheless according to [BGK+95]. Several scenarios are conceivable where runtime practically is of little importance, if it does not exceed certain bounds that, however, are predictably beyond the runtime that can possibly be expected by algorithms according to the GAILS method. To emphasis this expectation, the secondary goal in solving COPs is softened to finding a goal search state in a reasonable amount of time. Although often not stated, the goal in solving COPs in this discourse is of long-term nature.

2.3 Local Search

For many COPs such as *NP*-hard ones there are no methods that can guarantee to find a goal search state or a feasible search state at all in a provable and reasonable amount of time such as in polynomial time in dependence of the problem instance size. Nevertheless, methods that approximate goal or feasible search states can be contrived [Joh74, ACG+99]. These approximation methods mainly are constructed for optimization problems. They do not come as a free lunch, though. They trade off runtime for precision: the closer in cost a (feasible) search state is to be to a globally optimal search state, the longer they have to run to with certainty find such a search state. In the end, to come arbitrarily close to an optimal search state as concerns cost, they need time exponential with respect to the instance size again, but these approximation methods at least can guarantee a certain cost bound. Inventing such approximation methods is not an easy task, however, and involves a lot of mathematical understanding of the problem at hand. In particular, guaranteeing any bounds basically requires to find a proof for it. For some problems this cannot easily be done at all or cannot be done with a reasonable amount of effort. That's why so-called *search heuristics* or *heuristic search methods* (also called (*heuristics* for short) are invented and used.

Heuristic search methods are informed search methods [RN03, pp. 94ff] in contrast to uninformed search methods such as breadth-first or depth-first search [RN03, p. 73] which basically enumerate all possible search states of a given search space to find a goal search state. Informed search methods employ a *heuristic function* which sometimes simply is called *heuristic* for short also (as well as the resulting search methods). A heuristic function is a function that evaluates search states with respect to how promising they are to lead to a goal search state. A heuristic function effectively implements some rule of thumb drawn from experience. If not complemented with some exhaustive search mechanism (such as for A^* search [RN03, pp. 97ff]),the rule of thumb nature of heuristic functions entails that heuristics are not guaranteed to find a goal search state or a feasible search state at all. Generally, any time a decision between several alternatives has to be taken and this decision is solved following some procedure that does not compute all possible consequences, this procedure can be considered to be a heuristic. By not computing all consequences, suboptimal decisions are possible and often likely, yet, with the benefit of acceptable runtime. For this reason heuristics may not always achieve the desired outcome, but in practice they work well. Whereas approximation and exact methods try to heavily exploit certain mathematical laws induced by a problem specification, coming up with a heuristic function does not require such a deep insight into the specificities of a problem.

One group of heuristic search methods are called *local search heuristics* or *methods*, or, in a concrete instantiation, *local search algorithms* [AL97, Stü98, CDG99a, Stü99]). In its simplest form, local search methods start with an initial search state and advance to new search states by changing individual solution elements. The changes done in each step are relatively small affecting only a limited number of solution components. In the following, although in fact solution elements are exchanged, such changes will be termed "changes or modifications to solution components" to stress that the assignment of solution elements to solution components is changed, i.e. that there rather is a change *of* solution elements and *to* solution components. Each such change, possibly subject to fulfilling the constraints to maintain feasibility, will yield a new search state and is called *basic search state transition*. Local search methods in principle work on candidate search states from S. They can visit infeasible search states also but need not. Anyway, they should do this only temporarily, since any goal search state must be feasible. Several such basic search state transitions can be grouped arbitrarily. A transition from one search state to another in general, independent of its granularity, is called *search state transition*. If the cost during the search state transition increases or decreases, it is also called *improving* and *worsening search state transition*, respectively. All local search methods have to make decisions during the search concerning the search state transitions to do next. These decisions in the simplest form are about basic search state transitions, but also can comprise decisions about arbitrary complex search state transitions. In general, these decisions are called *search decisions*. A basic search state transition from one search state to another is also called *local search step*, again perhaps with attribute improving or worsening.[2]

Each local search method in general will start with a search state and end in a search state, in the meanwhile doing one or more basic search state transitions. The sequence of all search states visited during the search is called the *search trajectory*. Abstracting from the details of a search trajectory and only considering the start and end search states, each local search method in fact induces a search state transition. When viewing a search state as an object, a local search method basically operates on a search state and manipulates it, thereby transforming it to another search state. Local search methods transform one search state to yield a successor search state at an arbitrary level of complexity possibly involving many basic search state transitions are also called *local search operator* to emphasis the transformational character. Inversely, *any* search state transition can be regarded as being induced by a local search operator.

The name local search stems from the locality of changes done to transform a search state into a successor search state. Methods that use this search paradigm are called local search based methods or local search for short, in contrast to exact methods. Figure 2.2 demonstrates a standard so-called 2-opt (exchange) local search step for the TSP. The 2-opt local search step simply changes two edges such that a valid tour results again. The dashed connections in the left-hand side of Figure 2.2 will

---

[2]In some texts, a basic search state transition or local search step is also called move. However, this notion is reserved within this text to mean something else and will be defined later in Section 2.8.

be deleted and substituted by the dashed connections on the right hand side resulting in a valid tour again. Most local search steps for the TSP are based on so-called n-opt exchanges ($n \in \mathbb{N}^{+}$). There, n edges are removed from a tour and replaced in such a manner as to yield a feasible tour again.

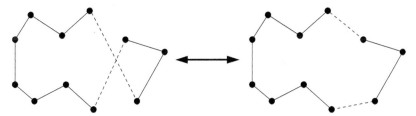

Figure 2.2: 2-opt local search step: Switch of two edges of a round trip of a TSP

The decision which n edges to replace with which other n edges can vary from random selection to a selection based on search state characteristics or other deterministic or semi-deterministic criteria. Looking at Figure 2.2, the same local search step could in principle be performed with two other edges. In fact, if some edge pairs did not lead to a successor search state with less cost, other pairs could be tried, in order to improve the cost of the successor search state. Local search steps not only denote actual basic search state transitions, the notion is also used in general to denote an implementation of a procedure for how to compute a successor search state from a current search state such that a basic search state transition results. This computation typically is based on a generic program how to incur systematic changes to the solution components of a search state to yield a successor search state. The generic nature results in several potential successors which as also called *neighboring search states* or *neighbors* for short. The notion of a neighbor again implies the notion of locality which gave rise to the name local search. To formalize the notion of a neighbor, consider the following definition according to [BR03]: A *neighborhood structure* $\mathcal{N}$ is a function $\mathcal{N} : S \rightarrow 2^{S}$ that assigns to every search state s ($s \in S$), a set of neighbors $\mathcal{N}(s)$ ($\mathcal{N}(s) \subseteq S$). The set $\mathcal{N}(s)$ is called the *neighborhood* of s. Neighborhood structures typically are specific to a certain problem type but independent of the instance size work for all instances of a problem type. Having a neighborhood structure, the set of all *local optima* can be defined as

$$S^{*} := \{s \in S' \,|\, \forall s' \in S . \mathcal{N}(s, s') \rightarrow f(s) \leq f(s')\} \subseteq S' \subseteq S$$

Analog to the argumentation in Section 2.2 and as often done in practice, infeasible search states are penalized strong enough such that no infeasible search state can possibly be a local optimum and hence the formal restriction for local optima to be a feasible search state can be avoided:

$$S^{*} := \{s \in S \,|\, \forall s' \in S . \mathcal{N}(s, s') \rightarrow f(s) \leq f(s')\} \subseteq S'$$

In principle, there is a one-to-one correspondence of local search steps to neighborhood structures. As has been argued, any local search step induces an neighborhood structure. Conversely, any neighborhood structure can be used to implement a local search step in principle. The local search step simply works by enumeration for any search state its neighborhood in any order and picking the successor search state for example be means of the first or best improvement heuristic.

Clearly, it holds $S^{**} \subseteq S^{*}$, i.e. the set of all global optima is a subset of the set of all local optima. Each local search step corresponds to a neighborhood structure and vice versa. The neighborhood of a search state s induced by a local search step simply comprises all search states that can be constructed from s by means of all systematic modifications to solution components that are possible according to the local search step's underlying generic program. In other words, for a given search state, a local search step enumerates the search state's neighborhood. Given a neighborhood of a search state s in turn, the local search step's underlying program simply can be implemented as an enumeration of the neighborhood of search state s. Any local search method is eventually constructed from local search steps and hence depends on a neighborhood structure. This implies that they all only work on a certain problem type. This respective problem type is called the *underlying COP type*. If a concrete problem instance is referred to, it is called *underlying COP instance*. With the notion of a local optimum, the goal in solving COPs can be reformulated. The goal then is to solve COPs as good as possible by finding a global or reasonably good *local optimum*. A goal search state then is a global or a reasonably good local optimum.

In the case of n-opt exchange local search steps for the TSP it holds that the greater n is, the *stronger* or more *powerful* is a local search step, because the major are the modifications to a current tour (and the smaller n, the *weaker* a local search step). This can be carried over to the general case: The more comprehensive search state modifications are, the stronger is a resulting local search step. Here, the number of edges exchanged, n, can be considered to be the *strength* of the local search step. Typically, changes made by stronger local search steps are not easily made undone by less powerful local search steps. The power of a local search step typically is correlated with the size of the neighborhoods induced by the neighborhood structure the local search step is based on. The more modifications can be incurred by a local search step, the more neighbors will result. In the general case, however, it need not always be this way.

The progress of local search methods is guided by the cost function of the COP taken on. It accordingly serves as heuristic function. The simplest local search methods are called *local search procedures*. Together with local search steps they are the commonly used local search operators. A local search procedure, also called *hill-climbing* [RN03] (stemming from handling maximization problems, in the context of this document it is rather a descent), tries to always proceed using an underlying local search step to a next search state whose cost is less than the cost of the current search state. Each local search procedure typically only employs one single and fixed local search step. This procedure will continuously improve search states and thus optimize them. It will stop in a local optimum when no neighbor with better cost can be found. Each chain of improving local search steps ending in a local optimum is called a *local search descent* and sometimes *local search* for short as well. Depending on the *start search state* where the descent begun, a local search procedure does not always end up in the same local optimum. The search space is subdivided into regions of so-called *attraction basins*. An attraction basin is a subset of the search space. Each search state in this subset will be transformed by a local search procedure to the same local optimum, if used as start search state for its local search descent. Attraction basins are dependent on local search steps and hence procedures: Different local search steps and procedures subdivide the search state space in different subsets of attraction basins. Along the same lines, attraction basins are dependent on the neighborhood structure that local search steps are based on. Attraction basins and the strength of local search steps are related. Changes made by stronger local search steps can lead out of attraction basins of less strong local search steps, leading a local search descent based on the less strong to e *new* local optimum.

Each neighborhood structure induces a so-called *connection graph* yielded by identifying search states as nodes and any two neighboring search states as edges. The progress of a local search method can be seen as moving from search state to search state in the connection graph along edges. Figure 2.3 illustrates such a progress. Nodes in this figure represent search states, edged represent a neighbor relationship. Depending on the neighborhood structure used, the graph not necessarily is bidirectional. The dashed arrows indicate the search steps done so far. The node with the dotted arrows originating is the current search state. The dotted arrows indicate the possible next steps (assuming that a local search method does never go directly back to a search state just visited). The question marks associated to the dotted arrows indicate that there is a search decision to make for the local search method where to step to next. A connection graph with respect to a fundamental local search step of the problem at hand can be used to measure *distance* between two search states in terms of the number of local search steps needed to transform one search state into another. A connection graph can also be viewed as topology. Extending this topology into another dimension, i.e. a cost dimension, yields, together with the cost function, a *cost surface*. The connection graph view of a search space also leads to the notion of a *region* of the search space. A region is a connected subgraph of the connection graph and hence a coherent subset of the search space.

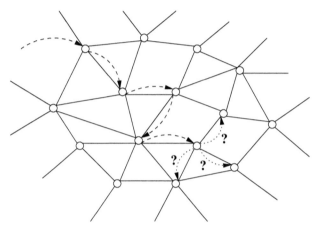

Figure 2.3: Progress of a local search method through a connection graph

Each neighborhood structure also induces a different set of local optima for a problem instance. Now, if several local search steps and procedures for different neighborhood structures are used alternately, the question arises which are the local optima for the combined neighborhood structures? In principle, any local optima for any neighborhood structure or only those that are common for all neighborhood structures can be regarded as a local optimum. That is, given a family of neighborhood structures $\mathcal{N}_{(m \in M)}$ ($\mathcal{N}_m \colon S \to 2^S$, $M = \{1, \ldots, k\}$, $m \in M$, $k \in \mathbb{N}^+$) let

$$S_m^* := \{s \in S \mid \forall s' \in S . \mathcal{N}_m(s, s') \to f(s) \leq f(s')\}$$

The *joined set of local optima* then is defined to be the union of all the sets of local optima from all neighborhood structures of the family:

$$S_{\bigcup \mathcal{N}_{(m \in M)}}^* := \bigcup_{m \in M} S_m^*$$

The *set of common local optima* is defined to be the intersection of all sets of all local optima from all neighborhood structures of the family:

$$S_{\bigcap \mathcal{N}_{(m \in M)}}^* := \bigcap_{m \in M} S_m^*$$

It holds:

$$S_{\bigcap \mathcal{N}_{(m \in M)}}^* = \bigcap_{m \in M} S_m^* \subseteq \bigcup_{m \in M} S_m^* = S_{\bigcup \mathcal{N}_{(m \in M)}}^*$$

Intuitively, the second definition for the set of all local optima is more appealing. Here, any local optimum from $S^*_{\cap \mathcal{N}_{(m \in M)}}$ cannot be improved by any local search procedure available. If it could, this would be contradictory to the principle of a local optimum. This, though, exactly happens for the first definition of the joined set of all local optima. Here, many local optima will exist which are not common to all neighborhood structures and accordingly can further be improved according to one or more neighborhood structures.

A union of all elements of a family $\mathcal{N}_{(m \in M)}$ of neighborhood structure, $\bigcup \mathcal{N}_{(m \in M)}$, called *united neighborhood structure*, can be defined straightforwardly as

$$\bigcup \mathcal{N}_{(m \in M)}(s) := \bigcup_{m \in M} \mathcal{N}_m(s)$$

The set of local optima according to the united neighborhood structure will be the same as the set of common local optima, $S^*_{\cap \mathcal{N}_{(m \in M)}}$, for the family of neighborhood structures. The other possibility of intersecting all neighborhoods of a search state cannot ensure that a search state is left with neighbors at all and therefore is useless for general contemplations. The definition of a united neighborhood structure and accordingly using a set of common local optima might not work very well in practice, though. It is cumbersome to always descent to a *common* local optimum since this effectively means to apply any local search procedure in round-robin fashion until no procedure can find an improvement anymore. So, even if the joined set of local optima contains local optima that can absolutely be improved according to other neighborhood structures, it might be more useful to be used in practice. It is larger than the set of common local optima so in any case will it contain all goal search states contained in the set of common local optima also.

Besides the choice of the corresponding neighborhood structure, other design decisions are to be made for a local search step. At first glance, it seems best to select the best neighboring search state to advance to next. This requires computing all neighboring search states otherwise one cannot be sure to have found the best neighboring search state. This greedy approach might become ineffective or even impractical, if the neighborhoods of individual search states are huge. In fact, experience has shown that for large neighborhoods the greedy so-called *best improvement* strategy is not useful [PM99]. Most of the computation time is used to enumerate neighborhoods while very few local search steps are actually made. To remedy this apparent waste of time one could step to the first cost-improved neighbor computed according to the local search steps induced neighborhood enumeration. The chances to find a better search state long before the complete neighborhood has been explored are higher and this approach usually decreased the time needed to make a local search step. This strategy is called *first improvement* and practically only explores a complete neighborhood in the case a local optimum has been reached. The rational why this strategy sometimes is better than the best improvement is that it might be better to do more local search steps or even any local search steps at all in any descending direction than to require to take the steepest descent if the latter takes too long. In other words, several small local search steps can lead farther in shorter time than one complicated huge local search step [PM99]. Other variations of the neighborhood exploration and neighbor selection decision strategies are conceivable. Note that these strategies are heuristics themselves with respect to the search decisions a local search methods has to do. Instead of systematically and deterministically exploring the neighborhood, random sampling can be performed. Also, the order in which a neighborhood is explored can be randomized in contrast to a deterministic enumeration or other elements of the local search steps can include some element of chance. Whenever randomization is involved in computing a local search step, notions *randomized local search step* and derived *randomized local search procedure* are also used to indicate their stochastic nature.

One variation of local search methods are so-called *construction local search methods* or *heuristics* [AL97, BR03]. These methods do not work on complete search state solution encodings. Instead, they construct a preferably good search state step by step from scratch by successively adding solution elements until a complete search state solution encoding has been constructed. Such methods can be viewed as local search methods, too, if in each local search step, only some solution elements are added by assigning them to solution components. Any search state with only partial solution encoding in the form of only a partial variable assignment of solution elements to solution components is called *partial search state*. A search state with complete solution encoding in the form of a complete assignment is called *complete search state*. The decision how to extend the current partial search state can be based on local information about the current partial search state and the possible (assignment) extensions. Neighborhood structures can be defined for partial search states here. The connection graph of conventional local search methods becomes a construction graph, since it is based on partial search states. It also can be thought of being directed, if partial search states can only be extended.

The advantages of local search methods over exact or approximate methods are mainly that they typically are quite easy and fast to contrive and to implement and that they find search states usable in practice quickly. These search states typically are not globally optimal, but close enough to the global optimum and thus reasonably good, i.e. good enough for practical purposes. The disadvantages of local search methods compared to exact or approximate methods comprise the abandonment of guaranteeing to find a globally optimal search state, a search state within a fixed range of the globally optimal search state, or even a feasible search state at all. For example if a problem type is strongly constrained and on account of this finding feasible search states already is a difficult task in itself. Local search methods might fail too often to find any feasible search state at all or reasonable good search state. Nonetheless, in practice they typically come within close range of the globally optimal search state.

2.4 Metaheuristics

The name local search stems from the fact that only immediate neighbors according to some neighborhood structure are considered by a local search step. Any other information, for example information about the last search states visited, are not stored and hence used per se. In [RN03, p. 111], the authors write that

> *this resembles trying to find the top of the Mount Everest in a thick fog while suffering from amnesia.*

A major problem in the context of search decisions of simpler local search methods such as local search procedures is that of getting stuck in local optima. One single local search descent is by no means guaranteed to find a global optimum. Instead, it almost certainly will end up in a local optimum that is not a global optimum or not even a relatively good local optimum, since there are typically lots of local optima in a search space [Sta95, Boe96, CFG$^+$96, SM99]. In a sense, a local search procedure easily is misguided by the heuristic function in the form of the cost function. Figure 2.4 illustrates a search space with several local optima. The y-axis represents the cost of a search state and the x-axis represents search states where neighbors are sequenced linearly according to a one-dimensional neighborhood structure. A local search procedure, illustrated by a dashed arrow, can get stuck in a local optima and has to overcome local maxima to eventually proceed to a global optimum.

To prevent from getting stuck in local optima, extensions to improve simple local search procedures to yield more foreseeing search decisions have been suggested. These improvements have taken two main directions. On the one hand, introducing some element of chance enables a local search

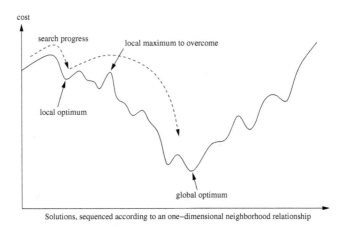

Figure 2.4: Model of a search space with several local optima

method to advance to otherwise unvisited search states by chance. This provides some kind of exploration which goes beyond a purely deterministic approach. On the other hand, strategies (or rather heuristics) for how to escape local optima have been devised. The first direction of improvement leads to so-called *stochastic local search methods* [Hoo98, Hoo99b, HS04], the other direction leads to the invention of *metaheuristics* [LO96, OL96, OK96, CDG99b, VMOR99, HR01, GK02, BR03, RdS03]. In practice, both approaches typically are combined. In particular, almost any metaheuristic is randomized in some parts.

Randomization of local search procedures can come in several flavors. One possibility is to randomize the construction of the first, initial search state. Additionally, randomized local search steps such that worsening local search steps are allowed can be introduced. In fact, both possibilities yield more robust and better performing local search procedures [Hoo98, Hoo99b, HS04]. To follow the idea of randomizing the construction of an initial search state, a snapshot metaheuristic called *random restart* can be contrived: Whenever a local optimum was found and a local search procedure got stuck the whole process is restarted by constructing a new random initial search state and subsequently improving it until the next local optimum is found. This effectively samples the space S^* of local optima. This procedure, however, is quite inefficient, since it is not directed in any way and experiments have suggested that the number of local optima of a problem with an exponential number of search states with respect to the instance size is exponential with respect to the instance size again [Sta95, CFG$^+$96, Boe96, SM99], rendering this approach as ineffective as a complete enumeration of the whole search space.

Altogether, the random restart technique is not very useful. Some strategic meta-component, also called *strategy component* (or *strategy* for short), in addition to pure chance has to be incorporated. This insight leads to the invention of metaheuristics. Any metaheuristic can be considered to consist of components that are local search steps or procedures and which are responsible for iteratively improving search states by improving its solution encoding, called *solution improvement components* and a strategy component responsible for avoiding to get stuck in local optima or for escaping local optima. The borders between these two types of components sometimes are blurred and blend in fairly smoothly. These two types of components can be found in any local search method in principle, although the degree of blending typically increases the simpler local search methods are. Yet, any local search method is heuristic in nature and for this reason certainly has to have a strategy component that finally makes search decision about the next search state transition. Therefore, the notions strategy component and strategy for short are carried over to *any* local search method.

Strictly speaking, the strategy component of a metaheuristic should be called meta-heuristic, since it controls other heuristics in the form of the improvement components which are heuristics in their own. The strategy component thereby decides about a coherent number of individual decisions at once, i.e. makes meta-decisions. It thereby is heuristic in nature itself. In so far, naming the combination of strategy *and* improvement components a metaheuristic is somehow misleading but nevertheless common parlance. Since metaheuristic themselves follow the local search paradigm for improving search states, they also belong to the broader class of local search methods. Local search methods in general, and simple ones such as local search procedures in particular, do not require an explicit strategy component acting at a meta-level. They have to make search decisions, though. Hence, any local search method and not only metaheuristics must have some parts that decide about which search state transitions to perform next. Regarding these parts together as a virtual strategy (component), the notion can be transferred to apply to local search method in general. The strategy (component) thereby guides and controls the search and effectively implements a search strategy or heuristic.

Local search methods are a superset of metaheuristics. Yet, metaheuristics are the most successful and mostly used subclass of local search methods with most research efforts devoted to. Examples for metaheuristics that use local search steps or procedures, either as a black box or as blended in subcomponents are:

- *Simulated Annealing* (SA) [Dow93, AKv97],

- *Tabu Search* (TS) [GTdW93, GL97],

- *Iterated Local Search* (ILS) [LMS01, LMS02],

- *Variable Neighborhood Search* (VNS) and *Variable Neighborhood Descent* (VND) [HM99, HM01, HM02],

- *Greedy Randomized Adaptive Search Procedure* (GRASP) [RR02], and

- *Guided Local Search* (GLS) [Vou97, VT02].

Metaheuristics that only secondarily use local search procedures are:

- *Ant Colony Optimization* (ACO) [DD99, DDG99, DS02, DS04] and

- evolutionary heuristics, the most popular ones among them being *Genetic Algorithms* (GA) [KP94, Mit96, RR03].

The latter two employ a *population* of simultaneously visited search states and are called *population-based* whereas the other metaheuristics only use one search trajectory and accordingly are called *trajectory methods* or *trajectory-based* [BR03]. More details and further references to metaheuristics can be found in [BR03].

Escapes from local optima or local search steps can be more or less comprehensive. If they are more comprehensive, a new search state will be more distant to the old local optimum or search state and will lead a local search method to another region of the search space. This way, the search space is explored, the search itself is diversified. If a local optimum escape or local search step is less comprehensive, a new search state will be in the vicinity of the last local optimum or search state which then rather is an exploitation of the current region of the search space and corresponds to an intensification of the search. During research in local search methods, and metaheuristics in particular, one decisive criterion for success of a method has been identified. This criterion is the balance between exploitation and exploration or, in other words, the balance between exploitation and exploration of the search progress [BR03]. Empirical experiments for several COPs [MGSK88, Müh91, BKM94, BKM94, Boe96] have revealed a tendency of good local optima to cluster in relatively close distance around the global optimum. That is, if a metaheuristic has found a good local optimum, it might be promising to search in the vicinity of this local optimum in order to detect another good, better, or even global optimum. On the other hand, an algorithm cannot know in advance whether a local optimum found really is good in absolute terms. For example, an algorithm might by chance have found only quite bad search states so far such that the best search state found during the search is only average among all local optima of the search space. In this case, the algorithm can easily wrongly regard such an average search state found as a high score one. Intensifying the search around this essentially average search state would be waist of time missing to almost surely find better search states someplace else. Instead, a diversification of the search towards other regions of the search space is more profitable in such a case. Additionally, if an intensifying search in the vicinity of a good local optimum has not found a better search state for some time, it might be a good idea to diversify the search again to settle down in more promising regions. In summary, a local search algorithm in action always faces the decision, whether to exploit a good search state locally hoping to find an even better one in the vicinity or whether to explore the search space to find another, more promising region of the search space that can be exploited in turn.

Generally, metaheuristics use other local search methods – not only local search steps or procedures – as almost or complete black-boxes and add an additional strategy component that guides the search beyond the guidance provided by the black-box local search methods. The goal is not only to escape local optima or avoid getting stuck in local optima, but in general to find a proper balance between exploitation and exploration and thereby or in general to improve performance further. Simple local search procedures are only exploiting or rather intensifying the search by doing a monotonically improving local search descent. The strategy component of a metaheuristic then is responsible for exploring other regions of the search space also in order not to miss promising ones; it is responsible for ensuring a certain diversification of the search. Solution improvement components of metaheuristics are problem type specific. In contrast, the strategy components are not and hence can and will be reused. Hence, summarizing, metaheuristics can be characterized as general-purpose heuristic methods that are intended to lead basic and problem type specific heuristics (e.g. local search procedure or construction heuristics) to more promising regions of a search space. Metaheuristics hence are general frameworks for the design of heuristic methods. The next section will give an example of a metaheuristic which will illustrate the notion of a metaheuristic.

2.5   Iterated Local Search

One of the simplest yet most appealing and successful metaheuristics is Iterated Local Search (ILS) [LMS01, LMS02, BR03]. The idea behind ILS simply is to do some kind of local search in the smaller (sub-) space of local optima S^* with $S^* \subseteq S$, since this subset must contain all goal search states. As simple as this idea is in theory, implementing it in practice cannot be done easily and straightforwardly. The problem is that no neighborhood structure for the space S^* of local optima can be given or computed easily: If there were efficient methods to quickly identify

Procedure *Iterated Local Search*

  s_0 = GenerateInitialSearchState

  s^* = LocalSearchProcedure(s_0)

  Repeat

    s' = Perturbation(s^*, *history*)

    $s^{*\prime}$ = LocalSearchProcedure(s')

    s^* = AcceptanceCriterion(s^*, $s^{*\prime}$, *history*)

  Until termination condition met

End

Algorithm 2.1:   Iterated Local Search

search states that are local optima, a problem itself possibly could be solved efficiently, in the case of *NP*-hard problems perhaps in polynomial time. Being able to solve a *NP*-hard problem in polynomial time implies that it holds $NP = P$ [GJ79]. This, however, is not expected by many researchers. Identifying local optima efficiently will at least require a deep understanding of the structure of a problem which has not been reached for any *NP*-hard problem yet. Up to now, the only practical approach known to find local optima is to sample them with the help of repeated local search descents, for example starting from randomly generated start search states. This was mentioned as being not very effective in the previous section, though (cf. Section 2.4). Instead, one can use a method which is intermediate between the random restart approach and a randomized local search procedure that always starts from the same start search state: repeatedly execute local search procedures from start search states that are connected more directly than just being two random samples of the search space. When a local optimum was reached, one could simply do a "jump" (called *perturbation*) to an "adjacent" search state by applying a stronger modification to the current search state than a typical local search step would do. The modifications done to a local optimum are not as comprehensive as constructing a new random complete search state, though. Certain parts of the search state that might represent good partial search states can be retained and reused, this way rendering the search more efficient. Repeatedly alternating a local search descent to a local optimum and incurring the stronger perturbation modifications to the local optimum will result in a biased, connected and directed walk through the set of local optima S^* building a chain of local optima in contrast to a random walk.

The general outline of the ILS metaheuristic following [LMS02] in pseudo code notation is given in Algorithm 2.1. Each cycle of the **Repeat** – **Until** loop thereby corresponds to one iteration. Although not shown explictly, the overall best search state found so far is stored also, of course. The other components of the ILS metaheuristic are described next:

- Function GenerateInitialSearchState simply generates the initial search state for the metaheuristic. The initial search state, s_o, typically is constructed straightforwardly, e.g. with a greedy approach, possibly randomized. Its quality is not decisive. Substantial difficulties do only arise, if it is hard to find a feasible search state at all.

- Function LocalSearchProcedure denotes a local search procedure that uses an underlying local search step to descend to the next local optimum. This function can be viewed as a black box. The only important feature it must possess is that it finds a new local optimum, denoted by s^* or $s^{*\prime}$.

- Function Perturbation represents the perturbation. A perturbation is a local search step and also modifies a current search state s^* to yield a successor search state s'. In contrast to a typical local search step, however, the modifications are more comprehensive, usually involving more solution components. As such, a perturbation is a stronger local search step compared to the local search steps employed by the local search procedure LocalSearchProcedure. The aim is to make a big step or rather a jump that first of all should lead to a new attraction basin with respect to the local search procedure employed in function LocalSearchProcedure and the local optimum s^* just found. Otherwise, the next call to function LocalSearchProcedure would undo the changes of the perturbation yielding the same local optimum as before. A perturbation therefore should be complementary to the local search procedure; it must not be too weak. If a perturbation, on the other hand, is too strong, i.e. "jumps" too far away, the whole search progress degenerates to a random restart like search. In the general case, the perturbation can base the construction of a new search state on a history of search states, too. This so-called *search state history* is labeled *history* in the pseudo code in Algorithm 2.1. The strength of a perturbation directly influences the behavior of an ILS in terms of the balance between exploitation and exploration that is induced by it: The stronger a perturbation is, the more amenable is an ILS for exploration. The weaker a perturbation is, the more likely will it concentrate on exploration.

- Function AcceptanceCriterion represents an acceptance criterion. An acceptance criterion has to decide whether to continue from the last local optimum $s^{*\prime}$ found or to back up to a previous one. More generally, an acceptance criterion has to produce a new starting search state s^* for the local search procedure LocalSearchProcedure. The acceptance criterion can base its decision on the current local optimum $s^{*\prime}$, the previous local optimum s^* and perhaps even on a complete history of search states visited previously (*history*). Typically, the search state history contains a selection of the local optima visited so far and is maintained by the acceptance criterion. The acceptance criterion has strong influence on the nature and effectiveness of the search. Together with the perturbation, it controls the balance between exploitation and exploration of the search.

ILS is more than just a variant of iteratively random restarting a local search procedure. Whereas random restart simply samples the set of local optima randomly, ILS can induce a certain bias with the help of the perturbation and acceptance criterion component which can guide the search. A perturbation reuses most solution elements which is particularly helpful when good search states have been found. Often, the better the search state at the begin of a local search procedure, the more likely is it that the local optimum found by the local search procedure is better; good local optima might cluster around the global optima or other good local optima [BKM94, Boe96]. Random restarts will not be able to reuse any good combination of solution elements and will most likely produce relatively poor start search states for local search procedures. An acceptance criterion that only accepts better search states to continue from will carry the reuse of good solution elements to the extreme. In fact, this acceptance criterion has quite successfully been applied [SH01]. By biasing the search to find new and better local optima the whole search process becomes strongly directed towards good search states. Additionally, an acceptance criterion can be implemented in such a way that it rejects any new local optima for a number of iterations but stores them in its search state history, this way effectively probing several applications of a perturbation by sampling a number of local optima (the perturbation must be randomized, of course). This sampling can be regarded as an enumeration of a neighborhood of the current local optimum s^*. This neighborhood is in fact based on the set of local optima S^*. In a sense, the search space then really is reduced to the much smaller (sub-) space S^* with $S^* \subseteq S$ of local optima, of course, at the expense of computation time that is needed by the local search procedure to find neighboring local optima. If the sampling acceptance criterion only accepts better samples, this will yield an ILS which effectively performs a local search descent in the set of local optima S^*.

ILS is considered to be a general-purpose metaheuristic, because those parts of ILS that organize the search in the form of an interplay of perturbation and acceptance criterion and, secondary, of the local search procedure as well, are generic in nature. The components are used as black-boxes and can be exchanged independently. Of course, for each new type of problem, adapted components specialized to the problem at hand have to be implemented. The pattern of ILS, i.e. a local search procedure embedded in a perturbation and an acceptance criterion, remains unchanged, though.

ILS has several other advantages as will be given next:

- A local search procedure started after a perturbation applied to a local optimum typically will require substantially less local search steps to find a new local optimum than the same local search procedure started from a search state constructed randomly and completely anew. This is because a perturbation changes only parts of a search state and does not build it from scratch anew. ILS can involve many runs of a local search procedure, i.e. visit lots of local optima and sometimes, it is better to examine many local optima, even if most of them are inferior, instead of trying only a few good ones (cf. Subsection 4.1.2).

- By means of an acceptance criterion and a history of search states, stagnation of the search can be detected (cf. Subsection 4.3.2), e.g. by detecting that the search circled around in the same region of the search space for a while. If an algorithm spends too much time searching for a goal search state in a certain region of the search space, it might miss to search in other regions which might contain better and more easy to find goal search states. Clearly, in such a case, the effort is wasted: it would be better to explore more of the search space instead of exploiting the currently visited region. The acceptance criterion can then decide to continue the search from a different region.

- The steering mechanism balancing exploitation vs. exploration is quite evident in the case of the ILS metaheuristic, which might be one of the reasons for its success. The two ingredients perturbation and acceptance criterion are responsible for where the search continues. The strength of the perturbation directly determines the degree of diversification. The acceptance criterion, too, has a strong influence on the nature and effectiveness of the search in that it, too, controls the balance between exploitation and exploration: an acceptance that only accepts a new local optimum $s^{*\prime}$ as the next current search state s^* if it is has better cost than the previous local optimum implements an extreme exploitation. The other end of the scale is an acceptance criterion that always accepts a new local optimum as the new current search state. Such a strategy will lead to extreme exploration.

- The conceptual simplicity of the ILS metaheuristic makes it easy to implement it with few parameters that have to be adjusted. The ILS parameters such as the kind of acceptance criterion to use or the strength of the perturbation have an intuitive and reconstructible effect and hence a meaning to the human user. In almost the same manner, the reduction of the search space to local optima reduces the size of the search space and makes it more amenable for humans.

The balance between exploitation and exploration is considered very important [BR03]. It seems natural to improve the decision when to do what. One possibility is to let an algorithm learn this decision by experience. In the case of ILS, this can be done easily by providing several perturbations, local search procedures, and perhaps acceptance criteria that represent different exploitation and exploration strategies to choose from. Accordingly, the topic of learning is discussed in the next section (cf. Section 2.6), and the topic of machine learning in the context of local search will be covered in the following section (cf. Section 2.7).

2.6   Regularities and Learning

The world as we know it is highly regular. Regular according to [Url05d] means to be

> *formed, built, arranged, or ordered according to some established rule, law, principle or type.*

A regularity is

> *the quality or state of being regular, something that is regular*

[Url05d]. Regularities are induced by some *underlying* law or rule and laws induce a regular structure. This is in contrast to states of chaos and implies the ability to predict, if the underlying laws are known. Prediction in turn can be used to improve behavior. This way, regularities can be exploited. Finding and exploiting regularities then means to find the underlying laws and to exploit them by using them for prediction. Regularities arise at any level of granularity and abstraction. In the context of combinatorial optimization, search spaces also exhibit regularities. There are laws implied by any generic problem specification or any concrete problem instance specification.

Examples of Regularities in COPs For example, the occurrence of local optima and the ability to find them via monotonic local search descents according to some neighborhood structure and a given cost function is a common regularity of COPs. As another example, consider the representation of a railway system. Different types of trains will yield different types of connections, some are faster, some are slower. The system of connections can be represented as a multi-graph. Each edge represents one specific train connection, an edge weight indicates travel time. Now, if the task is to find a route with shortest travel time between any two points, connections belonging to faster trains are to be preferred in this respect. Anyhow, it is clear that not only the train type of a connection but also its direction is of importance when computing such a route. Taking a train in the opposite "direction" is not advisable; the two generic problem characteristics edge weight and "direction" have to be considered together.

Laws underlying regularities and hence regularities themselves can be expressed by mathematical theorems given a formal specification such a formal problem (instance) specification. As the previous example has shown, these theorems then are formulated based on arbitrary combinations of generic problem characteristics. In general, the field of mathematics can be viewed as an attempt to formalize problem specifications and recognize and utilize regularities by describing the underlying laws in so-called theorems. Mathematics can be regarded as a formal language for describing certain aspects of the world with the aim to formulate general laws that can be used for prediction. Theorems describing (underlying laws of) regularities can be used to improve searching a COP search space, too. Basically all exact and approximate search methods that do not simply systematically enumerate and test all search states of a search space rely on exploiting regularities to achieve their speed up compared to pure enumeration schemes. The effectiveness of these exact search methods almost always stems from their factoring out large regions of the search space that need not be searched, since no better search states can provably be found there. Approximation search methods rely on proofs for approximation bounds. If there were no regularities exhibited by problems, then an improvement of methods would be impossible in principle. So the only potential for improvement of *any* exact and approximate search methods lies in the regularities of the search space and hence the performance and usefulness of exact and approximation search methods centrally depends on whether useful regularities can be discovered and exploited. The tendency is that the more regularities can be exploited, the better a method is performing. Ideally, all regularities will be exploited but this certainly is illusive. Nevertheless, considering that regularities vary in complexity, it can be striven to exploit all regularities up to a certain complexity or granularity, anyway many more than usual methods do. The argument that the only potential for improvement lies in the regularities can be transferred to local search methods also.

Some regularities will be of general nature and accordingly will be problem independent such as the monotonicity of local search descents. But different problem types also have different problem specification. This will result in different regularities or rather problem type specific invariants as indicated by previous research and experience with local search ([BKM94, Boe96, MSK97] discuss this topic, for example. For example, it has been found out that in almost any good tours for a TSP instance, more than 90% of all cities are connected to their two nearest neighbors [Wal98]. It can be expected that most differences in the performance of local search methods essentially stem from different characteristics for different problem types. The *no free lunch theorem* [WM97, DJW02] for example states that for any two optimization algorithms (hence including local search and random search algorithms), if one outperforms the other on a class of problem types, there will be another class of problem types where the opposite holds. In consequence, averaged over all problem types, no local search method will do better in the limit than any other, even not better than random search. Nevertheless, any subset of all problem types is solved better by some local search methods than others. The varying performance of different metaheuristics on different problems types is an indication for systematic sources for performance differences which can only result from the differences in the problem type specifications and the resulting varying regularities. This has empirically been proved by hitherto experience in local search research in fact showing varying performance of local search methods across different problem types [RDSB+03, SBMRD02, BR03].

The conjecture is that the differences in performance are due to the fact that some local search methods exploit regularities better than others. It can be argued that in order to further improve local search methods they *must* be able to exploit regularities induced by problem type specific regularities. Because of the no free lunch theorem, finding better general local search method strategies working for all types of problems is not feasible. Either, for each new problem type tackled a best local search methods has to be found empirically and hence circumstantially, or, it can be attempted to devise local search method that outperform other methods on a relevant subclass of all problem types such as a number of problem types occurring in practice. Such local search methods then must be tailored towards this subclass as a whole. If the subclass is too big, it is doubtful whether this will work. In any case, it means to adapt local search methods manually as before. Alternatively, one can strive to enable the local search methods to adapt to each new problem type automatically, basically yielding a "new" and "suitable" method for each problem type. This way, the theoretic restrictions of the no free lunch theorem are evaded effectively in practice. This adaption can only work by flexibly identifying and exploiting individual regularities for each new problem type. Therefore, regularities have to be identified or "learned about".

The process of finding and describing the underlying laws of regularities and hence the regularities themselves can be regarded as *learning*. One decisive trait of intelligent behavior is the ability to adapt to (dynamically changing) environments. The capability to adapt to an environment can be achieved by recognizing and exploiting regularities, hence one decisive trait of intelligent behavior is the ability to learn. This can only work, of course, if there are regularities present. Adaptation to chaotic environments is futile. Adaptation is not only crucial for good performance of an individual in dynamically changing environments but also in static environments if they exhibit a certain search space size and complexity. Even simple static and deterministic environments such as a COP can be too large and exhibit such an immense inherent complexity as to allow a systematic and comprehensive treatment, e.g. the search for a constrained search state in the form of a search state with a constrained solution encoding. For example, the TSP problem instance shown in Figure

2.1 with 532 cities has 2^{532} different search states. Consequently, first regularities are identified (and the underlying laws are proven) for a COP at hand and next methods are built using the underlying laws. This procedure is omnipresent in human behavior as exemplified with next example.

Mountain Crossing Example Local search has been labeled hill-climbing also [RN03], so consider the task of a local searcher in the form of a human trek leader crossing mountains such as the Alpes or the Rocky Mountains. Humans do not cross mountains using a straight-line path, even if the mountains have never been passed before. Instead, a human trek leader will look at the mountains before entering them to learn and to find and use passes, instead of going the direct way via ridges, peaks, steep slopes, cliffs or canyons. Clearly, the effective cost including the expectation for accidents and the estimated traveling time for following the direct straight-line path is too high, even though the straight-line distance is the shortest on the first side, neglecting the ups and downs. Any individual that is about to cross mountains will almost immediately, but latest after the first bad experiences during the search, learn to take passes and to go from one valley to the next. He or she will learn that taking a pass and following valleys to the next pass might be longer in distance, but is more safe and with respect to traveling time actually shorter and therefore better as concerns the overall cost measure.Someone crossing mountains will additionally learn that following valley after valley in the general direction of his or her target will eventually lead out of the mountains and to the target. Figure 2.5 illustrates such a path through a mountain ridge. The regularities revealed after a mountain search space analysis can be described by notions such as valleys, peaks, ridges, and so on. Humans utilize such regularities to become more effective in their task. In general, humans can search effectively even with their limited computational resources in terms of speed and memory because they learn and adapt. They get a "feeling" or simply know where to look first.

Figure 2.5: Crossing mountains via valleys and passes

Learning is based on experience and will produce knowledge about the regularities that are inherent to the experience made. The knowledge will be about the laws underlying the regularities which are responsible for producing the experiences and possible effects that can be used for prediction. The knowledge hence represents an understanding of how the experienced things work. An actual representation of the laws or rather regularities is called a *model*. A model is the outcome of a learning effort. This holds true for human learning as well as for machine learning. A model describing some law can then be used for prediction. In a given situation, the experience that is to be made can be predicted without actually having to make the experience. This is highly appreciable. The aim in learning is to generalize beyond the experience already made by trying to predict in new, unseen circumstances using a learned model. The assumption is that the learned model is valid beyond the circumstances that induced the model. A model consequently is also called a *hypothesis*. It can be seen as a hypothesis of how things work, a hypothesis about the underlying laws (of regularities). The experiences used for learning often are also called *training examples* or *examples* for short. The whole learning process thereby is always guided by some kind of a priori given performance measure. Typically, this is some kind of error measurement measuring the predictive quality of the learned hypothesis for new, unseen circumstances. These new experiences used for performance assessment are also called *test examples*. These should not be used for further learning, but for evaluation.

In machine learning, any hypothesis must by encoded by data structures. In principle, the data structures can be arbitrary complex, but because of limited resources in practice in terms of memory consumption and computation time, their principle form must somehow be restricted. Consider the task of learning a function from experience in the form of input-output examples. In principle, any functional form could be used to instantiate a hypothesis in the form of a functional description for computing function values from input values that fits the examples i.e. that will predict for the examples with little error. This involves two steps: First, an appropriate functional form has to be selected, next its generic parameters have to be adjusted. In terms of computation, first an appropriate generic data structure has to be found and next its parameters in the form of the generic and changeable data structure components have to be adjusted. For reason of efficiency and to prevent from over-fitting due to too exact function descriptions, in machine learning practice, the functional form can be committed to a concrete method and one generic data structure in advance. Or, the general functional form is restricted somehow. The drawback is that not all conceivable hypotheses – in the example functions – can be encoded any more. The set of all hypotheses that still can be represented is called the *hypothesis space* of a learning method. In the case of function approximation for example, the hypothesis space for example can be restricted to comprise only polynomial functions. A data structure for storing polynomial functions basically has to store a variable number of coefficients. The process of learning then consists of finding parameters in the form of a polynomial degree and of finding coefficients.

Often a function has to be learned. This is done by function approximators. Function approximators want to approximate an unknown function f ($f : X \to Y$) for which only training examples in the form of input-output pairs $(x, f(x))$ $((x, f(x)) \in X \times Y)$ are given. Here, x is the input and $f(x)$ is the output of unknown function f on input x. The task is to produce a model $\tilde{f}$ ($\tilde{f} : X \to Y$) that approximates f given only a set of training examples which can be thought of representing unknown function f. This task is also called *pure inductive learning* or simply *inductive learning*; no a priori knowledge is available or used standardly. Approximating function $\tilde{f}$ is a hypothesis stemming from the hypothesis spaces employed by the function approximator. Often, function approximators are parameterized representations of functions such as simple linear functions or polynomial function. Their hypothesis space then is defined by a generic functional form. The models are parameterized with a parameter vector $\vec{\theta}$ ($\vec{\theta} \in \mathbb{R}^n$, $n \in \mathbb{N}^+$) of real values and work on real-valued input vectors $\vec{s}$ ($\vec{s} \in \mathbb{R}$). For a hypothesis space containing all polynomial function, the parameter vector $\vec{\theta}$ represents the coefficients. Typically, all function approximators in some form employ generic models that are based on a real-valued parameter vector $\vec{\theta}$ working on real-value vectors $\vec{s}$ are common, too (cf. Subsection 3.3.4).

In practice, a function approximator implementation will not return a hypothesis $\tilde{f}$ in the form of a formal description that can be evaluated afterwards, but it will compute an internal representation in the form of appropriate data structures which then can be used to compute values for new, potentially unseen points of the domain X. An actual internal representation of $\tilde{f}$ as approximation for f is also called *internal model* or shorter simply *model*. Inductive learning in the form of function approximation is used by many reinforcement learning methods presented later in Chapter 3. It belongs to the class of *supervised learning* and can also be called *learning by instruction*. A teacher has to give instructive feedback that indicates *which* was the correct value. The correct values here are the $f(x)$ of the test examples.

2.7   The Role of Machine Learning in Search

As was mentioned before, exact and approximate search methods rely on insights into regularities that have been learned in advance by the method constructor (or someone else). Thus, exact and approximate search methods utilize a priori learning. In a sense, they adapt statically and a priori. As the mountain crossing example on page 26 has shown, adaptation by means of learning can also take place dynamically during the search. Learning during search requires automation; no efficient search method is likely to be interactive. In general, it is only consequent to automate the learning process of identifying and exploiting regularities by means of machine learning techniques. These can be built into search methods so they can identify and exploit regularities by themselves rendering them dynamically adaptive. In contrast to a priori learning of mathematical laws for method construction, methods that learn thereby do not need to find and formulate explicit laws in a specific representation (language); most importantly is to achieve an improved behavior. Laws detected by humans are represented in the brain as neural networks, but the exact working of knowledge representation of neural networks is still unknown. Nevertheless, being able to extract human understandable laws for regularities is even more appreciated. This then can be considered computer supported human learning.

The regularities that are exploited by exact and approximate search methods have been found and investigated by human beings explicitly. Considering the complexity of many search spaces and the rather restricted human brain resources, it is likely that the regularities identified so far by humans rather belong to the simpler ones or have been identified by chance; many others are still waiting to be found. The complexity of exploited regularities is increasing continuously and those regularities whose utilization can make a difference in performance might be arbitrary complex. Furthermore, in the case of humans, the insights into regularities typically are generalized over a number of problem types or all instances of a problem type at once and do not consider potential differences in regularities across individual problem types or even individual instances. This might make them harder to find and proof compared to regularities only valid for one problem type or instance. Accordingly, the effort to find such new laws that can be utilized successfully is substantially and increases. This argument is also supported by the observation that many new regularities are disguised by a huge amount of *noise*. Noise is irrelevant or meaningless information occurring along with desired relevant information. Noise can be regarded as one of the main obstacles that prevent humans from finding complex regularities. Considering their complexity, also completely deterministic environments such as COPs can exhibit pseudo-noise to humans. From a certain point on, there are probably too many regularities for humans to find them all with a reasonable amount of time and resources. As a conclusion, it can be assumed that the computational power of human beings is increasingly not able to deal with the this task of identifying and investigating regularities of complex search spaces to further improve search methods. By their pure computational power, computers are likely to be able to handle (pseudo)-noise better. Machine learning techniques might be able to discover even infinitesimal regularities efficiently, exploring them beyond human capacity. This maybe is the only means to detect and exploit new regularities at all. In this respect, it is only natural and even mandatory to automate the process of detecting and exploiting regularities.

Although not stated explicitly, these arguments hold true for local search methods solving COPs, in particular metaheuristics, also. Current practice in local search method research is that any modification or construction of local search methods mostly is done in the design and implementation phase. Currently, most new strategies for local search methods that are proposed cannot be considered to be learning. All these proposals are based on evolving insights into the working of known local search methods and are based on accumulating success in the analysis of individual search spaces [Wei90, JF95, Sta95, MF99, SS03, HS04]. Yet, during searching the search space, the strategies of local search methods typically remain fixed. Any insights in the working of local search methods enter the local search method construction a priori, e.g. in the form of improved strategies and tuning of parameter settings. In fact, the tuning of parameters has been identified in practice to have a huge influence on the success, perhaps as important for performance as the method architecture itself: A method with an inadequate architecture cannot be tuned to perform well, but even a method with the most adequate architecture will perform poorly, if it is not tuned properly, i.e. even the best architecture can be "thwarted" by poor tuning. Tuning basically is learning to choose from several settings of some flexible, but during the search nevertheless fixed strategy. Again, the process of getting more insights in the inner working of local search methods including any subsequent invention of more suited search strategies can be regarded as learning. This learning typically takes place on the side of the local search method constructor and/or tuner a priori in between running algorithms. According to [BP97],

> *it is often the case that the user is a crucial learning component of a heuristic algorithm, whose eventual success on a problem should be credited more to the human smartness than to the algorithm intrinsic potentiality.*

Tuning is an optimization problem and in aces of discrete parameter value sets also a COP. Learning itself is an optimization problem and for the same reasons as for tuning perhaps even a COP. So basically, the problem of tuning is transferred into another; one COP is transferred into another, hopefully easier to solve one. As a natural extension and improvement of local search methods, learning should be automated and should take place during search, preferably including a transfer of the learned across several search processes to enable accumulation and refinement of the learned. Machine learning techniques incorporated into local search methods as a new basic ingredient can be used to identify and exploit regularities, e.g. to derive rules for a proper balance between exploitation and exploration. The hope is that local search methods will learn about and thereby identify the factors which truly influence performance as autonomous as possible or at least help human researchers in doing so. As before, a proof of any laws is not necessary. Additionally, by means of the stochastic nature of most local search methods their behavior cannot be proved exactly anyway. It suffices, if the performance measures of the methods increase substantially, i.e. significantly in terms of statistical testing. Experience in practice will eventually rule out all approaches that do not work.

Recall that learning implies building and using a model to make decisions. In this context, adaptation is considered to be the result of learning. Learning is intended to discover knowledge and represent it in a flexible model. If the hypothesis space is too constrained, no true new knowledge can be represented and hence discovered. Another decisive trait of learning is the ability to generalize and transfer what was learned. Now, learning can be differently comprehensive. Technically, adaptation can also be seen as a weaker form of learning. Here, only some but not all model parameters are adjusted, i.e. the hypothesis space is very constraint. In the function approximation example from the previous section, Section 2.6, fixing the polynomial degree and requiring to find only some coefficients while most of the coefficients are already determined can be viewed as adaptation. It can also be seen as some form of tuning. The learning method is case of adaptation in the technical sense mostly is already determined in advance.

The notion of adaptation can also be used in the sense that something is fit. In the context of local search, in this sense, any local search method is adaptive, because their behavior is dependent on and fit to the search trajectory. Trivially, the information needed to decide on the next search state transition is computed from the current search state. Each local search method must necessarily do it when guided by the cost function. To exclude trivial cases, a local search method typically is labeled adaptive as soon as it derives and memorizes some information from the search trajectory with the intention to create a sensitivity of its behavior to the specifics of the current search process and problem instance at hand. The intention is to direct the search in a more straight manner to goal search states. The behavior supposedly is not following a fixed strategy independent of the peculiarities of the current search process as is done by conventional local search methods but adapts itself to the peculiarities. Adaptive local search methods must incorporate some kind of memorization of information in the form of a model which influences the decisions made during the search. For example, consider a TS method that can set certain search states temporarily tabu, e.g. because they have recently been visited before. This prevents them from becoming a new search state again. Such a TS method accordingly employs a model which consists of the tabu list and the corresponding decision mechanism and consequently can be labeled adaptive.

Comparable to the TS example, the behavior of many local search methods is strongly influenced by information gathered, but is not directly and crucially controlled by it. In the case of meta-heuristics, central parts of the strategy (component) remain fixed and cannot be transferred across problem instances. The TS here cannot be considered as performing learning, since in the context of local search generalization of the learned not only should mean the ability to generalize over the search states within one problem instance, but also over several problem instances as well. This presupposes that the models can be transferred across several instances and that they are independent of a specific instance, in particular of the instance size. Transfer of the learned implies that the strategy (component) of a metaheuristic can be transferred also, which in turn presupposes that the search decisions basically exclusive are determined by the learned model: The strategy component basically consists of the learned model. The learned model then is representing and centrally implementing the search strategy, not only modifying it. If such a strategy component guides the search and results in adaptive behavior, it is called *learning strategy component* or shorter *learning component*. A learning component controls the search and modifies its underlying model. For example, consider a number of different local search steps that can be applied in each step. A local search method can be considered to learn and employ a learning component, if it learns a model in the form of a function in sufficiently generic form which maps search states to available local search steps. Each current search state then is input into the function and the output local search step is executed. If the function representation is problem instance independent, it can be transferred and used for other problem instances as well.

Several local search methods, mainly in the form of metaheuristics, have been proposed that are labeled adaptive and are supposed to be adaptive [BT94, BKM94, Bey95, DS02, PR00, PK01, Hoo02]. Such local search methods cannot be considered to incorporate learning according to the view just described, though (cf. Section 4.4). None of these make use of the full potential of machine learning techniques and do not have a central learning component (cf. Section 4.4). That's why the following nomenclature is used within this document. All local search methods that do not incorporate some form of information gathering and memorization during search to influence its search decisions are called *non-adaptive*. Their strategy (component) remains fixed. Simple best and first improvement local search procedures fall in this category. Those methods that do extract and memorize information from the search trajectory in a model and which thereby use the technically weaker form of learning, adaptation, by adjusting only certain model parameters using a very constrained hypothesis space are called *adaptive*. They do not employ a learning strategy component, but they do employ some form of model nevertheless. Those methods that incorporate a true learning strategy component which centrally makes the search decisions, which employ a sufficiently large and powerful hypothesis spaces, and which can transfer models across problem instances are called *true learning* or simply *learning* local search methods. In the context of local search methods then, learning implies being adaptive but not the other way round.

2.8 Agents

The motivation for incorporating learning to improve local search method performance is appealing and easily reconstructible. Evidently, learning is a survival strategy proven to work in nature many times. The question that arises is how to actually and practically accomplish the application and incorporation of machine learning techniques? An answer to this question can be given by means of the notion of an agent.

One of the objectives of the field of artificial or computational intelligence can be regarded as building entities that behave intelligently. These entities are called *agents* [RN03]. Agents are entities that act within an *environment* autonomously. They perceive the environment by means of *sensors* and carry out *actions* to interact with the environment. The *perceptions* need not encompass all aspects of the environment but can be partial; an agent might only partially observe an environment. The behavior of an agent is evaluated with respect to some kind of *performance measure* and the goal of an agent is to maximize its performance with respect to the given performance measure. The actual value of a performance measure is determined by the current situation or *state* of the environment, also called *environment state*, including the agent, and possibly the environment states encountered during the course of interaction so far. The performance measure is a function with all possible environment snapshots, i.e. environment states an environment as a whole can be in, or any sequence of environment states, as its domain. An agent interacts and hence changes its environment by taking actions. In each state, an agent can choose among several actions. The set of applicable actions might differ from environment state to environment state. An agent's behavior is induced by its action selection strategy. In each current environment state, an agent finally has to select and apply one of its available and applicable actions. This selection strategy is called *policy* and can hence be modeled as a (possibly randomized) mapping of environment states as an agent can observe it to actions. An agent can be a robot, but it can also be some program that for example searches the Internet independently. Summarizing, the decisive traits of an agent are [RN03]:

- It is autonomous within an environment (may it be real or virtual),

- it senses/perceives the environment,

- it interacts with the environment by means of applicable actions thereby changing it, and

- it does so in order to maximize a given performance measure.

In principle, the environments agents interact with are dynamic in nature. They involve changes over time, for example triggered by actions done by the agent or *events* external to the agent. Any dynamic system inevitably involves the aspect of time. The dynamics of an environment over time often are modeled as a sequence of discrete time points. For each time point an environment can be regarded as being in a current state that statically describes its state of affairs. This environment state includes agents, since an agent itself is also part of the environment. Agent actions and other events will occur at these discrete time points and will change the overall environment state leading to a new current environment state at the next time point. Modeling an environment in the form of environment states, actions, and events that lead from one environment state to another is a very popular and successful model of how the world works and has extensively been used. In particular, in the field of artificial or computational intelligence the world often is modeled this way. This model typically is adopted for agents, too [RN03].

An agent environment can be split into two parts: The environment outside an agent is called *external environment*, the internal state of belief of an agent is called *internal environment*. Anything that cannot be controlled directly by an agent is considered to be part of the external environment outside the agent. The boundary between agent and environment is where an agent's absolute control ends, but this is by no means the boundary of its knowledge as well. An agent's internal environment is an internal representation of the environment as a whole which is needed by the agent in order to reason about its environment as a whole. It has to contain a description of the current environment state in the form of description of the current state of affairs of the external environment as the agent can observe it and a description of its own state of affairs, for example summarizing some aspects of its own computation up to now. This internal representation of the environment as a whole as can be observed by an agent is denoted by *agent state*.   Since any internal representation of an agent belongs to its internal state of belief and can be considered to be an agent's state also, agent state and internal environment are identical.

An agent environment has dynamic and static parts. Since an agent acts in an environment and, together with other events, changes it over time, some representation of the environment's and agent's dynamics is needed, too in the internal representation of the environment as a whole. This dynamic component is modeled in the form of actions that can be applied by an agent and other events that happen independent from an agent in addition to model describing potential outcomes of actions and events. Applying actions will have some effect to an agents environment in that the environment will react by changing its environment state and accordingly the agent state changes also when an agent adjusts its inner environment representation. Each environment state transition and hence agent state transition triggered by an action is denoted by *agent move* or *move* for short. The environment or agent state an agent moves to is denoted *move target* or *target of a move*. Both agent actions and other agent external events entail changes to the environment state and hence agent state that have to be modeled, too, in so-called *transition models*. The changes or rather transitions need not be deterministic, the same action in the same environment state can have different results, perhaps simply because other external events occur independently at the same time. A popular way to model such nondeterministic outcomes is by assigning probabilities to them. Having a model of the environments dynamics in the form of transition models as part of its agent state enables an agent to reason about its interaction with the environment. Such a model enables an agent to predict to some extent the outcome of different applicable actions with respect to the expected resulting environment state and hence with respect to the performance measure to maximize. Since the whole environment any agent interacts in is in general not completely observable for the agent, it can only reason about its inner representation of the environment state, i.e. its agent state. Hence any discussion about agents and agent design henceforth will be with respect to agent states. A policy for example, more precisely is a (possibly randomized) mapping of agent states to actions.

Many methods have been proposed to design autonomous, intelligent, adaptive, and learning agents incorporating lots of machine learning techniques. These methods could be reused to incorporate machine learning techniques into local search methods if some resemblance to the concept of an agent could be find in these local search methods. Fortunately, this is the case straightforwardly. Local search methods can be viewed to work on environments and environment states. In their case, the environment states are search states consisting of a solution encoding for the underlying COP type (cf. Section 2.3). They carry out transitions from one search state to another. In a sense, a local search method advances from search state to search state by applying actions in the form of local modifications to the solution encoding of the current search state, i.e. in the form of local search operators such as local search steps or local search procedures. In the special setting of local search methods solving COPs, the whole environment typically is fully observable for a local search method, i.e. the sensors of a local search method are perfect. This means, the external environment *is* the current search state. Consequently, the external environment can be mapped completely to an inner representation in the form of the search state solution encoding. This part of an local search method equivalent to an internal environment of an agent is also called *search state*. The other part of such an equivalent internal environment for a local search method can comprise additional information about the search history in the form of move, step or iteration counters, and so on, and is called *heuristic state part*. Generally, it contains any necessary information about the progress of the local search method or rather heuristic that is not a solution encoding but which is needed or computed for its operation. This kind of information is called *heuristic information*.

The mentioned characteristics of local search methods coincide with many traits of an agent as stated before. A local search method is autonomous, it perceives the environment as represented by its search state, it interacts with environment by means of the local search operators it employs, and it does so in order to maximize a performance measure in the form of the cost function of the COP to be solved. The search and heuristic state parts of a local search method together can form an agent state. If a local search method furthermore can choose from several local search operators to induce search state transitions and hence agent state transitions, there is no principle difference to the notion of an agent as just described and used in the field of artificial intelligence. Accordingly, a local search method can easily be viewed as a (virtual) agent as well. Let such an agent be called *local search agent* (LSA). An LSA can be regarded as modeling and being equivalent to a local search method. Executing an LSA means executing a local search method. As such, a concrete LSA instantiation can also be regarded as being an algorithm. The strategy component of a local search method then becomes that strategy component for an LSA which basically boils down to action selection and hence is equal to the policy of an LSA. The performance measure for an LSA is given or rather induced by the cost function of a COP to be solved such that the goal of an LSA becomes the goal in solving COPs. The state of an LSA, now called *LSA state* exactly has two parts in the form of a search state which simply comprises the solution encoding and a heuristic state part which stores any other information. Since the environment in the case of a COP is completely observable and also controllable by an LSA – no LSA external events happen – an LSA state basically is the state of the COP environment as a whole, the overall environment state. Since accordingly any change to an LSA environment is triggered by the LSA itself, actions an LSA can do can be defined to be *anything* that change an LSA state. Such actions are called also *local search actions*. Figure 2.6 illustrates an LSA moving. The picture is an extension of Figure 2.3. Nodes now represent LSA states. Edges still represent the neighbor relation, now in terms of actions that can bring an LSA form one LSA state to the other. The dashed arrows indicate the search progress of the LSA labeled "LSA" in this figure so far. Dotted arrows marked with a question marked indicate that several action are applicable and that the LSA's policy has to select one for application in the current LSA state the LSA is residing at the moment. They will yield to the LSA state the dotted arrows point to, which might be unknown to the LSA, though, if it does not have a transition model available.

Local search actions basically are local search operators as defined in Section 2.3. However, they originally only induce state transitions in the context of search states. As such, they change an LSA state as well but do not affect the heuristic state part of an LSA state. In contrast, any change to an LSA state as a whole, perhaps affecting only one of the two LSA state parts, search or heuristic state part, can happen, too, according to the just given definition of a local search action. Therefore, any change to an LSA state as a whole resulting in a state transition to the LSA state is called an *LSA move* and each LSA move is triggered by a local search action. Local search operators hence are not local search actions directly, but typically local search actions centrally consist of local search operators and only extend them in that they additionally collect and update heuristic information. That's why the nomenclature for search states from Section 2.3 is transferred: An LSA move that basically consists of a basic search state transition is called *basic LSA move* The special case of LSA moves that move an LSA from one local optimum to the next is denoted by *local optima LSA move*.

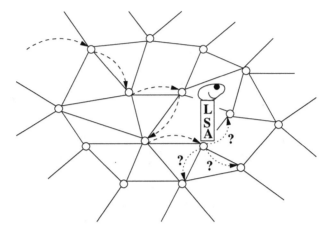

Figure 2.6: Progress of a local search agent

Local search actions can be based on any kind of local search operator, including local search methods and thus local search actions implement local search methods. The local search methods come in the form of local search steps or complete local search methods such as metaheuristics. Most of these local search operators use other ones for their operation. For example, a local search procedures uses a local search step, an ILS metaheuristic uses a perturbation local search step, a local search procedure, and an acceptance criterion for its operation. In general, local search operators can use or rather can consist of any composition or combination of other local search operators thus forming arbitrarily complex hierarchies of local search operators. As an example, consider an ILS metaheuristic (cf. Figure 4.1 and its discussion in Subsection 4.1.3). Also, one iteration of an ILS comprising a perturbation application with a subsequent local search descent and an acceptance criterion application can be regarded as a local search operator. Such a local search operator is special in so far as it only visits local optima and thus only operates on the subspace of local optima S^*. Since local search operators can easily be extended to be local search actions, the same arguments just given hold true for local search actions as well. Consequently, any hierarchical composition of local search actions can be considered a local search action again. The resulting local search action as well will exhibits the central trait of local search actions, namely that they produce a new LSA state. In principle, the level of granularity of a local search action is irrelevant as long as its runtime is finite and as long as it yields a new successor LSA state.

The most important part of an LSA state is its search state. As a consequence, concepts and attribute defined and used so far for search states can be transferred to the LSA state notion and basically mean the same but only regard the search state part of an LSA state. The attributes for example comprise good, bad, best, worst, average, better, worse, improving, worsening, minimal, maximal, suboptimal, inferior, quality, cost of and so on and always refer to the cost of a search state part of an LSA state. Concepts for example comprise *candidate* and *feasible LSA states*, *goal LSA state*, *start LSA state*, *neighboring LSA states*, *partial LSA state*, *complete LSA state*, *LSA state history*, local optimum, and so on. Analog, attributes for local search steps and local search operators and methods such as weak, strong, weaker, stronger, randomized, and so carry over also. In the following, discussions often only concern the search state part of an LSA state. In case of an LSA, state can be thought of as being its search state without harm, since this part is mostly affected. Except for the next chapter, Chapter 3, only LSA agents are concerned.

Therefore, the notion LSA state by default is abbreviated to state, also names for other concepts which contain a part "LSA state". The same abbreviation transfer to composite notions holds true for the abbreviation of the notion of LSA move to simply move also. The notion of LSA state is only used for explicit emphasis. The notion local search action is abbreviated also to action in the following and is used in full name only for emphasis purposes. In the next chapter, the notion state means agent state which as a special case comprises LSA state also. The same holds true for the notions move and action.

Trajectory methods [BR03] most closely resemble the LSA view on solving COPs. Trajectory methods are metaheuristics that start with an initial search state and proceed in discrete time steps from one search state to the next. According to [BR03]:

> *The algorithm starts from an initial state (the initial solution) and describes a trajectory in the search space.*

Typical trajectory methods are Simulated Annealing (SA), Tabu Search (TS), Iterated Local Search (ILS), Variable Neighborhood Search/Descent (VNS/VND), and Guided Local Search (GLS). Seen this way, a COP taken on with a trajectory method can consequently be viewed as an LSA that applies sequences of actions to reach the goal in solving COPs. Even other classes of metaheuristics following the view of [BR03] can be cast in the framework of an LSA. Population-based metaheuristics for example can be regarded as employing a population of individual LSAs. Here, however, some additional problems such as coordinating a population has to be dealt with which nevertheless does not seem infeasible. The notion of an LSA can also be transferred to population-based methods by extending the state of an LSA to comprise a population of search states. The further discussion will be with a trajectory-based LSA in mind, though.

This chapter has addressed several issues in the context of local search and machine learning as a preparation for enhancing local search methods with machine learning techniques. The notion of an LSA was introduced to capture certain aspects of local search methods in order to get a starting point for this venture. In the following chapter, Chapter 3, one class of machine learning techniques, Reinforcement Learning, will be presented which can be applied directly following the LSA view to local search methods this way enhancing them with machine learning techniques.

Chapter 3

Reinforcement Learning

The scenario of a local search agent (LSA) described in the previous chapter, Chapter 2, is a special case of a scenario that research in *reinforcement learning* deals with. Both scenarios are concerned with an agent that operates on agent states (form now on state for short) which in turn model an environment. Repeatedly in each state, the agent chooses from several applicable actions. Each action will then lead to another state and will give a feedback to the agent. This chapter covers all aspects of what reinforcement learning is and how to tackle reinforcement learning problems. Although this chapter is kept general to the reinforcement learning scenario, all insights can be transferred directly to the LSA scenario yielding a learning LSA which is enhanced with reinforcement learning called *learning LSA*. The next chapter, Chapter 4, then presents the Guided Adaptive Iterative Local Search (GAILS) method which can be regarded as a realization of a reinforcement learning based learning LSA.

The first section of this chapter, Section 3.1, is concerned with providing a precise problem definition of the reinforcement learning scenario and introduces means to formulate the goal to be achieved by an agent. The second section, Section 3.2, then discusses how to solve reinforcement problems presenting many different methods. The last section, Section 3.3, finally is concerned with issues in function approximation, since all reinforcement problem solution methods presented in the second section eventually will be based on supervised learning in the form of function approximation.

3.1 Reinforcement Learning Problems

Reinforcement learning problems consist of several parts which model the most important problem characteristics such as which actions are available, how the environment responds to agent interactions, and what the goal of an agent is. This section gives a precise and formal problem definition in Subsection 3.1.2 and covers further important aspects such as different goal definitions (Subsection 3.1.3) and problem variants such as Markov decision problems (Subsection 3.1.1).

3.1.1 Markov Decision Processes

Solving a reinforcement problem involves moving an agent from one state to the next altogether forming a sequence of states. Such a sequencing of states also is called *process* or *decision process*. In each state an agent has to decide which action to use next in order to trigger the next state transition. Each individual process actually undertaken depends not only on the actions chosen, but also on the environment dynamics and responses. Since an environment not necessarily is deterministic, an action can have several possible outcomes in the form of a successor state. The usual approach to model nondeterminism in action outcomes is to employ probability distributions over the possible outcomes, and possibly over feedback, too. If the environment dynamics as represented by transition and feedback probabilities do not change over time, such a process is said to be *stationary*. If these probabilities furthermore do only depend on a *finite* number of previously visited states, taken actions, and received feedback, the process is called to be *Markovian* or fulfilling the *Markov assumption*. Basically this means that each state can be finitely encoded such that any information contained in a finite sequence of states, taken actions, and received feedback can be encoded such that it can also be contained in the last state of the sequence. Each new state then contains the information from a finite number of previously visited states including taken actions and received feedback such that it provides any relevant information for determining the transition and feedback probabilities in each move. Formally, according to [SB98, p. 63] a state signal has the *Markov property*, if and only if it holds that

$$P(s_{t+1} = s, r_{t+1} = r \mid s_t, a_t, r_t, s_{t-1}, a_{t-1}, r_{t-1}, \ldots, s_1, a_1, r_1, s_0, a_0) = P(s_{t+1} = s, r_{t+1} = r \mid s_t, a_t)$$

where t ($t \in \mathbb{N}$) is the current time point, s_t ($s_t \in S$) is the current state, a_t ($a_t \in A(s_t) \subseteq \mathcal{A}$) is the next action to take at the current time point t, the s_i ($s_i \in S$, $i \in \{0, \ldots, t-1\}$) are the previously visited states of the process, a_i ($a_i \in A(s_i)$) is the action that was taken in state s_i, r_j ($r_j \in \mathbb{R}$, $j \in \{1, \ldots, t\}$) is the feedback obtained during the state transition from s_{j-1} to s_j (also denoted by $s_{j-1} \rightarrow s_j$), s ($s \in S$) is the potential choice for the new state at time $t+1$, and r ($r \in \mathbb{R}$) is the potential new feedback obtained for the next state transition. Processes that fulfill the Markov property are called *Markov decision processes* (MDP). Although in practice the Markov assumption does not necessarily hold, it typically can be assumed, since the violations are relatively small and practically ignorable; they mainly stem from principle theoretical contemplations about the problem characteristics. In fact, the formal definition of a reinforcement learning problem in this document presupposes that the Markov property holds. The definition of reinforcement learning problems is given with this presupposition in mind.

3.1.2   Reinforcement Learning Problem Definition

The standard scenario of reinforcement learning consists of an agent that can interact with an environment via perceptions and actions. The perceptions may, but need not, completely characterize the current state of the environment. In the case of an LSA, the environment basically consists of a search state of a COP and accordingly the perceptions of an LSA are complete (cf. Section 2.8). An agent iteratively interacts in discrete time steps with the environment by executing one of a set of applicable actions which may differ from state to state. The actions applied by an agent in each step or rather move change the state of the environment which in turn gives rise to new perceptions of the new environment state. Seen this way, actions yield new successor states. One of the perceptions an agent receives is a feedback signal in the form of a scalar numerical *reinforcement*, also called *immediate reward* or simply *reward*. The law that governs each immediate reward is called *reinforcement signal*, *reward signal*. The reward signal assigns a scalar value to each move. The actual scalar value is also called *reward*.

An agent's behavior is induced by its action selection strategy which is modeled as a policy (cf. Section 2.8). A policy is a possibly randomized mapping of states to actions. The aim of an agent is to behave as to maximize a given performance measure representing its goal. The measure thereby is computed from the rewards obtained for each agent's move. The computing function is called *return function* or *return model*. Its resulting value is called *return value* or *return* for short. Typically, the performance measure in the form of the return model is a possibly discounted sum of the rewards. Since in general environments are nondeterministic, the aim of an agent more precisely is to maximize the expected return. Figure 3.1 depicts the standard reinforcement learning scenario (t is the current time point, s_t are the perceptions describing the state at time t, a_t is the taken action at time t, while r_t is the reward observed after execution of action a_{t-1} corresponding to state transitions $s_{t-1} \rightarrow s_t$). The policy component of the agent in this figure includes all elements that together implement the behavior of the agent.

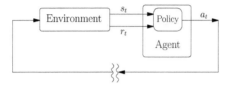

Figure 3.1: Model of the reinforcement learning problem

In addition to these basic elements of a reinforcement learning scenario, several issues are of importance:

- Rewards can also be negative, comparable to punishments in nature. The individual rewards received can be viewed as pleasure and pain an agent senses during its interaction with the environment and which drives its actions the same way living beings are fundamentally driven by pleasure and pain [Tho91].

- As has already been mentioned, in general the environment will be nondeterministic. The same action applied in the same state will not necessarily lead to the same successor state or yield the same reward. Since many local search methods and local search steps are randomized, this also holds true for LSAs. Nevertheless, it is assumed that the environment will be stationary, i.e. the probabilities of state transitions and rewards do not change over time although it is conceivable and typically the case in real-world applications.

- The reward signal and the return model together model the goal of an agent; they must be designed properly in order to correctly reflect the intended goal and make the agent learn to achieve this goal. The return model together with the reward signal can be viewed as a model of optimal behavior, since the aim is to maximize the expected accumulation (i.e. return) of immediate rewards ([KLM96], p.240). The return model thus is also called *optimality model*. The design of the return model and the reward signal are one of the most important design issues in reinforcement learning. In particular, the design *must* be part of the external environment and outside the scope and control of an agent. If the reward signal or the return model were manipulable by an agent, it could achieved its goal of maximizing the return trivially by simply changing the reward signal or return model instead of having to interact with its environment. According to [SB98, pp. 56f],

  [...] the reward signal is your way of communicating to the agent what you want it to achieve, not how you want it achieved.

  In particular, the reward signal is not the place to impart prior knowledge about how to achieve what we want it [the agent] to do.

Considering the LSA scenario, a reward signal which is based on a cost function of a COP such as the delta in cost per move is outside the control of an LSA. Of course, different return model and reward signal designs can be tried in order to find out which are best suited to truly reflect the goal in solving COPs (cf. Section 4.3), but during the run of an LSA, they must kept fixed.

- A complete specification of an external environment including state representation, applicable actions and its effects, rewards signal, and return model then is called *reinforcement learning problem* or *task*.   A reinforcement problem is an instance of the general reinforcement learning scenario.

- Reinforcement learning is defined as a problem, not as a class of algorithms. Any algorithm that solves a reinforcement learning problem is a *reinforcement learning algorithm*. Hence, reinforcement learning rather is a framework for learning. Any methods proposed to solve reinforcement learning problems are called *reinforcement learning methods* or *reinforcement learning techniques*.

Reinforcement learning problems typically are modeled as MDPs. This document follows this practice and formally defines a *reinforcement learning problem* or *task* as follows:

It consists of a 7-tuple

$$(s_0, S, \mathcal{A}, A\colon S \to \mathcal{P}(\mathcal{A}), r'\colon S \times \mathcal{A} \times S \to \mathbb{R}, T\colon S \times \mathcal{A} \times S \to [0,1], R\colon (\mathbb{N} \to \mathbb{R}) \to \mathbb{R}),$$

where

- S is a (possibly infinitely) set of states,

- $\mathcal{A}$ is a set of actions,

- s_0 ($s_0 \in S$) is the start state,

- A ($A\colon S \to \mathcal{P}(\mathcal{A})$) is a function that assigns to each state s ($s \in S$) a set $A(s)$ ($A(s) \subseteq \mathcal{A}$) of applicable actions which typically simply is the same set A, $A := A(s) = \mathcal{A}$, for all $s \in S$,

- r' ($r'\colon S \times \mathcal{A} \times S \to \mathbb{R}$), is the reward signal which varies with states, applied actions, and reached successor states,

- T ($T\colon S \times \mathcal{A} \times S \to [0,1]$) is the transition model, i.e. a state transition function in the form of a probability distribution for each pair (s,a) ($(s,a) \in S \times \mathcal{A}$) over S that assigns a probability for making a transition from a state s to a successor state s' ($s' \in S$) when applying action a ($a \in A(s)$) (since a reinforcement learning problem is modeled as an MDP here, the transition model T that models the environment behavior takes only the current and the next state into account), and

- R ($R\colon (\mathbb{N} \to \mathbb{R}) \to \mathbb{R}$) is the return model that computes for a sequence of rewards the long-term performance measure an agent tries to maximize representing its goal. It typically does so by accumulating the individual rewards.

For the course of an agent interaction let r ($r\colon \mathbb{N} \to \mathbb{R}$) be a function that records the rewards an agent has observed during its interaction with $r(0) := 0$. Function r often is also written as $(r_t)_{t\in\mathbb{N}}$ ($(r_t)_{t\in\mathbb{N}}\colon \mathbb{N} \to \mathbb{R}$) with $r_t := r(t)$ being the reward an agent obtains for state transition $s_{t-1} \to s_t$. The form $(r_t)_{t\in\mathbb{N}}$ is more vivid and accordingly used as a short cut for the sequence of rewards received by an agent just before time points $0, 1, 2, \ldots$. Sometimes, only a subset of a sequence of rewards starting at some time point t is considered. Such a subset of a sequence $(r_t)_{t\in\mathbb{N}}$ is denoted by $(r_{t+k})_{k\in\mathbb{N}}$ ($(r_{t+k})_{k\in\mathbb{N}}\colon (\mathbb{N} \setminus \{0, \ldots, t-1\}) \to \mathbb{R}$). It is the chain of rewards an agent will observe when starting in state s_t with $((r_{t+k})_{k\in\mathbb{N}^+})(0) := 0$ and $((r_{t+k})_{k\in\mathbb{N}^+})(n) := r_{t+i}$ for $n \in \mathbb{N}$.

Some reinforcement problems naturally break down in *finite episodes*, also only called *episodes*. For a given n the function recording the observed rewards only returns zero: $\forall n' \in \mathbb{N} . n' > n \to r_{n'} = 0$.

Each episode ends in a special state called *terminal state*. If an agent, after reaching a terminal state and ending an episode, is reset to tackle the problem anew from some new start state (possibly chosen randomly), the task is also called *episodic task* [SB98, p. 58]. If a search trajectory potentially is infinite, a task is also called *continuous*. In order to simplify notions, continuous tasks are also called *continuous episodes* in contrast to typical finite episodes. Episodic learning or using finite episodes then is denoted *episodic learning approach* or *episodic approach*, while learning with continuous episodes is also called *continuous learning approach* or *continuous approach*. Depending on the stopping criterion for LSAs (or rather its represented local search methods) and how the search trajectory of an LSA is split, all types of episodes can occur in the LSA scenario as well (cf. Subsection 4.3.2).

3.1.3 Return

The return model determines how an agent should take the future (rewards) into account in the decisions that are to be taken now. It thereby determines the goal an agent will follow. The return model accumulates future rewards. Its summation can take on different forms along several dimensions. It is important to determine whether the task is episodic and has a finite-horizon or whether it potentially is continuous and nonterminating and has an infinite-horizon. Note that infinite-horizon does not mean that all sequences necessarily are infinite, only that there is no fixed deadline. Any (immediate) reward can be incorporated with equal weight in the return model accumulation or rewards can be assigned different weights, hence influencing the total return differently. This way, the long-term nature of a goal can be adjusted. The latter approach is useful in order to give rewards in the nearer future more weight compared to rewards that are located farther in the future. This makes sense in, if it is important to reach the goal quickly or in a reasonable number of moves. If an agent only takes actions to get a final huge reward in the far future this reward might come to late and the agent might not "survive" long enough to obtain the huge future reward. It might be more practical to greedily obtain some positive rewards in the nearer future. One popular method to assign decreasing weights for future rewards is by geometrically discounting future rewards with a *discount factor* or *rate* γ ($\gamma \in [0, 1]$). This discount factor can be interpreted [KLM96, p. 240]

> *[...] in several ways. It can be seen as an interest rate, a probability of living another step, or as a mathematical trick to bound the infinite sum.*

It makes a difference with respect to an agent's behavior if the return model is based on the summation of exact rewards as observed or whether it is based on some kind of averaged reward encountered during the progress.

Based on the just described design options – episode length and episode horizon, reward weights, and reward accumulation vs. averaging – to form a return model, the following principle return models or rather optimality models have been identified as being useful:

- *Finite-horizon return model*:
$$R((r_t)_{t\in\mathbb{N}}) := \sum_{t=0}^{T'} r_t$$
with T' ($T' \in \mathbb{N}$) being the time horizon (and $\forall\, t' \in \mathbb{N}.\, t' > T' \rightarrow r_{t'} = 0$, of course).

- *Infinite-horizon discounted return model*:
$$R((r_t)_{t\in\mathbb{N}}) := \sum_{t=0}^{\infty} \gamma^t\, r_{t+1}$$

with γ being the discount factor. If $\gamma < 1$ and $(r_t)_{t \in \mathbb{N}}$ is bounded, $\sum\limits_{t=0}^{\infty} \gamma^t r_{t+1}$ has a finite value.

- *Average-reward return model*:

$$R((r_t)_{t \in \mathbb{N}}) := \lim_{h \to \infty} \frac{1}{h} \sum_{t=1}^{h} r_t$$

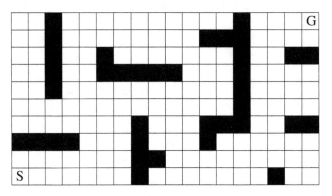

Figure 3.2: Example grid maze

Grid Maze Example   As an example for how reward signals and return models can be designed and used consider a grid maze like the one in Figure 3.2. The maze consists of a number of fields, some of them are walls, the rest is free space (walls are black in the figure). The state of an agent within the maze is taken to be the field it currently occupies. Hence the state space is finite. The actions an agent can take are "Up", "Down", "Left", and "Right" which will move the agent to the respective neighboring field of its current state. An agent cannot move to a wall field. Instead, an action that would lead to such a field has no effect and lefts the agent in place. One field is special in the maze, since it is the goal field (marked "G" in the picture). The task of the agent is to find a shortest path to the goal from its current position (marked "S") . When the goal field is reached, one episode has finished, since the goal state serves as the terminal state. To ensure that on the one hand reaching the goal state really is a goal for the agent in that it will maximize its long-term return, the transition to the goal state is assigned a high positive reward (pleasure). In order to ensure a quick movement to the goal state, any other transition costs a negative reward (pain) to the agent. Given a return model that does a simple accumulation of the rewards, an agent trying to learn to maximize its long-term return within this maze will have to learn to move as quickly and directly as possible to the goal state using the least moves possible, i.e. finding the shortest paths to the goal from any field it has started in, since each move adds to the costs. To complicate things in the maze, actions could be made non-deterministic beyond the rule not to move through walls. For example, each action might with a probability greater than zero just do the opposite move.

The setup of the reward signal and the return models in the grid maze example was done as to reflect the true goal the agent is supposed to achieve. Problems where a substantial reward is only received when entering a terminal state are called *delayed reward problems*. The respective reward is called *delayed reward*. If all other moves have a reward of zero the problem is called *pure delayed reward problem.* The respective reward is called *pure delayed reward.* The special difficulty in delayed reward problems for an agent is that it must learn which moves are desirable to take based on a reward, i.e. information, that might only be obtained in the far future. Since global optima or reasonable good local optima of COPs are not known in advance and hence goal states are not known in advance, LSAs will not operate in delayed reward problems. If a second type of terminal state such as a pit is present that gives rise to a huge negative reward, the problem becomes an *avoidance problem* which, nevertheless, is also not applicable for LSAs. For some tasks time plays a vital role. In such tasks each move is assigned a penalty in the form of a negative cost. In this case, the agent must avoid making too many moves, i.e. take too much time, to reach its goal. Move costs now are quite relevant for LSAs again, since the secondary goal in solving COPs is to find goal states quickly (and will be discussed for an LSA in detail in Subsection 4.3.3).

3.2 Solving Reinforcement Learning Problems

Knowing what a reinforcement learning problem is, the next question is how to solve such a problem? How to build an agent or a learning LSA that can solve a reinforcement learning problem or a COP, respectively? Clearly, to solve a reinforcement learning problem an agent in the end has to come up with a policy that achieves its goal of maximizing the expected long-term performance measure, i.e. the return. Recall that in the case of an LSA, this goal is the goal in solving COPs which is to find a goal states and, secondary, to find them as soon as possible (cf. Section 2.2). One possibility to come up with such a policy is to learn it.

As was discussed at the beginning of this chapter, an agent in the context of reinforcement learning has no teacher that can tell which action to choose in which state. The only feedback available is in the form of a reward, for example a delta in cost in the case of an LSA, and has two disadvantages. First, it is of evaluative nature only. A reward is an indication of how good the just now taken action in the previous state was but not which was the best action to carry out in this previous state. In both scenarios, reinforcement learning and learning LSA, generally no supervisor is available that tells whether a taken action was correct or wrong or even good or bad. Least of all, an agent is told which action would have been best, independent of the actually taken action. This would be instructive feedback. Second, it is short-term in nature, whereas the goal of an agent is of long-term nature in principle. The feedback an agent receives is merely a scale of goodness of the immediate result of an action which does not directly reflect the goodness with respect to the goal an agent pursuits. This is especially apparent for a learning LSA where no one can tell the absolute goodness of local optima found in advance, for example (cf. Section 2.4). The problem resulting from the short-term nature of rewards contrasting to the potentially long-term nature of an agent goal that has to be solved by reinforcement learning methods is called the *temporal credit assignment problem.*

An agent does not know how good a taken action now will be in the long run when later and rather delayed rewards may have huge effects on the overall return. An agent does not know which consequences are entailed by a current action selection in the far future. Current choices can have very bad effects in the future and no teacher is available to tell. Accordingly, supervised learning in the form of given input-output pairs for states and actions cannot be applied directly to solve a reinforcement learning problem. Instead, an agent must learn through trial-and-error interaction with the environment to achieve its goal. While an agent interacts, it also learns. In contrast to supervised learning, an agent must explicitly explore the environment in order not to miss important opportunities to learn thereby catching a fundamental trade-off between exploration and exploitation of the environment. When it interacts it must also invest in learning the important regularities guiding its behavior thoroughly. The same trial-and-error interaction is needed in principle for a learning LSA trying to solve a COP independent from whether it learns or not. The trade-off between exploration and exploitation directly translates to a necessary balance between exploration and exploitation which was identified as one key to successfully solving COPs with local search methods (cf. Chapter 1 and [BR03]).

The necessary trial-and-error interaction of an agent and the problems faced by it because of delayed rewards are the most distinguishing features of reinforcement learning problems. This section deals with how one can design learning methods that solve reinforcement learning problems and overcome these difficulties. It will start with defining the notion of a policy that controls an agent in Subsection 3.2.1. Next, problems with large search spaces, the notion of a value function that can be used to actually implement a policy and what an optimal policy is are discussed in subsections 3.2.2 and 3.2.3, respectively. Learning value functions and policies with various methods is addressed in subsections 3.2.4 until 3.2.7. Finally, some variants of the presented solution methods and other approaches are discussed in Subsection 3.2.8.

3.2.1 Policies

Several methods have been proposed to solve a reinforcement learning problem. Common to all approaches is that they all try to learn and implement a policy that in turn implements an agent's behavior. A *policy* π is a stochastic mapping from states to actions:

$$\pi \colon S \times \mathcal{A} \to [0, 1] \text{ with } \forall s \in S . \forall a \notin A(s) . \pi(s, a) = 0 \land \sum_{a \in \mathcal{A}} \pi(s, a) = \sum_{a \in A(s)} \pi(s, a) = 1$$

If it holds that $\forall s \in S . \exists a \in A(s) . \pi(s, a) = 1$, the stochastic policy will become deterministic and degenerates to a simple mapping in the form of $\pi \colon S \to \mathcal{A}$.

When learning policies, several aspects have to taken into account. First, it makes a difference, whether an agent can use the models of the environment's dynamics. It makes a difference whether the transition model and/or the reward signal is known to an agent or not. If these models are not given explicitly to an agent, it can learn them from the experiences gathered during the search. Obviously, having such models, learned or given a priori, provides useful information that might greatly enhance learning a policy. *Model-based* methods either learn a model of the environment or get one a priori to use it to either learn and implement a policy, or both. *Model-free* methods learn and implement a policy without using a model of the environment. In general, it cannot be expected that an environment model is available or it is the case that learning such a model takes too much effort not justifying the gain in learning quality. In the context of an LSA and combinatorial optimization where actions are local search steps, procedure or even more complex state transitions that typically are randomized, a model of the environment lies beyond reach, even for estimation. Several methods to learn and implement policies without a model have been proposed, too. In fact, reinforcement learning mostly is concerned with the question how to obtain policies when models are not known.

The next issue relevant to policy learning is how an agent represents a policy. An agent can learn a policy directly in the form of a mapping from states to actions or it can learn so-called *utility functions* that can be used to implement policies indirectly. Directly learning policies means that complete policies have to be executed, evaluated, and compared. This approach is described more closely in Subsection 3.2.7. Before, the approach of acquiring policies via learning and using utility functions will be discussed.

Utility functions assign numerical values to states. A utility of a state expresses the desirability of an agent to be in this state, of course with respect to its long-term goal. Because of the potential stochastic nature of environments as holds true in particular for a learning LSA, the utility of a state for an agent is the expected return starting its interaction in this state. Since agent interactions are behavior and accordingly policy dependent, state utilities are, too. A utility function U^π ($U^\pi \colon S \to \mathbb{R}$) for a given policy π an agent follows then is defined as:

$$U^\pi(s) := E\left(R((r_{t+k})_{k\in\mathbb{N}^+}) \mid \pi, s = s_t\right) = E_\pi(R_t \mid s = s_t)$$

where s ($s \in S$) is a state, s_t ($s_t \in S$) is the state at time t ($t \in \mathbb{N}$), sequence $(r_{t+k})_{k\in\mathbb{N}}$ $((r_{t+k})_{k\in\mathbb{N}} \colon (\mathbb{N} \setminus \{0, \dots, t-1\}) \to \mathbb{R})$ is the chain of rewards an agent will observe when starting in state s_t (cf. Subsection 3.1.3), R_t ($R_t \in \mathbb{R}$) is a short cut for $R((r_{t+k})_{k\in\mathbb{N}})$ ($R\colon (\mathbb{N} \to \mathbb{R}) \to \mathbb{R}$), E ($E \in \mathbb{R}$) is the expectation, and E_π ($E_\pi \in \mathbb{R}$) denotes the expectation with respect to an agent following policy π. $R((r_{t+k})_{k\in\mathbb{N}})$ will be given by one of the return models from the previous section and hence:

$$R((r_{t+k})_{k\in\mathbb{N}}) \quad = \quad R(r_t, r_{t+1}, r_{t+2}, \dots) \quad = \quad R(0, r_{t+1}, r_{t+2}, \dots)$$

$$= \quad \sum_{k=0}^{\infty} \gamma^k r_{t+k+1} \ \text{ or } \ \sum_{k=1}^{T'} r_{t+k} \ \text{ or } \ \lim_{h\to\infty} \frac{1}{h} \sum_{k=1}^{h} r_{t+k}$$

where $r_{t+i} = ((r_{t+k})_{k\in\mathbb{N}^+})(i)$ ($i \in \mathbb{N}^+$) (cf. Subsection 3.1.3) is the reward received for state transition $s_{t+i-1} \to s_{t+i}$, γ is the discount factor, and T' ($T' \in \mathbb{N}^+$) is the finite horizon (cf. Subsection 3.1.3).

If one assumes that an agent's preferences are based on state utilities which in turn are based on state sequences and that these sequences are stationary, i.e.

[...] if you prefer one future to another starting tomorrow, then you should still prefer that future if it were to start today [...]

[RN03, p. 617], only two ways to assign utilities to state sequences are possible: Additive reward accumulation functions and discounted reward accumulation functions corresponding to the finite-horizon undiscounted and infinite-horizon discounted return model from Section 3.1, respectively. Following the principle of maximum expected utility (MEU) a rational agent should carry out an action that maximizes its expected utility of being in the resulting next state. The justification for the MEU according to [RN03, p. 585], is:

If an agent maximizes a utility function that correctly reflects the performance measure by which its behavior is being judged, then it will achieve the highest possible performance score if we average over the environments in which the agent could be placed.

If an agent has access to transition and reward signals and is given a utility function U ($U\colon S \to \mathbb{R}$), the expected utility $E(Q)$ ($Q\colon S \times \mathcal{A} \to \mathbb{R}$) of any action a applicable in a state s can be computed:

$$E(Q(s, a)) = \sum_{s'\in S} T(s, a, s')\left(r'(s, a, s') + \gamma U(s')\right)$$

where T ($T\colon S \times \mathcal{A} \times S \to [0, 1]$) and r ($r'\colon S \times \mathcal{A} \times S \to \mathbb{R}$) are the transition and reward signal from the reinforcement learning problem definition (see Subsection 3.1.2), respectively. The equation for an undiscounted return model essentially is the same, only factor γ ($\gamma \in [0, 1]$) is omitted.

Different types of policies can be implemented using utility functions. Utility functions can be used to implement stochastic policies that approximately fit to the MEU principle induced by a given utility function. A stochastic policy can be represented by a *preference structure* that assigns each action per state a preference value. The expected utility of the outcome of an action a applied in state s, $E\left(Q(s,a)\right)$, acts as the preference value. Preference values can be used to compute the probability of choosing an action in a state by using some kind of probability distribution, e.g. a Boltzmann distribution (also called Gibbs distribution or *softmax policy*, if $\tau = 1$) [Luc59, Bri90]:

$$\pi_t(s,a) = P(a_t = a \mid s_t = s) = \frac{e^{\frac{p_t(s,a)}{\tau}}}{\sum\limits_{a' \in A(s)} e^{\frac{p_t(s,a')}{\tau}}}$$

where a_t ($a_t \in A(s_t) \subseteq \mathcal{A}$) is the next action to take at time point t in current state s_t, $p_t(s,a)$ ($p_t \colon S \times \mathcal{A} \to \mathbb{R}$) represents the preference of choosing action a in state s at time t and τ ($\tau \in \mathbb{R}^+$), is called *temperature*. Subscript t of p_t indicates that preferences can change over time. The higher temperature τ is, the more closely each action's probability values lie together, even if the preferences are quite different. The more τ approaches zero, the higher the probability for the actions with higher preferences. In the limit $\tau \to 0$, the policy behaves as a *greedy policy* selecting the maximally preferred action with probability 1 or, in the case of ties, each of n ($n \in \mathbb{N}^+$) equally maximally preferred actions with probability $\frac{1}{n}$. The greedy policy represents the MEU described before.

Using utility function values directly as preferences, one can implement policies by means of a look-ahead search also. For example, a greedy policy is implemented by searching for the successor state with the highest utility by first making a virtual transition to all successor states and next computing their utilities. If ties are broken systematically, a greedy policy easily can be made deterministic. Sometimes, however, it is more useful to use a policy that is only approximately greedy. As was already mentioned, many reinforcement learning methods learn during interaction, in particular they learn or rather improve utility functions and thereby improve the policy. If the agent does only follow one fixed policy based on a given utility function, both utility function and policy will come to a fix point where they do not change anymore, since the policy perfectly fits the utility function and vice versa. This fix point needs not necessarily be a good policy, it can be rather suboptimal. The agent then failed to explore the search space to find opportunities to improve both utility function and deduced policy, rather sticking with one policy only. For this reason, small nondeterministic variants of policies are used. One example is the softmax policy from before, another exploring policy is called ϵ-*greedy policy*. It selects any of n greedy actions with probability $\frac{1-\epsilon}{n}$ each and any of m ($m \in \mathbb{N}^+$), non-greedy actions with probability $\frac{\epsilon}{m}$ each for $\epsilon \in (0,1]$.

3.2.2   Large Search Spaces

Another important issue in policy learning and learning a utility function is the size of the search space tackled. Huge or even infinite state spaces have to be taken into account, also. When solving COPs as is the aim here, the size of the search space is finite, but nevertheless exponential in size with respect to instance size. The grid maze example of Subsection 3.1.3 has a finite and relatively small state space. The states of such a state space can be stored as a table. Any policy or utility function then can be implemented as a lookup table. If the state space, gets too large after all, condensed representations of states and state spaces have to be chosen. Such a condensed representation typically is a vector $\vec{s}$ ($\vec{s} \in \mathbb{R}^n$) of n so-called *features* that are extracted from a state s. The extraction process can be regarded as a function $feat$ ($feat \colon S \to \mathbb{R}^n$) with $\vec{s} := feat(s)$. The derived condensed space of all features is called *feature search space* (or *feature space*) and denoted by $\vec{S}$, $\vec{S} := feat(S) := \bigcup_{s \in S}\{feat(s)\} = \bigcup_{s \in S}\{\vec{s}\}$. If no confusion is possible, the feature vector $\vec{s}$ itself also is referred to as *features* or *state features*. Individual feature will be denoted

by *individual feature* when an explicit distinction is necessary. The value will be called *individual feature values* or *feature values* for short. The feature values typically are from $\mathbb{R}^n$ since there are function approximators available that can be used to learn and represent a function $fa\colon \mathbb{R}^n \to \mathbb{R}$, for example and with a utility function $U_{fa}\colon S \to \mathbb{R}$, accordingly as a composition of first extracting features from a state and next applying the function approximator to the extracted feature vector:
$U_{fa} := fa \circ feat$

A feature is a characteristic, a distinguishing trait, quality, or property of a state that is encoded as a real value. As such it can be viewed as an attribute. Each state then has a specific feature vector. For example, the cost of a search state in any COP can be regarded as a feature. In the case of the Traveling Salesman Problem (TSP), the average edge length of a tour can be a feature. All features of a state feature vector together somehow have to extract the important characteristics of states of a problem while neglecting the unimportant. All information cannot possibly be used because of the size of the search space, but superfluous and non-relevant parts can be neglected. Features can be used to represent utility or direct policy functions. Instead of the whole search space, the condensed feature space acts as domain for utility or policy functions. Features need some predictive power so correct or at least useful utility or policy functions can be found. Certainly, the feature "color" of a car is not qualified to predict its maximum speed, even if most Ferrari cars are red. Abstractly, features must be representative not only of a search space as a whole but also of the regularities of the search space that are to be exploited for learning. They must capture the representative and predictive traits of a problem instance and perhaps of all, or subsets of, problem instances of a problem type. Of course, the regularities are not known in advance, so a certain trial and error procedure will be involved in feature construction and selection. Basically, all reinforcement learning methods presented later can be extended to work with features, but they will only work, if the features selected have some predictive power.

The rational to introduce features was to condense the search space representation because the original search space is too large. The condensed search space, R^n, however, is also infinite. The decisive difference is that its instance size n remains fixed, regardless of the instance size of the underlying problem instance and that n therefore can be kept relatively small. This enables to represent utility and policy functions by function approximation. All methods presented in this section were originally intended for lookup table representations of the value functions, but they can be transferred straightforwardly to using function approximations over a condensed search space using features. The special type of function approximator is irrelevant in principle, it potentially only affects performance (cf. subsections 3.3.4 and 3.3.3). Adjusting the methods for solving reinforcement learning problems works as follows. Instead of providing a state s for the lookup table, its representative in the form of its features $\vec{s}$ is used. The utility or policy function representations simply have to be adjusted to work as function approximators over the feature space. Any learning method described later will be presented using update rules. The righthand side of these rules simply is the new function value for the given feature vector. Feature vector and the new value then are a typical input-output pair forming a training example for a function approximator yielding a typical supervised learning scenario. These inputs typically come in an incremental fashion and updates of the utility or policy functions are supposed to happen immediately, so learning has to be done online and preferably in an incremental fashion. If a function approximator cannot work online in an incremental fashion, it can somehow be simulated by collecting input pairs into sets also called *batches* and processing batches of samples instead of each sample individually. If a task is episodic, such batches can be formed quite naturally by collecting the training examples until the end of an episode.

The advantage of using feature-based utility functions is that they can generalize: since features are not a one-to-one mapping from the actual state space, any change of a utility or policy function with respect to a certain feature vector will affect all states with similar feature vectors, too. That is, learning not only takes place for states visited by an agent during search but also for states not visited during the search. This can speed up learning tremendously. On the other hand, using features always involves some kind of approximation, originating in the condensing nature of the feature representation. Unintentional generalizations can occur with the result that a function approximator fails to find an appropriate or even stable functional relationship for the utility or policy function. This in particular can happen, if the features selected are not appropriate with respect to what is supposed to be learned. Successful generalization also crucially depends on the power of the function approximator used. If its hypothesis space does not even faintly contain a hypothesis that might fit the true utility or policy function to represent, it will fail and accordingly the whole learning effort.

In general, learned utility or policy functions cannot be correct for the whole feature space and hence not for the whole search space. Most supervised learning function approximators try to minimize the *mean-squared error* (MSE):

$$MSE(\vec{\theta}_t) := \sum_{s \in S} P(s)(U^{\pi}(\vec{s}) - U(\vec{\theta}_t)(\vec{s}))^2$$

where U^{π} is supposed to be the true utility function for policy π and $U(\vec{\theta}_t)$ is the approximated utility function at time t dependent on the model parameter vector $\vec{\theta}_t$ ($\vec{\theta}_t \in \mathbb{R}^n$). A probability distribution P ($P : S \rightarrow [0, 1]$) over S weights the errors. Since the error cannot be reduced to zero for all states, P becomes important. Reducing the error for some state s, and induced feature vector $\vec{s}$, by increasing its weight $P(s)$ can only be done when it is decreased for other states which then are likely to have a larger error. This makes sense, if these other states are less important. Distribution P in fact has a big impact on how generalization takes place, e.g. in which region of the state space. For most methods described in the previous subsection, P is the distribution with which states are visited and hence training examples are collected. The weights then basically are the frequencies with which states were encountered. Other distributions are conceivable but harder to implement. Here again, the fundamental trade-off between exploitation and exploration emerges because only the parts of the search space that were visited can influence the error measure.

Theoretical research in reinforcement learning addresses the issue of convergence of utility and policy functions either in the form of lookup tables or in the form of function approximation over a condensed state space. The relevant results for lookup tables and finite state spaces will be covered in the next two subsections. Unfortunately, when generalization by means of features and function approximation comes into play, most convergence results no longer hold. In practice, however, convergence frequently occurs or the learned utility or policy functions work reasonably well. Further issues in generalization using function approximators are discussed in Subsection 3.3.1.

3.2.3 Value Functions and Optimal Policies

In the context of greedy or ϵ-greedy policies, any special meaning of the numerical value provided by utility functions is irrelevant; they only need to properly reflect rankings among states. Even in the general case of policies implemented by means of a preference structure, the exact values might not be so important, e.g. because the used probability distribution smoothes out differences with respect to the final probabilities. If the numerical values provided by a utility function do not have a special meaning beyond rankings among states or if only the ranking ability is required, utility functions are also called *value functions*. From now on, only value functions will be considered, since for this discourse mainly rankings between utilities are of importance. Value functions are denoted with a V instead of a U.

Value functions can come in different flavors. Which type of value function is useful depends on whether transition models and reward signals are available to an agent or not. For example, if no transition models are available, which can be assumed to be the case for LSAs solving COPs with its actions based on randomized local search steps and procedures, value functions cannot directly be used to implement a policy. A look-ahead search in this case is not possible, because an agent has no means to compute the expected values of successor states; it even does not know which action might lead to which successor states. One popular remedy to this problem is to use value functions for action-state pairs instead of only for single states. The latter kind of value function is called *action-value function*, while value functions for states only are called *state-value functions*. The meaning of an action-value function Q^π ($Q^\pi : S \times \mathcal{A} \to \mathbb{R}$) with respect to a policy π is defined as the expected return when taking action a in state s:

$$Q^\pi(s, a) := E_\pi(R_t \,|\, s_t = s, a_t = a)$$

If s is a terminal state, its value is zero for all a, of course. State-value and action-value functions for a policy π are closely connected. Bellman in the 50's [Bel57] first described this connection via the following equations which accordingly are called *Bellman equations* (the original equations only considered state-value functions, though):

$$
\begin{aligned}
Q^\pi(s, a) &= E_\pi(R_t \,|\, s_t = s, a_t = a) \\[2mm]
&= E_\pi(r_{t+1} + \gamma \sum_{k=0}^{\infty} \gamma^k \, r_{t+k+2} \,|\, s_t = s, a_t = a) \\[2mm]
&= \sum_{s' \in S} T(s, a, s') \, (r'(s, a, s') + \gamma \, E_\pi(\sum_{k=0}^{\infty} \gamma^k \, r_{t+k+2} \,|\, s_{t+1} = s')) \\[2mm]
&= \sum_{s' \in S} T(s, a, s') \, (r'(s, a, s') + \gamma \, V^\pi(s')) \\[2mm]
&= E_\pi(r_{t+1} + \gamma \, V^\pi(s_{t+1}) \,|\, s_t = s, a_t = a)
\end{aligned}
$$

and

$$
\begin{aligned}
V^\pi(s) &= E_\pi(R_t \,|\, s_t = s) \\[2mm]
&= \sum_{a \in A(s)} \pi(s, a) \, E_\pi(R_t \,|\, s_t = s, a_t = a) \\[2mm]
&= \sum_{a \in A(s)} \pi(s, a) \, Q^\pi(s, a) \\[2mm]
&= \sum_{a \in A(s)} \pi(s, a) \, (\sum_{s' \in S} T(s, a, s') \, (r'(s, a, s') + \gamma \, V^\pi(s'))) \\[2mm]
&= E_\pi(r_{t+1} + \gamma \sum_{k=0}^{\infty} \gamma^k \, r_{t+k+2} \,|\, s_t = s)
\end{aligned}
$$

where again r_{t+1} is the reward received at time point t after applying action a_t to state s_t yielding the transition to the new state s_{t+1}. Also, R_t is a short cut for $R((r_{t+k})_{k \in \mathbb{N}})$. T and r' are the transition and reward signal from the reinforcement learning problem definition (cf. Subsection 3.1.2), respectively, and V^π and Q^π are the state- and action-value functions for policy π, respectively. The equations for an undiscounted return model essentially are the same, only factor γ is omitted (or is set to 1). Since this holds true for all following equations for value functions, only the equations for the discounted return model will be presented. The advantages of using action-value functions are obvious. An agent does not need a model for action selection. In the case of an (ϵ-) greedy policy a one-step or rather one-move look-ahead search simply maximizes over the applicable actions in a state. Note that steps of an agent are called moves (cf. Section

2.8). Nevertheless, in combination with the notion look-ahead search, n-step is common parlance as will be throughout this document.

Whether a policy is learned and applied directly or whether it is derived via utility or value functions does not make a difference in principle. What matters is that a behavior of an agent is implemented. All approaches have in common that they eventually implement behavior by mapping states or state representations to actions by some kind of internal function representation. This function in the end represents the behavioral strategy that governs an agent and is hence called *strategy function*. In this discourse it denotes any mapping an agent employs to guide its search.

Policies can be evaluated with respect to what they are supposed to achieve: maximizing the expected long-term return of an agent. Accordingly, value functions can be used to order policies. Value functions define a partial ordering $\geq (\setminus, \geq \subseteq (S \times \mathcal{A} \times S \rightarrow [0,1]) \times (S \times \mathcal{A} \times S \rightarrow [0,1]))$ over policies by the following equivalent definitions ($\pi, \pi' \in S \times \mathcal{A} \rightarrow [0,1]$ are policies):

$$\pi \geq \pi' \quad :\Leftrightarrow \quad \forall s \in S . V^{\pi}(s) \geq V^{\pi'}(s)$$

$$\Leftrightarrow \quad \forall s \in S . Q^{\pi}(s, \pi(s)) \geq V^{\pi'}(s)$$

with (since π, π' are stochastic):

$$Q^{\pi}(s, \pi(s)) = \sum_{a \in S(s)} \pi(s, a) \, Q^{\pi}(s, a)$$

If the state space is finite, there will be at least one optimal policy with respect to the partial order $\geq$ thus defined. Any such optimal policy will be denoted with π^* ($\pi^* \colon S \times \mathcal{A} \rightarrow [0,1]$). All optimal policies have the same associated state-value function called *optimal state-value function* which is denoted by V^* ($V^* \colon S \rightarrow \mathbb{R}$). For a state s it can be defined as:

$$V^*(s) := \max_{\pi} V^{\pi}(s)$$

Any optimal policies also share the same *optimal action-value function* which is denoted by Q^* ($Q^* \colon S \times \mathcal{A} \rightarrow \mathbb{R}$) and which for a state s and an action a accordingly is:

$$Q^*(s, a) := \max_{\pi} Q^{\pi}(s, a)$$

Informally, for a state-action pair (s, a) ($(s, a) \in S \times \mathcal{A}$), $Q^*(s, a)$ represents the expected return if taking action a in state s (a is not necessarily the greedy action for this state) and subsequently following the optimal policy. According to the relationship between state-value and action-value functions stated before, Q^* can be written in terms of V^* as follows yielding the *Bellman optimality equations* (again, the original equations were only stated for V^*):

$$
\begin{aligned}
Q^*(s, a) &= E_*(R_t \,|\, s_t = s, a_t = a) \\[2mm]
&= E_*(r_{t+1} + \gamma \sum_{k=0}^{\infty} \gamma^t \, r_{t+k+2} \,|\, s_t = s, a_t = a) \\[2mm]
&= E_*(r_{t+1} + \gamma V^*(s_{t+1}) \,|\, s_t = s, a_t = a) \\[2mm]
&= E_*(r_{t+1} + \gamma \max_{a' \in A(s_{t+1})} Q^*(s_{t+1}, a') \,|\, s_t = s, a_t = a) \\[2mm]
&= \sum_{s' \in S} T(s, a, s') \, (r'(s, a, s') + \gamma \max_{a' \in A(s')} Q^*(s', a')) \\[2mm]
&= \sum_{s' \in S} T(s, a, s') \, (r'(s, a, s') + \gamma V^*(s'))
\end{aligned}
$$

The other direction looks as follows:

$$V^*(s) = \max_{a \in A(s)} Q^*(s, a)$$

$$= \max_{a \in A(s)} E_*(r_{t+1} + \gamma V^*(s_{t+1}) \mid s_t = s, a_t = a)$$

$$= \max_{a \in A(s)} \sum_{s' \in S} T(s, a, s')(r'(s, a, s') + \gamma V^*(s'))$$

For finite state spaces, the Bellman equations and the Bellman optimality equations have unique solutions [Bel57, How60, Wat89].

The partial order over policies can be used to compare and improve policies to eventually wind up with an optimal or at least good enough policy. Now, when the value function for a policy π has been found it can be used to improve this policy to a new and, with respect to the partial order among policies $\geq$, better policy π':

$$\pi'(s) := \arg\max_{a \in A(s)} Q^\pi(s, a)$$

$$= \arg\max_{a \in A(s)} \sum_{s' \in S} T(s, a, s')(r'(s, a, s') + \gamma V^\pi(s'))$$

for all $s \in S$. This works due to a theorem called *policy improvement theorem* [Bel57, How60, Wat89] which states that for each pair π and π' of policies it holds:

$$\forall s \in S . Q^\pi(s, \pi'(s)) \geq V^\pi(s) \quad \Rightarrow \quad \pi' \geq \pi$$

$$\Rightarrow \quad \forall s \in S . V^{\pi'}(s) \geq V^\pi(s)$$

Improving a policy using the policy improvement theorem is called *policy improvement*. It is also referred to as learning *control* in the form or action selection. In order to be able to perform policy improvement, a policy π has to be evaluated first, i.e. its corresponding value functions Q^π or V^π have to be computed or rather learned first. This process is called *policy evaluation* or *prediction problem*. Together, policy evaluation and improvement yield a straightforward approach to find the optimal policy by simply alternating these two steps (an arc labeled with ev denotes a policy evaluation step, an arc labeled with im denotes a policy improvement step):

$$\pi_0 \xrightarrow{\text{ev}} V^{\pi_0} \xrightarrow{\text{im}} \pi_1 \xrightarrow{\text{ev}} V^{\pi_1} \xrightarrow{\text{im}} \pi_2 \xrightarrow{\text{ev}} \dots \xrightarrow{\text{im}} \pi^* \xrightarrow{\text{ev}} V^*$$

This process is called *policy iteration*. It will converge to the optimal policy, as long as each policy evaluation step is guaranteed to converge to the true value function or at least produce a better approximation thereof, and if the policy improvement strictly improves the old policy based on the evaluated value function. It works because policy iteration in fact constructs a sequence of monotonically improving policies which will reach the optimal policy in the limit [Bel57, How60, Wat89]. Policy improvement includes a look-ahead search. Again, this can be based on a model or it can be model-free. As will be seen next, not only policy improvement can be model-based or model-free, policy evaluation independently can be model-based or model-free, also. Policy iteration approaches are called model-free only, if both the policy evaluation part *and* the policy improvement part of an iteration are model-free.

3.2.4 Learning Value Functions

Policies are often implemented via value functions and evaluating value functions is one crucial part of policy iteration which basically governs any method solving reinforcement learning problems.

The question how to learn value functions naturally arises? In particular, learning optimal value functions V^* and Q^* is required. There are several approaches which will be presented next. They can be divided in model-based and model-free methods. Model-based methods are covered first. The methods presented next are described for the case that value functions are represented as lookup table. The most important methods nevertheless can straightforwardly be enhanced by using a condensed feature state space and function approximators (cf. Subsection 3.2.2). Convergence guarantees, might thereby be lost, though.

The dependence of the state utilities on each other according to the Bellman equations can be exploited in order to learn value functions. The Bellman equations, however, are based on the environment models, so an agent has to be provided with the models somehow. But given the models, methods called *dynamic programming* (DP) [Bel57] can be used. DP denotes a collection of algorithms that use transition and reward signals in order to learn value functions which are represented as lookup tables. They use the important property of value functions that these can be rewritten recursively in the form of the Bellman equations (cf. Subsection 3.2.3). If functions π, T and r' are known and the state space is finite, Bellman equations can be instantiated for each state of the finite search space yielding a number of equations that can be used to solve for V^π yielding a unique solution [Bel57]. Hence, by solving the Bellman equations, one can obtain the value function for a certain policy or for the optimal policy directly when using the Bellman optimality equations.

A first glance disadvantage of this approach is that for large state spaces such as for COPs solving the resulting equations is not practical. If the state space is even infinite, solving the equations for V^π and Q^π, or V^* and Q^*, exactly is not feasible at all. Instead, the recursive rewrite rules of the Bellman equation can be used as update rules in an iterative procedure. In each iteration the value of each state is updated by a fresh, more exact value as computed by the update rules. Since the equations are guaranteed to have a unique solution, a fix point to this iterative process will exist and the unique solution will be the fix point of the iteration [Bel57]. A process like this is also called *fix point iteration*. To be more precise, for evaluating a policy π, the Bellman equations yield the following update rule for state-value functions where $V_t^\pi(s)$ denotes the value function value at time point t for state s:

$$V_{t+1}^\pi(s) \leftarrow \sum_{a \in A(s)} \pi(s,a) \left(\sum_{s' \in S} T(s,a,s') \left(r'(s,a,s') + \gamma V_t^\pi(s') \right) \right)$$

For each state in the state space, in each iteration the respective function value will be updated yielding a sequence of value functions $V_0^\pi, V_1^\pi, V_3^\pi, \ldots$. The name of this policy evaluation method is *iterative policy evaluation*. Under the condition that $\gamma < 1$ or that eventual termination under policy π is guaranteed from all states, a unique fix point will exist and it will hold for all $s \in S$ [Bel57, How60]:

$$\lim_{t \to \infty} V_t^\pi(s) = V^\pi(s)$$

Of course, for large state spaces this approach is infeasible since complete sweeps over the search space are necessary. When solving COPs this is always the case, since their search space is exponential in size dependent on the instance size. The sweeps can, nevertheless, be abbreviated and accelerated in several ways without loosing the convergence guarantees. One kind of speedup can be achieved by asynchronously updating the function values in two ways. Such methods are called *asynchronous DP* [Ber82, Ber83]. First, updates of a function value for a state immediately replace the old value and can be used for the next updates of the current sweep instead of accumulating updates until the end of a sweep and applying them between two sweeps. Fresh, more precise values will be used sooner and hence will accelerate convergence. Second, the order of state updates in a sweep does not influence convergence in principle but can increase the rate of convergence as long as each state repeatedly and infinitely often is updated in the limit [Ber82, Ber83]. *Prioritized sweeping* [MA93, PW93] is one method that does intelligent asynchronous sweeps thereby accelerating convergence substantially.

Value function evaluation is needed for the policy evaluation part of policy iteration (cf. Subsection 3.2.3). Policy iteration can be carried out on an episode by episode basis or it can be finer grained which is useful for continuous tasks. The alternating parts of an iteration, policy evaluation and improvement, can be abbreviated to the point where only updates for one state or state-action pair are done at each iteration and for each part. The general approach of alternating policy evaluation and policy improvement at any level of granularity is called *Generalized Policy Iteration* (GPI) in [SB98, p. 90]. The authors state:

> Almost all reinforcement learning methods are well described as GPI. That is, all have identifiable policies and value functions, with the policy being improved with respect to the value function and the value function always being driven towards the value function for the policy.

Convergence still is guaranteed at even the finest GPI granularity – i.e. if only one policy evaluation and improvement update per iteration is done – under conditions stated before: Each state is updated infinitely often and $\gamma < 1$ or eventual termination is guaranteed [Ber87, Ber95, BT96, SB98].

One special case of an abbreviated policy iteration is to do only one update for each state in the policy evaluation step in each iteration. Convergence guarantees will still hold under the conditions just mentioned [Ber87]. This kind of abbreviated fix point iteration for the policy evaluation step can be combined directly with the maximizing operation of the policy improvement step yielding update rules induced by the Bellman optimality equations. In this case, the alternating steps of policy improvement can be condensed almost completely. This procedure is called *value iteration*. Its update rule is:

$$V_{t+1}(s) \leftarrow \max_{a \in A(s)} \sum_{s' \in S} T(s, a, s') \left(r'(s, a, s') + \gamma V_t(s') \right)$$

Under the same conditions stated before it holds for all $s \in S$ [Ber87]:

$$\lim_{t \to \infty} V_t(s) = V^*(s)$$

One important issue with respect to policy evaluation and policy iteration is when to stop it. Clearly, the real value function and the truly optimal policy will only be found in the limit. But in practice this seems not to be needed; sufficiently close suboptimal policies and value functions will do as well. Since it is difficult to estimate how far a policy is from the optimal policy, its value function is consulted instead. The typical approach is to compare successive value function snapshots in the sequence of value functions and compute a delta, potentially for all states. If these deltas or rather some kind of statistic over them do not exceed a certain given bound ϵ ($\epsilon \in \mathbb{R}^+$) the iteration can be stopped, since further changes will even be smaller. For example, the maximum operator can be used yielding the following stopping criterion:

$$\max_{s \in S} |V_{t+1}(s) - V_t(s)| < \epsilon$$

3.2.5 Estimation Methods

As was pointed out, the dynamic programming methods crucially depend on models for transitions and reward signals. If no such models can be given a priori to an agent, they can be learned during the search. Dynamic programming methods following these lines are called *Adaptive Dynamic Programming* methods. Again, policy improvement can be done via any GPI instantiation. Typically, when learning models as well, the whole state space has to be gone through more often than when given models a priori. This often takes too much time in practice and the speed of convergence is slow. In general, providing models is a very strong prerequisite usually not met in practice or not feasible, for example in the case of an LSA solving a COPs based on randomized local search steps and procedures. Fortunately, there are other methods that do not need to learn models explicitly. They also use the Bellman equations. They can be viewed as methods that solve the Bellman equations approximately by using actual experience obtained during the search of an agent. These experienced actual transition feedbacks replace the knowledge about the expected transition feedbacks which only can be computed, if the transition model and reward signal are known. The methods presented next so to say use samples to estimate the expected values that show up in the Bellman equations and accordingly are named *estimation methods*. Their convergence and correctness not surprisingly depends on obtaining representative samples, e.g. that they cover the relevant parts of the search space and these thoroughly enough.

In examination of the need for estimation methods to cover the proper parts of a search space, the challenge of exploration vs. exploitation truly and clearly becomes apparent. The typical remedy to this problem is to allow exploration by not always greedily following the policy that is just being evaluated, but sometimes use suboptimal, exploring actions. This can, for example, be implemented by an ϵ-greedy policy. In this case, the value function will in fact be learned with respect to the ϵ-greedy policy used, which, differs only slightly from the greedy policy, though. Methods that do estimate the policy they are following during learning are called *on-policy* methods. Other approaches use two different policies. One (explorative) policy is used to take the samples and guide the search during a learning phase, whereas a second policy actually is learned and optimized. The policy guiding the learning phase is called *behavior policy*, the policy that gets optimized is called the *estimation policy*. Methods that employ two policies and learn a policy different from the one producing behavior during the learning phase are called *off-policy* methods [SB98, p. 122 and p. 126]. Often, however, the distinctions are blurred in that after each update the behavior policy is based on the estimation policy, thus no separation into learning and execution phase is needed anymore. Those methods presented next which are off-policy are mentioned explicitly when they are introduced, all other methods are on-policy.

The most straightforward method to learn value functions is by simple estimation from examples. Such methods are called *Monte-Carlo* (MC) estimation methods. Sometimes they are also called *direct utility estimation*. These methods do only work in episodic tasks. They record the rewards observed by an agent during an episode. At the end of the episode they compute a sample return for possibly any state or state-action pair encountered during the episode from the collected rewards using the return model given. This reward is an estimation of the true return model's value for the states and state-action pairs it was computed for, of course, with respect to the policy followed. This is because any search of an agent reaching a terminal state following a fixed policy during the episode can be seen as a sample of possible behavior of the agent. By the law of large numbers, the averaged estimation will converge to the true value, i.e. will correctly approximate the real state- or action-value function in the limit. In effect, in each update each new sample will shift the average value a little bit into the direction of the new sample which is also called *target* of the update. The actual value of a target of an update operation is also called *update* for short and typically used a training example for function approximator update. Therefore, updates are also called update or training examples throughout this document. Let $V_t^\pi(s)$ be the value function at time t for state s with respect to policy π, and let $k_s^{(t)}$ ($k_s^{(t)} \in \mathbb{N}^+$) be the number of samples already obtained for state s until time t. Then, after the update the new value function at time $t+1$ for state s can be written as:

$$V_{t+1}^\pi(s_t) \leftarrow V_t^\pi(s_t) + \alpha_{s_t}^{(t)} \left(R_t - V_t^\pi(s_t) \right)$$

with $\alpha_{s_t}^{(t)} := \frac{1}{k_{s_t}^{(t)}}$ $(\alpha_{s_t}^{(t)} \in \mathbb{R})$. Factor α is the *learning rate* use to smooth and control learning.

The update rule can be modified. For example, α $(\alpha \in \mathbb{R})$ or rather the $\alpha_{s_t}^{(t)}$ can be decreased over time. The averaging method will converge for arbitrary α if α is properly decreased over time [BT96, SB98] or if $\alpha < 1$ is sufficiently small. The update target R_t can be varied, too, as will yield to other methods presented in the next subsections.

Although MC estimation is appealingly simple, it has severe disadvantages. Waiting until the end of an episode before an agent can update value functions might take too long or is impractical or even infeasible altogether, e.g. for continuous tasks as is the typical case for an LSA which is based on trajectory method local search methods such as the Iterated Local Search (ILS) metaheuristics. Also, convergence might be rather slow, since many episodes have to come to an end. The disadvantage of this approach is that it does not take into account that utilities are not independent from each other. Among other things, this is reflected by the fact that all states or state-action pairs along the search trajectory of the episode are eligible to have return estimates to be computed for them. Consequently, improvements of the pure MC sampling methods have been proposed. These methods are called *temporal difference methods* because they work based on so-called *temporal credit assignments*.

3.2.6 Temporal Difference Learning

Instead of waiting for producing a suitable update target until the end of an episode, temporal difference methods use an approximation of R_t. In fact, $r_{t+1} + \gamma V_t^\pi(s_{t+1})$ is a sample for R_t and hence for $V_t^\pi(s_t)$ which supposedly is more correct than the old approximation $V_t^\pi(s_t)$ because it incorporates a really experienced value r_{t+1}. This variation, together with the Bellman equations, leads to the following new update rule:

$$V_{t+1}^\pi(s_t) \leftarrow V_t^\pi(s_t) + \alpha \left(r_{t+1} + \gamma V_t^\pi(s_{t+1}) - V_t^\pi(s_t) \right)$$

The requirements for α with respect to convergence stay the same as for the simple MC method presented in the previous subsection. The method just described is denoted TD(0) [Sut88]. It does not need a model for learning as can be seen immediately, but it still employs a state-value function and therefore needs a model for policy implementation. If policy π is fixed and the step-size parameter α is sufficiently small, TD(0) will converge to V^π in the mean. If the step-size parameter α is decreased over time in a sequence $(a_k)_{k \in \mathbb{N}}$ $((a_k)_{k \in \mathbb{N}} \in \mathbb{N} \to \mathbb{R})$ and it holds

$$\sum_{k=1}^\infty a_k = \infty \text{ and } \sum_{k=1}^\infty \alpha_k^2 < \infty$$

TD(0) will converge to V^π with probability 1 [Sut88, Day92]. Using a temporal-difference method for value function evaluation enables an agent such as a learning LSA to obtain a training example for the function approximator that represents the value function in each move and to directly use it for the next one.

Transferred to using action-value functions, the TD(0) algorithm becomes a method called *Sarsa* [RN94, Sut96]. Since it learns an action-value function, it can be used to directly combine policy evaluation and policy improvement in each iteration. The action-value functions can be used directly to derive the behavior policy, [1] e.g. by means of an ϵ-greedy policy. In each iteration an

---

[1] The behavior policy is also the estimation policy, since Sarsa is on-policy.

improved policy is used, if in each iteration, the new action is (almost) greedily selected according to the improved old policy as represented by the improved action-value function. The update is:

$$Q_{t+1}^{\pi}(s_t, a_t) \leftarrow Q_t^{\pi}(s_t, a_t) + \alpha\left(r_{t+1} + \gamma\, Q_t^{\pi}(s_{t+1}, a_{t+1}) - Q_t^{\pi}(s_t, a_t)\right)$$

Sarsa can be transformed into an off-policy algorithm easily by using both a behavioral and estimation policies simultaneously (cf. Subsection 3.2.5), then being called *Q-learning* [Wat89, Wat92]. It does not need a model for neither learning nor control (i.e. action selection) and hence is completely model-free. The behavior again is generated via an ϵ-greedy policy-based on the action-values for each state. The update rule for the simplest form called *one-step Q-learning* is [SB98, p. 148]:

$$Q_{t+1}^{\pi}(s_t, a_t) \leftarrow Q_t^{\pi}(s_t, a_t) + \alpha\left(r_{t+1} + \gamma\, \max_{a \in A(s)} Q_t^{\pi}(s_{t+1}, a) - Q_t^{\pi}(s_t, a_t)\right)$$

To further refine the update rules for TD methods, other targets for update can be considered. Instead of sticking with just looking in the future one step or rather move, several moves could be carried out and the accordingly observed rewards could be collected. This will give an even more precise approximation for the return, since more actually experienced rewards enter the estimation. Let the target for such an n-step look-ahead be:[2]

$$R_t^{(n)} := r_{t+1} + \gamma\, r_{t+2} + \gamma^2\, r_{t+3} + \gamma^{n-1}\, r_{t+n} + \gamma^n\, V_t^{\pi}(s_{t+n})$$

In the limit $n \to \infty$, $R_t^{(n)}$ becomes R_t:

$$\lim_{n \to \infty} R_t^{(n)} = R_t$$

Yet another alteration of the target can be considered by not using a single n-step target but a weighted sum of several or even all targets:

$$R_t^{\lambda} := (1 - \lambda) \sum_{n=1}^{\infty} \lambda^{n-1} R_t^{(n)}$$

This last generalization is denoted TD(λ). By using so-called *eligibility traces*, the update rule for state-value function evaluation becomes:

$$V_{t+1}^{\pi}(s) \leftarrow V_t^{\pi}(s) + \alpha\left(r_{t+1} + \gamma\, V_t^{\pi}(s_{t+1}) - V_t^{\pi}(s_t)\right) e_t(s)$$

where for all $s \in S$, independent of s_t and s_{t+1}, eligibility $e_t(s)$ $(e_t \colon S \to \mathbb{R}_0^+)$ is defined as:

$$e_t(s) := \begin{cases} \gamma\, \lambda\, e_{t-1}(s) & \text{if } s \neq s_t; \\ \gamma\, \lambda\, e_{t-1}(s) + 1 & \text{if } s = s_t. \end{cases}$$

Each state s of the search trajectory gets assigned an eligibility value $e_t(s)$ that signals how much influence the reward obtained currently at time t should have when updating the state. Eligibility traces work by decaying in each move all former state eligibilities by the product of discount factor and weight λ. Only the eligibility of the current state s_t is increased to reflect its new weight. Eligibility traces work by distributing the necessary updates portion-wise backwards along the search trajectory. The update for each state of the search trajectory of target R^{λ} is not done immediately, which is not possible anyway, since most parts of R^{λ} can only be computed in the future. Instead, it is split into pieces, one piece for each future move that will influence the update target for the current state according to the weighted sum of n-step return targets and taking into

---

[2]The end of the line for $n \to \infty$ would be R_t.

account its proper weight. Eligibility traces do exactly this. For $\lambda = 0$, TD(λ) becomes a simple one-step target update method, for $\lambda = 1$, TD(λ) becomes an MC method.

Eligibility traces and the update rule just described can efficiently be implemented, since only eligibilities for the states of the search trajectory have to be considered (all others are zero) and accordingly only for these states the value function has to be updated. Additionally, since the weights are decaying geometrically, eligibility values for only a limited part of the search trajectory have to be kept in memory and consulted for update. Eligibility traces are rather implemented as a list of state or state-action pair eligibilities. Eligibility traces as just presented are called *accumulating* eligibility traces, since if a state is encountered twice in a short period of time, the old, not yet decayed eligibility for this state will be further increased, possibly beyond 1. *Replacing* eligibilities instead do not add a 1 to the eligibility of the current state, but simply set the eligibility for the current state to 1 exactly.

Q-learning following the TD(λ) approach in its simplest form is called Sarsa(λ) [RN94]. Its behavioral policy is implemented as for Sarsa and the updates work as follows:

$$Q_{t+1}^{\pi}(s,a) \leftarrow Q_t^{\pi}(s,a) + \alpha \left(r_{t+1} + \gamma \, Q_t^{\pi}(s_{t+1}, a_{t+1}) - Q_t^{\pi}(s_t, a_t) \right) e_t(s,a)$$

where for all $s \in S$ and $a \in \mathcal{A}$, independent of s_t, a_t, s_{t+1}, eligibility $e_t(s,a)$ ($e_t \colon S \times \mathcal{A} \to \in \mathbb{R}_0^+$) is defined as: a_{t+1} and

$$e_t(s,a) = \begin{cases} \gamma \lambda e_{t-1}(s,a) + 1 & \text{if } s = s_t, a = a_t; \\ \gamma \lambda e_{t-1}(s,a) & \text{otherwise.} \end{cases}$$

Adjusting this eligibility update rule for replacing eligibility traces is straightforward. Again, the update rule does not require to update all action-state pairs in practice, but only those whose eligibility is significantly greater than zero which typically is a relatively small number. Sarsa(λ) is an on-policy algorithm. One off-policy version of it is called Watkins's Q(λ) [Wat89, Wat92], [SB98, pp. 182ff]. It also tries to compute a target of a weighted sum of n-step returns as Q-learning does, also using eligibility traces. It also uses some kind of behavioral policy, e.g. an ϵ-greedy policy that guides the search, but tries to make use of the maximum operator max to directly learn the action-values for the optimal policy. However, care has to be taken for this approach. Not all n-step returns can be incorporated but only those that correspond to a state reached via a greedy action. As soon as a non-greedy action is executed, the eligibilities are obsolete. In general, if a_{t+n} is the first non-greedy action taken, the backup is done towards the following target:

$$r_{t+1} + \gamma \, r_{t+2} + \gamma^2 \, r_{t+2} + \ldots + \gamma^{n-1} \, r_{t+n} + \max_{a \in A(s_{t+n})} Q_t(s_{t+n}, a)$$

The eligibility traces work as for Sarsa(λ), only they are reset to zero as soon as the first non-greedy action has been taken. Let $\mathcal{I}_{xy} := 1 \iff x = y$, then they become:

$$e_t(s,a) := \mathcal{I}_{s\,s_t} \times \mathcal{I}_{a\,a_t} + \begin{cases} \gamma \lambda e_{t-1}(s,a) + 1 & \text{if } Q_{t-1}(s_t, a_t) = \max_{a' \in A(s_t)} Q_{t-1}(s_t, a'); \\ 0 & \text{otherwise.} \end{cases}$$

The update rule for Watkins's Q(λ) is:

$$Q_{t+1}^{\pi}(s,a) \leftarrow Q_t^{\pi}(s,a) + \alpha \left(r_{t+1} + \gamma \max_{a' \in A(s_t)} Q_t^{\pi}(s_{t+1}, a') - Q_t^{\pi}(s_t, a_t) \right) e_t(s,a)$$

for all states s and actions a. A complete sweep over state-action pairs is not necessary in practice for the same aforementioned reasons with respect to eligibility decay. Convergence rules for $0 < \lambda < 1$ have not yet been proven for Watkins's Q(λ).

Recall that one major disadvantage of using generalization and function approximation is that the convergence proof for the methods introduced in the previous section do not hold anymore (cf. Subsection 3.2.2). In the case of an infinite state space convergence guarantees cannot be given either. Sometimes, stable value functions cannot be learned, e.g. because successive updates periodically undo themselves or even mutually increase errors that build up. Looking for new convergence proofs for these cases is still an active area of research. Methods that do compute their target for update using an old estimate, i.e. DP and TD(λ) methods for $\lambda < 1$, are called *bootstrapping*. MC methods are not bootstrapping. Bootstrapping methods combined with function approximation tend not to minimize the MSE (cf. Subsection 3.2.2) but only find near-optimal MSE solutions. Furthermore, if they are off-policy methods, the distribution P with which the samples are drawn is not the same as the one used to train the function approximator. This can lead to divergence and even infinite MSE in some cases [Bai95]. To prevent this from happening, more robust function approximators, that do not extrapolate such as nearest neighborhood methods, and local weighted regression can be used. Alternatively, the learning goal can be changed. Instead of trying to minimize the MSE, the so-called mean-squared *Bellman error* (MSBE) can be tried to be minimized for policy π:

$$\text{MSBE}(\vec{\theta}_t) = \sum_{s \in S} P(s) \, (E_\pi(r_{t+1} + \gamma \, V^\pi(\vec{\theta})(\vec{s})(s_{t+1}) \mid s_t = s) - V^\pi(\vec{\theta})(\vec{s})(s))^2$$

Although bootstrapping methods can diverge, in practice they have been found to typically perform better than methods that do not bootstrap; convergence occurs often enough [SB98, p. 220].

3.2.7   Learning Policies Directly

All policy learning introduced so far was essentially based on learning value functions and thereafter inducing policies from them. This subsection briefly illustrates the approach of directly learning policies. Direct policy learning methods are also called *direct policy search* methods. There, the policy is not implemented via value functions that assign utilities to states or state-action pairs that in turn are used for look-ahead or for maximizing search. Instead, the agent becomes more of a reflex agent that adjusts its reflexes. The policy function π can either be represented directly or via a preference structure (cf. Subsection 3.2.1. Each approach uses function approximation over the search space, perhaps based on features, for representing the policy directly or for representing preference level functions for the preference structure. The function approximation can be in the form of a lookup table, but typically a parameterized function approximator or a direct functional form such as a polynomial of a fixed degree is used. If the functional form or the function approximator is differential in dependence on a parameter vector $\vec{\theta}$ ($\vec{\theta} \in \mathbb{R}^n$), gradient descent can be used. Policy search will adjust the parameter vector $\vec{\theta}$ until a policy is found that performs well.

Gradient descent search can be done straightforwardly for deterministic policies and environments. The policy is executed and the observed reward is accumulated yielding a new desired policy value $\rho(\vec{\theta})$ ($\rho(\vec{\theta}) \in \mathbb{R}$). Since the policy and the environment is deterministic, the policy value ρ in effect is only dependent on $\vec{\theta}$. Next, the gradient of the policy or preference structure is computed, perhaps empirically by local search of a neighborhood structure for $\vec{\theta}$ on $\mathbb{R}^+$, and the actual parameter for the policy or preference structure representation is adjusted in direction of the gradient. In general, if a neighborhood over policies can be defined, local search can be used to improve policies. When using a preference structure to represent a policy, such local search like policy improvement scheme is to apply systematic probing updates to the preference structure that can be undone. Such, candidate policy preferences can be obtained and can be tested. The best among the candidate policies then gets selected and the respective updates are made permanently, this way implementing an empirical policy improvement procedure.

In any case, multiple policies have to be executed. Unfortunately, executing a policy may be costly, since in the worst case it involves waiting until the end of an episode or, in the case of using the discounted return model from Subsection 3.1.3, as long as the weights are sufficiently small such that the resulting rewards can safely be omitted. Yet, directly learning policies essentially means that complete policies have to be tested, i.e. executed, in order to be able to evaluate and compare them. If the environment or the policy is truly stochastic, which unfortunately is the case when solving COPs with a learning LSA, then empirical gradient descent as just described is difficult to perform since it is difficult to get good estimates of neighboring states in the policy space and hence an unbiased estimate of the gradient. Direct policy search is quite a new approach where research still is in its infancy [Wil92, NJ00, SMSM00, SM01, Kak02, BS03, BKNS04].

3.2.8   Other Approaches and Variants

Sometimes, the policy is represented explicitly, separately, and independently from the value function in contrast to the look-ahead search methods that were described hitherto. These methods implement a policy indirectly by deriving it from the value function. The approaches that explicitly represent a policy in addition to value functions are called *actor-critic* methods. The policy representation is the *actor*, while a value function is serving as *critic* that assesses or rather criticizes the action selection of the actor. After each taken action, and reward and new state encountered, the critic establishes a new, potentially more exact estimation of the value of the previous state or state-action pair, as described before. Again, this new, potentially more correct estimate is used to compare it to the old estimation by computing a delta. This delta then is not only used to update the value function (as an approximation) but is also used as feedback input to the actor that accordingly can update its policy representation, too. This architecture is illustrated in Figure 3.3.

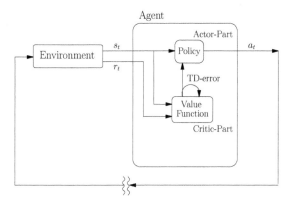

Figure 3.3:   Model of an actor-critic architecture

If a stochastic policy is used, it can be represented by a preference structure that assigns to each action per state a preference value. This preference value can be used to compute the probability for each action to be chosen in a state by using some kind of probability distribution, e.g. a Boltzmann distribution (cf. Subsection 3.2.1). The preference structure can be represented as a lookup table or using a function approximator. In either case, after each action selection a_t at time t in state s_t, the delta $\Delta_t V$ ($\Delta_t V \in \mathbb{R}$) from the critic can be used to compute a new target for the preference structure at position (s_t, a_t) $((s_t, a_t) \in S \times \mathcal{A})$ by:

$$p(s_t, a_t) \leftarrow p(s_t, a_t) + \beta \Delta_t V$$

where β $(\beta \in [0,1])$ is a step-size parameter comparable to α from the value function estimation used to smooth the approximation of the preference structure. Other update rules for the preference structure are conceivable, too [SB98, p. 152]. The advantages of explicitly representing the policy are that less computation is needed for finding the proper actions in each state – possibly expensive evaluations of one or more value functions can be omitted – and that stochastic policies can be represented and learned directly. Also, it may be easier to incorporate domain dependent constraints on action selection (which can, for example, be done implicitly by restricting applicable actions in dependence of a state s (cf. set $A(s)$ in the definition of a reinforcement learning problem in Subsection 3.1.2). Eligibility traces can be incorporated in the actor critic approach as well (cf. Subsection 3.2.6). The critic part uses any of the afore mentioned methods to evaluate value functions using eligibility traces, e.g. TD(λ). The actor part needs two eligibility traces, one for each state and one for each state-action pair ([SB98, p. 185].

A method for undiscounted continuous tasks different from all other hitherto presented methods has been proposed. This approach is called *R-learning* [Sch93, BT96]. It works by breaking down the long-term goal to a short-term goal by trying to maximize the obtained reward per time step averaged over *all* time steps (rather than just greedily the next step). Given a policy π, the value function is defined as:

$$\rho^\pi := \lim_{k \to \infty} \frac{1}{k} \sum_{t=1}^{k} E_\pi(r_t)$$

This directly reflects the average-reward optimality model from Subsection 3.1.3, since

$$E\left(R((r_t)_{t \in \mathbb{N}}) \right) = E\left(\lim_{h \to \infty} \frac{1}{h} \sum_{t=0}^{h} r_t \right) = \lim_{h \to \infty} E\left(\frac{1}{h} \sum_{t=0}^{h} r_t \right) = \lim_{h \to \infty} \frac{1}{h} \sum_{t=0}^{h} E(r_t)$$

according to the theorem of Lebesgue, since $\frac{1}{h} \sum_{t=0}^{h} r_t$ is integrable for all h $(h \in \mathbb{N}^+)$ and all $(r_t)_{t \in \mathbb{N}}$. As a prerequisite, the process must be ergodic, i.e. the probability of reaching any state from any given state under any policy must be non-zero. In other words, ρ^π must not depend on the start state. In the long run, the average-reward of course will be the same, but during the search, there will be temporary differences so-called *residuals* that can be exploited and that form the value of a state or a state-action pair:

$$\widetilde{V}^\pi(s) := \sum_{k=1}^{\infty} E_\pi(r_{t+k} - \rho^\pi \mid s_t = s)$$

$$\widetilde{Q}^\pi(s, a) := \sum_{k=1}^{\infty} E_\pi(r_{t+k} - \rho^\pi \mid s_t = s, a_t = a)$$

In each state s_t at time t an action a_t is chosen. The update rule for an action-value function representation then is:

$$\widetilde{Q}_{t+1}^\pi(s, a) \leftarrow \widetilde{Q}_t^\pi(s, a) + \alpha\left(r_{t+1} - \rho + \max_{a' \in A(s_{t+1})} \widetilde{Q}_t^\pi(s_{t+1}, a') - \widetilde{Q}_t^\pi(s_t, a_t) \right)$$

If action a_t was the greedy action, i.e. if $\widetilde{Q}_t^\pi(s, a) = \max_{a' \in A(s)} \widetilde{Q}_t^\pi(s, a')$, the average expected reward representation per time step is updated, too:

$$\rho_{t+1} \leftarrow \rho_t + \beta\left(r_{t+1} - \rho + \max_{a' \in A(s_{t+1})} \widetilde{Q}_t^\pi(s_{t+1}, a') - \widetilde{Q}_t^\pi(s_t, a_t) \right)$$

Since the long-term goal of maximizing the average return received at each time step is not exactly the same as other long-term goals in the form of maximizing the return for an interaction, this approach will yield other optimal policies. Nevertheless, in most practical cases these two long-term goals are correlated strongly enough.

3.3 Function Approximation

Any reinforcement learning techniques presented so far in this chapter will base the actual implementation of a policy on some kind of strategy function. Learning a policy then boils down to computing and providing proper training examples for strategy functions. These examples can then be used as training examples and input into function approximators that approximate the function as exemplified by the training examples. In the easiest case, the function approximator used will be as simple as a lookup table, but such use cases are seldom in practice. Consequently, "real" function approximators have to be employed.

This section will discuss several issues concerning function approximation for reinforcement learning including generalization and normalization (Subsection 3.3.1), strategy function representation (Subsection 3.3.2), hazards and pitfalls when using function approximators (Subsection 3.3.3), and what types of function approximator techniques are suitable to be applied in the context of reinforcement learning (Subsection 3.3.4).

3.3.1 Generalization and Normalization

Generalization in the context of reinforcement learning combined with function approximation so far was only concerned with generalization across the states of a single search space (cf. Subsection 3.2.2). Generalization, nevertheless, can also take place very fruitfully across several problem instances, typically across the instance of one problem type but perhaps even beyond across several problem types. As is the case for COPs, all instances of a problem type have the same principle problem definition, but they differ in the details, most of all their instance and their search space sizes. Lookup table representations of strategy functions therefore cannot be transferred across multiple problem instances. In general, the existence of a suitable strategy function is dependent on the existence of a problem instance independent state representation. Such an independent state representation must abstract from the actual search space and provide a condensed characterization catching traits of states independent of any particular instance size. Such a representation has been presented before in the form of features (cf. Subsection 3.2.2).

Trivially, all problem instances must have the same fixed set of features independent of the instance size. The number of features can easily be kept fix independent of instance size, though. Next, predictive features must be chosen that capture traits valid for all problem instances and which nevertheless have predictive power (cf. Subsection 3.2.2). This crucially entails that feature values must be transferable across problem instances, too, in order to mean the same for all instances and enable function approximators to predict correctly – typically only for one problem type. Finally, not too many features should be employed in order not to overstrain any function approximator used. If all these criteria are fulfilled, there are no formal objections against generalization via function approximation across problem instances (of a problem type typically). In the contrary, transferring a strategy function across problem instances is appreciated and even necessary. Reuse of what was learned before by building upon previous algorithm runs can save effort. Then, a strategy function is repeatedly improved over multiple instances in order to use it for future guidance on new, unseen instances. Once a strategy function has been learned it might remain fixed but an algorithm can as well keep go on learning thus improving the strategy function combining knowledge learned in previous runs in the form of the strategy function with more relevant experience from the current run.

Normalization of features has just been identified as one prerequisite for successful transfer of strategy functions. Normalization of features including most of all the cost of a state and, derived, rewards obtained during search (cf. Subsection 4.3.3) is not an easy task. It seems to be feasible for individual problem types, though, since all instances of a problem type, in particular those for COPs, share some common basic structure to base normalization on. Normalization for a problem type means to transfer any concrete feature values for any problem instance and state to the same interval. The hope is that the resulting values in the common interval in essence have the same meaning for all instances of a problem type and hence learning a strategy function based on these normalized features is valid for all instances. One possibility to normalize a feature is to derive upper and lower bounds, UB and LB ($UB, LB \in \mathbb{R}$), respectively, for the values of a feature. With the help of upper and lower bounds, an actual feature value x ($x \in \mathbb{R}$) can easily put in relation to the upper and lower bounds. As an example consider a normalization function $norm$ ($norm \colon \mathbb{R} \to [0, 1]$), called *linear normalization* here, as follows:

$$norm(x) := \frac{x - LB}{UB - LB}$$

This kind of normalization has the advantage that the result will always be in the interval $[0, 1]$. The tighter the bounds are, the better this and other kinds of normalization work (cf. [BM98, BM00]). If the bounds are too loose, this might not work at all, because only a small fraction of the interval $[0, 1]$ might be used which requires a too high resolution on the side of the function approximator. Also, if the bounds for different problem instances are of different quality, the same feature normalizes to different parts of the interval $[0, 1]$. For example, in one problem instance it always normalizes to $[0, 0.5]$, in another one only to $[0.5, 1]$ and in yet another problem instance to the complete interval $[0, 1]$. Clearly, generalization in such a case across two or more instances is handicapped or might be futile altogether.

If normalization is not possible, another possibility to transfer a learned strategy function across different problem instances is to learn different strategy functions for different instance sizes hoping that by eliminating the influence of the instance size the features will have the same meaning. Also, one could collect a number of learned strategy functions from n ($n \in \mathbb{N}^+$) different instances, apply them to a new one regardless, and base any actual decision on a voting among the n strategy functions previously learned. This approach, as proposed in [BM98, BM00], is robust with respect to outliers. The actual values of the features and hence of the strategy functions are only used for a ranking and accordingly a final decision about which state or action is best becomes less sensitive to the problem instance size.

Summarizing, transferring anything that was learned before such as a strategy function in the case of a learning LSA solving COPs can help to gain efficiency. In order to transfer a learned strategy function to other problem instances, a general feature-based state space representation in form of instance (size) independent, hence normalized state features must be devised and function approximation must be employed (cf. [ZD97, pp. 34 – 36] and [MBPS98]).

3.3.2 Strategy Function Representation

Function approximators are used in reinforcement learning methods to represent strategy functions. The type of the strategy functions is a core component in designing a reinforcement learning method and hence a learning LSA and strongly influences learning. The basic types of strategy functions are state-value function, action-value function, and direct policy function implementation, typically in the form of a preference structure. All types of strategy functions have their advantages and disadvantages. When using an action-value function or a strategy function directly implementing a policy, the next action can be chosen without actually having to apply any action, i.e. that are model-free. Using a state-value function requires at least a one-step look-ahead search potentially involving several quite costly probing moves before one is selected, if no transition model is available. This holds true even more if actions are strongly randomized and the same action might have to be applied several times to build a representative average. If actions are local search steps or procedures, as is the case for a learning LSA, this can easily happen.

A particular problem arises in the context of action-value function representation. The problem is how action-value functions can best be represented using function approximators. When using one single function approximator for n-dimensional real-valued input vectors for representing an action-value function based on features, not only the features of a state have to be input to the function approximator but also the action. The feature vector $\vec{s}$ for a state s ($s \in S$), typically is from $\mathbb{R}^n$, i.e. $\vec{s} \in \mathbb{R}^n$, while the set of actions applicable in a state s, $A(s)$ ($A(s) \subseteq \mathcal{A}$), is a discrete set of abstract objects. There are two principle solutions to this problem. One solution is to simultaneously input the feature vector and actions encoded as real-valued vectors also in one combined vector to a function approximator. The other solution is to represent action-value functions using one state-value function per action.

For the first solution of simultaneously inputing actions combined with the feature vector $\vec{s}$ of a state to one function approximator, actions have to be encoded somehow into a real-valued vector $\vec{a}$ ($\vec{a} \in \mathbb{R}^m$, $m \in \mathbb{N}^+$). The function approximator now works on input vectors ($\vec{s}, \vec{a}$) from $\mathbb{R}^{n+m}$, i.e. ($\vec{s}, \vec{a}$) $\in \mathbb{R}^{n+m}$. This is possible but cumbersome and requires a lot of learning overhead on the side of the function approximator since it essentially has to learn to decide first which action was input before it can decide about the value for this action in a state. The aim when using action-value function is to find this action with the best value in a state in order to apply this action. It requires in the case of this first solution to find the maximum over a continuous action-value function Q ($Q \colon \mathbb{R}^{n+m} \to \mathbb{R}$) with a subspace fixed to $\vec{s}$. This, in turn, is an optimization problem by itself that might be infeasible to be solved quickly enough since it requires to compute the partial derivation of Q at feature vector $\vec{s}$ with respect to the action encodings: $\frac{\partial Q(\vec{s}, \vec{a})}{\partial \vec{a}}$. Looking at it this way, a big problem strikes: The solution to this optimization problem needs not, and most probably will not, correspond to any action encoding, since there are only finitely many actions encoded, while the solution space for actions is continuous in $\mathbb{R}^m$. One can settle with using the actions that are "nearest" to a solution found, if the action encoding is designed accordingly. First, this introduces the additional overhead to define and actually find "nearest" actions and, second, designing an action encoding with equal spaces with respect to some defined distance measure in between any two actions is not easy. Such an encoding is necessary, though, because otherwise an a priori bias towards some actions is introduced. A drawback of this approach is the need to define an adequate distance measure. One possibility is to encode actions the same way as categorical variables in multiple linear regression in binary code or even with one binary dimension per action [Mye90, pp. 140f]. The drawback then is that it increases the dimension of input vectors for function approximators and will certainly impede learning.

The second solution to represent an action-value function is to employ for a predefined set of applicable actions $\mathcal{A}$ a family $Q_{(a \in \mathcal{A})}$ ($Q_a \colon \vec{S} \to \mathbb{R}$, $a \in \mathcal{A}$) of function approximators. Here, for any action selection, a learning LSA just has to evaluate one function approximator for each action and then use the action which suits it best, e.g. which valued best. The action-value function as a whole is then composed from the so-called *individual action-value functions* $Q_{(a \in \mathcal{A})}$. On the one hand, employing one function approximator per applicable action introduces some overhead both in terms of memory consumption and computation. All function approximator internal models (i.e. data structures) must be kept in memory and computation of the next move will affect several function approximators. Learning for most reinforcement learning solution methods such as Q-learning on the other hand, does not affect all function approximators, but at most some few actions that were recently executed. This solution to representing action-value functions altogether seems more suitable, at least to begin with.

In contrast to value functions, learning a policy directly involves evaluating several policies and comparing them. This can be done by means of an empirical policy optimization over the space of all policies by means of yet another local search cf. Subsection 3.2.7). This, however, involves maintaining and executing several policies in parallel to choose from. Several probing policies that are very similar to the current one, i.e. that are neighbors to the current policy, have to be constructed. If policies are represented by function approximators, this is not supported per se by them, though. Using function approximators for representing policies directly seems hardly applicable. Functional representations can be used. These are parameterized and policies are optimized by gradually changing the parameters according to an empirical gradient descent in terms of the parameters (cf. Subsection 3.2.7). Another possibility to represent policies directly is by means of a preference structure (cf. Subsection 3.2.7). There, direct policy learning by means of a local search is easier. Also, this can be done using an action-critic architecture (cf. Subsection 3.2.8). Altogether, a preference structure representation in direct policy learning approaches seems to be favorable.

3.3.3   General Function Approximation Issues

Using a condensed feature-based search space and function approximation for representing and learning strategy functions is quite appealing. Nonetheless, care has to be taken and several pitfalls have to be avoided in order to ensure successful learning. Hazards and other issues involved in function approximation in the context of reinforcement learning are discussed in this subsection.

Most reinforcement learning techniques per default trigger learning online and incrementally. Here, function approximators representing strategy functions are provided with training examples after each move. In contrast to *online* learning, so-called *batch* learning   (cf. Subsection 3.2.2), waits either longer until a batch of training examples have been accumulated or until the end of a search, only collecting training examples during the search and triggering learning afterwards (cf. Subsection 4.2.1). One way to relate online and batch learning is by viewing batch learning as a means to simulate online learning, but it is not true online learning. For efficiency reasons, in online learning a function approximator's internal model should not be computed completely anew after each new training example presented to it. Instead, the internal model should be only partially modified and adapted while at the same time yielding the same internal model as if all training examples so far were given at once. This kind of procedure is called *incremental learning*. Not all function approximation techniques fulfill this requirement of incrementally adjusting internal models. Many function approximator techniques are only batch oriented. Given a set of training examples, the function approximator's internal model is computed for these examples completely anew, discarding any old internal model. If a machine learning technique cannot work incrementally per se, some solutions to this problem are conceivable to adjust them for online learning. A reinforcement learning method can be made batch oriented by accumulating updates instead of triggering learning after each new training example has been computed. If a reinforcement learning method used is based on episodes, the updates collected during an episode offer themselves to be single batches of training examples. Still, across batches such as episodes, the old learned representation will be discarded, if the function approximator employed does not work incrementally. The unpleasant side-effect is that anything which was learned by previous batches is forgotten. Accumulating the training examples over all existing batches is prohibitive because of the overwhelming number of training examples that will result. To remedy this obvious disadvantage, a collection of representative training examples of old batches can be maintained that are added to each new batch. In this case, splitting up the search progress into episodes is not necessary anymore; learning simply is triggered in regular periods, reducing the number of actual function updates. This can be done by a function approximator directly or in a wrapper completely transparent for a reinforcement learning method and the function approximator, respectively.

Another design decision to make concerns how often and how many updates and hence training examples should be computed. In principle, training examples can be computed after each move of an agent. In the case of an LSA, moves can be the result of applying local search actions consisting of an arbitrary hierarchical local search action composition. Accordingly, a move can be of any granularity including basic moves in the form of one local search steps or local optima moves which comprise several local search steps (cf. Section 2.8). One drawback of assigning rewards based on a too basic agent move might be that during a search then masses of training examples for a strategy function will be computed. The individual training examples obtained might differ only infinitesimally between consecutive moves: For consecutive training examples, both the feature vector values and the target values might be too close. Too many training examples then can perhaps not be handled by function approximators. The computational overhead increases substantially without significantly improving learning while overstraining a function approximator. It would be better to use fewer but sufficiently different yet representative training examples. Therefore, it might be advisable to store and/or use only parts of the search trajectory such as every n-th transition or to use the next training example that is sufficiently different from the previous training example.

Related to the problems occurring when providing too many training examples is the issue of memory consumption of function approximators. The memory requirements for a function approximator should not be underestimated. During the course of a search, virtually thousands of training examples will be input to a function approximator even if obtained economical. They probably cannot all be held in memory although some function approximator techniques require so (cf. Subsection 3.3.4). Some kind of forgetting might be necessary to implement. This can result in derived problems such as the problem of concept drift, though [SLS99b, Rüp01, Rüp02, Ban04]. This happens when old examples are forgotten about and new examples are not representative for the old, forgotten examples, too. That what is learned, the concept, drifts from the concept that is representing the old examples to the concept representing the new ones. Even if all training examples stem from the same basic population, there might be a temporary shift during the sampling that effectively shifts the concept that is representing the currently used set of training examples.

Clustering of training examples is a topic that needs to be considered, too. It can well be that an agent or a learning LSA will only visit some very limited region of the whole search space which will result in training examples whose feature vectors are also only from some limited region of the space of all possible feature vectors $\mathbb{R}^n$. In this case it might be necessary to adjust the function approximator to this kind of clustering. Even if the search space gets explored thoroughly, the resulting feature vectors might still cover only a small subset of all possible feature vectors. The function approximators should focus on a proper approximation for the regions of the features space that are of interest as supplied by the training examples (cf. Subsection 3.2.2).

Error measurement for function approximators has to be considered. Typically, the approximation quality of function approximators is assessed by some kind of mean-squared error over a set of validation examples. For any validation example, an error value is computed. For example, the supervised error [WD00] is the difference between the target value provided by the validation example and the actual value as computed by the function approximator. These errors are squared, accumulated and averaged. Beside the supervised error, other error measurements are conceivable in the context of reinforcement learning such as the Bellman error (cf. Subsection 3.2.6 and [WD00]) and the advantage error [WD00]. Perhaps even a combination of several errors can be used to assess the performance of a function approximation (cf. [WD00]). Note that any errors computed can also be used as a termination criterion for function approximation and for a reinforcement effort such as a learning LSA run as a whole. If the error falls below a certain threshold, it is considered to be sufficiently small and learning is deemed to having been successful.

Finally, care has to be taken concerning the training and application settings of strategy functions. First of all, a strategy function is strongly connected to the long-term goal it represents. It has to be ensured that the strategy function application is valid. A strategy function only supports the use case for which it was trained. Training a strategy function based on episodes and using it in a continuous setting, for example, might not guide the search towards the aspired long-term goal the strategy function was trained for. Besides being representative, features also must not be too numerous. The more features are input to function approximators the longer the computation, possible exceeding its means at some point. Too many features increase the danger of having too many irrelevant features that only disguise the real relationships and complicate things for function approximators.

3.3.4   Function Approximator Types

Concluding this section about function approximation in reinforcement learning, several type of function approximators are briefly presented and discussed concerning their suitability in the context of strategy function approximation. The choice of an appropriate function approximator is crucial and comes with distinct advantages and disadvantages. Many different function approximation techniques have been proposed. It is not clear at once which technique can be fruitfully used in the context of approximating a strategy function in particular in the context of solving COPs by means of a learning LSA. Clearly, incremental learning techniques are favorable for the online use case of function approximators in reinforcement learning, but as was indicated before, incremental learning technique is not required necessarily. The following list will itemize the most common function approximation techniques that seem to be useful in the context of reinforcement learning. Important characteristics are described briefly as well:

- *Regression*: Regression tries to find parameters for a parameterized function such that some error measurement is minimized. The parameterized function can be a linear, quadratic, an arbitrary polynomial, or any other computable function in principle. Regression typically is non-incremental and processes all training examples at once. Forgetting of unnecessary training examples is not intended originally, but should be possible. In the simplest form, representative training examples can be obtained by evaluating the current approximation at representative data points. Memory consumption is relatively low, only the parameters of the parameterized function that is represented need to be stored. Evaluation is quick, too, since only a function description has to be evaluated. Regression will inherently focus on possible clusters of training examples, since more error measurements will occur there increasing the urge to be as exact as possible in those regions. The most frequently used functional forms are:

  - *Linear*: Linear regression has successfully been for value function approximation already in the context of reinforcement learning even in an incremental fashion [Boy99, Boy02]. However, it is debatable whether the hypothesis space is big enough to enable learning of adequate functions for complex COPs, for example.

- *Quadratic*: Quadratic regression is more powerful, but also far more complex to compute the model parameters compared with linear regression. It is doubtful as well, if the quadratic model is powerful enough.

- *Polynomial*: The general case of polynomial regression might be very powerful in terms of representable functions but the computation of model parameters can be arbitrarily complex.

- *Local Learning*: Local learning, also called *lazy learning* or *memory-* or *instance-based learning* [RN03, pp. 733 – 736] basically first stores all training examples and computes an approximation value for a given input based only on training examples that are near to the input. The nearness relation is weighted according to some distance measure. Therefor, local learning can naturally handle the clustering of training examples. Learning typically takes place when computing a value, hence training such a function approximator generally is fast while using it to approximate a value for a certain input vector can be very costly. In the context of learning LSAs the opposite is more needed, though. The memory consumption can be very high, since per default all training examples accumulated so far will be kept in memory. The memory consumption scales up with the number of training examples. Since the whole computation of an approximation value is postponed until it is requested and based on all training examples stored in principle, too, the time needed for evaluation scales up with the number of training examples stored also. Together, this probably quickly becomes intolerable, so if the number of training examples increases, some kind of forgetting needs to be implemented. Some forms of local learning are:

  - *Nearest Neighbor*: Here, only the nearest neighbors of an input vector are used for performing a regression, for example linear regression. Each neighbor enters the regression with equal weight [RN03, pp. 733 – 735].

  - *Locally Weighted Regression*: So-called kernels in the form of probability distributions determine dependent on some distance measure the weight of each training example and do compute as approximation actual value as the weighted sum of all the values of the stored training examples [AMS97]. This simplest case is computing an average value, i.e. in effect a constant function, and then returning its function value at the requested input vector. This approach can be extended to compute more complex regression models with the weighted training examples such as:
    * Linear regression models,
    * Quadratic regression models,
    * Polynomial regression models, or
    * Radial-Basis Functions (RBFN) [CLPL03].

- *Support Vector Machine Regression*: Support vector machines (SVM) [SS02, SS04], [RN03, pp. 749 - 752] compute as approximation value for a given input vector grounded on a kernel-based hyperplane induced by stored training examples. However, they do not store and use all training examples but only those that are most representative. Accordingly, learning by identifying these so-called support vectors takes more time, while function value computation is comparably fast. Recent developments have also introduced incremental versions of SVMs [SLS99a, Rüp01, Rüp02, EMM02, TPM03, Ban04]. Since only the support vectors are stored whose number can somehow be adjusted in many implementations [Joa99, CB01, Rüp04, Url05g, CL04, CB04, Joa04], memory consumption problems stemming from the need to store many training examples are not to be expected. Yet, too many training examples will impede the computation of the hyperplane and the support vectors, too. But if used in incremental mode, this will only happen once and in the training phase only. Clusters will be focused on by having an increased density of support vectors in such regions. Altogether, this kind of function approximator has been shown to be computationally efficient while having great predictive accuracy, thus being able to handle lots of training examples.

- *Regression and Model trees*:   Regression trees divide the input space into partitions by splitting the input space at each node of a so-called regression tree (in the form of a tree shaped hierarchy of nodes) into two exclusive parts [BFOS93, Tor99]. Each input vector will be tested once at a node for each level of the regression tree. The test of each node will indicate which child node is used for testing at the next level. Finally, the input vector will reach of leave of the tree. Each such leave representing one partition of the input space. The function value then is computed according to a regression of any form (typically constant or linear) over the training examples that are located in the region of the leave. The regression tree, i.e. the input space partitioning, is established according to the training examples. By keeping only representative training examples per leave, i.e. those that describe the regression in the leaves sufficiently, the problem of memory consumption can be mostly evaded. If too many training examples are collected, this will substantially slow down the computation of the regression, too, but adequate handling of clustering of training examples is inherently supported. This method is not incremental. It has the advantage of computing function values quickly while having good prediction accuracy, though. Regression trees are similar to local learning methods without the potential need to store all training examples. They can be adopted to support incremental learning and are potentially faster than pure local learning methods concerning the computation of an approximation.

- *Neural networks*: Neural networks are modeling the neurons and the neuron network-like connection structure of human brains where neurons are structured in layers [Roj96], [RN03, pp. 736 - 748]. Neural networks can express very complicated functions and can compute function values efficiently. On the other hand, the more complex a function representation has to be, the more neurons and layers have to be incorporated, and the more difficult and time consuming learning will become. Incremental learning is possible – in fact done exclusively – but only in a very slow manner. One hazard with incremental learning and hence learning in general of neural networks is the problem of forgetting (cf. [ZD97]). The number of training examples input to a neural network does not affect the memory consumption or the time needed to compute approximation values. It does, however, affect the time needed for training substantially. In particular, if large numbers of training examples are to be learned, re-training of old training examples have to be performed, otherwise their influence is not appropriate. This might require to keep a substantial number of training examples in memory for re-training. How neural networks handle clustering of training examples cannot be assessed per se.

Chapter 4

The GAILS Method

In Section 2.7 it has been recognized that it is almost mandatory to seriously begin with augmenting local search methods with machine learning techniques. Several basic ideas and ingredients for how to accomplish this venture such as the notion of a local search agent (LSA) and reinforcement learning have already been sketched before in Section 2.8 and Chapter 3. This chapter will bring together these basics with the necessary details to make the approach work for the task of solving COPs. It will present the idea and concepts of the Guided Adaptive Iterated Local Search (GAILS) method.

In the following, only LSAs are concerned. Accordingly, the notions action and move concern an LSA and are used instead of the notions local search action and LSA move. Recall from Section 2.8 that the notion state denotes an LSA state. This abbreviation is valid for composed concept names also. An LSA state's search state part will be denoted by search state. Recall also that concepts and attributes defined and used for search states can be transferred to the LSA state notion and basically mean the same but only regard the search state part of a state. Any derivation from these rules serves for emphasis purposes. For example, the state space of all LSA states will also be denoted by S, the set of all LSA states with a search state that is a local optimum will then accordingly be denoted by S^*.

The concepts of the GAILS methods will be presented in the first section, Section 4.1. The following sections of this chapter then will further elaborate on the details of the GAILS methods such as learning opportunities, how to devise actions, and conceivable learning scenarios in Section 4.2. Next, the problem of how to make sure that an LSA in fact is pursuing a proper goal when solving a combinatorial optimization problem (COP) is addressed in Section 4.3. This chapter is concluded by Section 4.4 reviewing work related to the GAILS approach and brief discussion.

4.1 Concept

This section is intended to present the idea behind the GAILS method and to present and discuss its core concepts. The idea and some already invented core concepts are outlined in Subsection 4.1.1. A discussion about the universality of the GAILS method is carried out in Subsection 4.1.2. Finally, the concept of actions within the GAILS method are described in detail in Subsection 4.1.3.

4.1.1   Idea

The motivation for the invention of the GAILS methods is to incorporate machine learning techniques into local search methods. The question is how to accomplish this incorporation? First glance opportunities to learn mainly are aimed at improving the individual components of local search methods, and metaheuristics in particular. In the case of metaheuristics, learning to improve local optima escape strategies can be attempted. Experience in metaheuristic research has shown that most important in improving metaheuristics is to devise useful neighborhood structures to base local search steps and local optimum escape strategies on [HM02, HS04]. Useful neighborhood structures can also lead to better distance measures, since these typically are based on neighborhood structures and local search steps (cf. Section 2.3), and better distance measures can lead to a better tuning of the crucial balance between exploitation and exploration [BR03, HS04]. Accordingly, neighborhood structures can be learned. This can either be done by learning them completely anew or by adjusting some variable parameters such as the size or the strength, either in advance or during search. Given a number of different neighborhood structures, a straightforward possibility to learn is to flexibly choose from several given neighborhood structures in each step of the search. This can be viewed as an adaptive version of VND [HM99, HM01, HM02]. The cost function can also be subject to manipulation. Learning to flexibly modify the cost function and guiding the search this way further elaborates on GLS [VT95, Vou97, MT00, VT02, MTF03].

The list of learning opportunities can be prolonged much further, only some starting points have been exemplified. The major problem that is faced by all the previous proposals is the same: direct and supervised learning is not possible during local search. Recall from Section 2.6 that supervised learning is learning by instruction; a teacher has to give instructive feedback. Consider having the choice to carry out local search steps based on different neighborhood structures. In this case, instructive feedback indicates whether the local search step taken was wrong or right, and perhaps gives the correct or best local search step, *independent* of the actually taken local search step. This kind of feedback is not available in local search for combinatorial optimization. The primary goal in solving COPs is to find a goal search state in the form of a global or reasonably good local optimum (cf. Section 2.2). Since it is not known in advance what "reasonably" good quality is or even how to detect a global optimum, the real quality of a local optimum found during a search process can only be estimated relatively to what was seen so far. Fortunately, learning does not necessarily require instructive feedback. Learning scenarios where only evaluative feedback is available fall in the class of reinforcement learning; feedback there only has to indicate how good a taken local search step was instead of also indicating the best local search step possible and comes in the form or rewards (cf. Chapter 3). Learning techniques for such scenarios, as have been extensively discussed before in Chapter 3, only require some kind of partial ordering on the evaluative feedback in the form of rewards such as real numbers.

Because of the pure evaluative nature of the feedback available in the form of search state costs, it can be argued that reinforcement learning techniques are the only choice for incorporating machine learning in local search methods. As regards such an incorporation, note that COP environments LSAs operate on do not change during time but remain stationary and accordingly can be modeled with appropriate reward signals and return models as a Markov decision process (MDP). The states of a resulting reinforcement learning problem can be considered to be LSA states which centrally contain a search state (and hence a solution encoding) for the COP to be solved. Other prerequisites for applying reinforcement learning techniques are the notion of an agent and the notion of an action. All these ingredients have been proposed before in the form of LSAs and local search actions (cf. Section 2.8). Altogether, solving COPs by means of local search methods can be modeled easily as a reinforcement learning problem and so reinforcement learning techniques as presented before can directly be applied. This is the central idea behind the GAILS method.

One of the core concepts of the GAILS method presented here is the new concept of a virtual LSA representing the reinforcement learning agent. The cost function of COPs can be used to compute evaluative feedback in the form of rewards. The actions inducing moves from one state to a successor state are allowed to be arbitrary hierarchical composition of other actions which is the second core concept of the GAILS method. The third core concept of the GAILS method is solving a COP by means of an LSA and viewing this as solving an MDP. The problem of solving a COP by means of an LSA hence can be considered a reinforcement learning problem. This enables an LSA to apply existing techniques for solving such problems by learning and adapting its policy. The result then is a *reinforcement learning LSA* for solving COPs. In fact, any LSA that learns its policy in this context will have to do reinforcement learning in some form, so the notion reinforcement learning LSA is from now on abbreviated to *learning LSA*. In contrast to learning LSAs, the LSA notion originally only views a local search method as virtual agent. The local search method thereby does not learn but follows a fixed policy as determined by its strategy (component). In case of a learning LSA, the policy typically is implemented by means of a strategy function which accordingly in fact then is a learning strategy component or rather learning component, since the strategy function is subject to learning also. In the discourse of this document, all reinforcement learning approaches presented and used by a learning LSA employ a strategy function for implementation of a policy for action selection (cf. Section 2.7 and Subsection 3.2.3). The strategy functions typically are represented by function approximators and typically are value functions. The value functions and function approximators do not operate on the set of all states, but on a vector of real values $\vec{s}$ ($\vec{s} \in \mathbb{R}^n, n \in \mathbb{N}^+$) and map this vector to a real value (cf. Subsection 3.2.2). The real-valued components of the vector $\vec{s}$ – which is also called feature vector – are features of the current state that are intended to summarize the most important characteristics of the current state. The aim is to enable the function approximators that represent the value functions and that operate of real-valued vectors to learn a proper representation, i.e. a proper mapping and hence a proper value function. Summarizing, any GAILS method instantiation can be defined as a learning LSA that works with actions consisting of arbitrary hierarchies built from some set of basic actions and learns by means of reinforcement learning techniques.

4.1.2 Iterating Local Search

One potential hazard with directly applying reinforcement learning techniques in a learning LSA setting to solve COPs can be that the devised actions are too fine grained. For example, in order for a local search procedure to reach a local optimum a lot of local search steps have to be taken. Trying to learn on a too fine grained level such as on the level of local search steps might produce too many too close training examples, rendering strategy function approximation difficult (cf. Subsection 3.3.3). Furthermore, the very limited view of local search steps might not be beneficial for learning something about the long-term goal (cf. subsections 4.2.3, 4.3.2, and 4.3.3). Except for the last local search step of a local search descent, they are not directly connected to a local optimum from the relevant and goal state containing subspace of local optima. Ideally, local search is elevated to the next level of abstraction which is moving an LSA in the set of local optima, S^*, directly. This cannot be achieved based on actions that include local search steps such that mostly suboptimal states are visited. Unfortunately, no neighborhood structure for S^* is known or can be computed efficiently as yet. Nevertheless, local search on a neighborhood structure for S^* can be approximated by letting a learning LSA learn to choose among actions inducing local optima moves only. A learned policy then effectively implements an abbreviation of a neighbor enumeration and selection scheme, i.e. a local search step on a neighborhood structure on S^*.

The ILS metaheuristics effectively operates on the subspace of local optima, S^*. In each iteration a composition of local search operators in the form of a perturbation, local search procedure, and an acceptance criterion is applied. Let this coherent composition be denoted by *ILS local search operators* and its extension to be an action for an LSA by *ILS-action*. Then, by applying ILS local search operators, an ILS conducts a biased walk through the set of local optima S^* (cf. Section 2.5). The bias is mainly introduced by the perturbation and the acceptance criterion component of an ILS local search operator, but in the end the walk might still be a rather undirected and random trial-and-error interaction. Instead of probing the local optima search space S^* more or less by pure chance, a reinforcement learning variant of ILS in the form of a learning LSA can learn to choose a proper ILS-action for each next move.

Providing the necessary ILS-actions to choose from is not difficult. Many perturbations can be varied in strength or according to other parameters and typically several local search procedures are available for a problem type. Additionally, several acceptance criteria, potentially problem type independent can be invented. Any combination of a perturbation, a local search procedure, and an acceptance criterion results in a different ILS-action. One of the main advantages of ILS is that its components such as perturbation and acceptance criterion have interpretable meaning to humans (cf. Section 2.5). This makes ILS and hence an LSA using ILS-actions potentially easier to analyze and understand what was learned: Any rules learned for action selection might be interpretable for humans, too. For example, when using perturbations and local search procedures with different strengths to form ILS-actions, learning to choose the proper ILS-action essentially means to learn to choose the proper strength for the next action which directly translates to the problem of balancing between exploration and exploitation.

An LSA based on ILS-action is very appealing. If it is a learning LSA, it becomes adaptive and guided by the incorporation of a learning strategy component. It conducts a guided and adaptive walk through the set of local optima, S^*. The new concept of a learning LSA solving COPs accordingly is denoted *Guided Adaptive Iterated Local Search* (GAILS) method. Thereby, walking through the set of local optima is by no means a typical trait of the ILS metaheuristic only. As will be pointed out in the next paragraph, it is necessary for any local search method to ensure at least occasional visits to local optima to obtain the best results possible. This entails that any search trajectory of a local search method and hence of an LSA will repeatedly contain local optima. Any state sequence in between local optima can in some sense be regarded as escapes from the previous local optimum and a following descent to the next. Any local search method seen abstractly is iteratively visiting local optima and hence a variant of a very general view on ILS: Escape parts correspond to perturbations, the descent parts coincide with the local search procedures (that perhaps also can do worsening local search steps). Borders between escape and descent parts can become blurred. The acceptance criterion simply always accepts for this simple ILS variant. Any local search method that is enhanced with a learning component then is a guided adaptive iterated local search. Any learning LSA with a learning strategy component and appropriate actions defined in principle is a GAILS method instantiation, regardless of the learning method, such as reinforcement learning, employed. In this spirit, the GAILS method certainly is inspired by the concrete ILS metaheuristic and in a sense reinforcement learning but in its abstract concepts is far more general along the lines that ILS is far more general as a schema than the special metaheuristic instantiation.

But why is iteratively visiting local optima almost mandatory for any local search method to achieve good performance? In practice, the best search states found by local search methods almost always are local optima for one or more of the neighborhood structures used. If not doing local search descents, easy opportunities to almost surely improve a current search state will be missed. As a result, almost surely suboptimal search states will be found. Hence, an algorithm that does not periodical local search descents will behave sub-optimally. Unfortunately, this intuition cannot be proven and is not true in general. The intuition mainly stems from empirical observations where metaheuristics enhanced with additional local search procedures that where triggered periodically, for example applied to the best search states found by the original version of the algorithm, performed better [Mos99, SH00, HS04, DS04]. Local search descents will do no harm to the search state quality; all they do is improve the cost or simply do nothing. The downside of additionally applied local search procedures, though, can be the overhead of computation entailed by them. If limited resources are given, for example in terms of computation time, a local search descent enhanced variant of a local search method might miss to advance as far as the original version and exactly this way misses to find high quality search states that were found by the original variant at the end of its runs. Sometimes it simply might be better to visit as many search states as possible rather than to try to concentrate on finding fewer but hopefully better search states. This can be made vivid by looking at random local search where at each time a neighbor is picked randomly from a given neighborhood structure. It has been proved that random local search will find a global optimum in the limit with probability 1, if all search states of the search space are reachable by the local search step [HS04]. This holds as well for the special case where any two search states are neighbors. The result is easy to verify by observing that in the limit with probability 1, all search states will be visited at some time, since the probability to visit any search state is greater than zero. Regardless, in practice, visiting local optima periodically is almost surely improving performance. In the setting of GAILS the overhead of doing additional local search descents will be in addition to the overhead induced by the learning components of a learning LSA and might absolutely be negligible compared to the learning effort. Since additionally runtime is regarded as a secondary issue in this discourse (cf. Section 2.2), it is assumed that enhancing a local search method with occasional local search descent to form learning LSAs most likely will improve its performance in terms of search state quality.

As a consequence, any actions of an LSA should either wind up in a local optimum by making only local optima moves or they should ensure repeated visits of local optima during their execution. This entails that actions must not be constrained to be extended local search steps but in principle must be able to induce any complexity of search state transitions and hence arbitrary moves. Hence, any local search operator and therefore any local search method must be eligible to be extended to an action. From now on, it is assumed that any method and concrete algorithm periodically visits local optima. Any method or learning scenario that does not sample local optima at all is assumed to be (almost surely) inferior to a method or learning scenario that does so and will not be considered in the future discussion.

4.1.3 Action-Hierarchies

Recall from sections 2.3 and 2.8 that actions on the one hand are defined to be *anything* that changes a state. According to another definition, actions are extended local search operators in that they additionally also collect and update heuristic information and thus, in contrast to local search operators, potentially not only affect the search state part of a state but also the heuristic state part. Actions that only change the heuristic state part of a state are well conceivable. Recall from sections 2.3 and 2.8 also that local search operators implement local search methods and induce arbitrary search state transitions. Local search methods such as metaheuristics use other local search methods and thus can be regarded as consisting of arbitrary hierarchical composition of other local search methods finally based on basic local search operators such as local search steps. As a consequence, local search operators in principle as well consist of arbitrary hierarchical compositions of other local search operators. In the context of reinforcement learning, using actions that are built hierarchically is not new. *Hierarchical reinforcement learning* primarily is concerned with how to learn policy for hierarchically built actions (cf. Subsection 4.4.5, [PR97, Par98, HMK+98, HGM00, RR00, Die00, MMG01, Hen02, SR02, BM03, GM03a]). In contrast, the discussion presented next concentrate on how to build actions for LSAs hierarchically.

As an example for how local search operators can be built, consider a local search operator implementing a local search procedure. Local search procedures iterate a local search step until a local optimum has been reached. The lower part of Figure 4.1 illustrates such a local search operator. There, the local search operator represented by the box labeled "LsProcedure" implements a conventional local search procedure as can be seen from the pseudo-code inside the box. Variable s_0 in the pseudo-code provides for the start state. After the local search descent, the local optimum reached will be accessible via s_0 again. The implemented local search procedure needs for its operation another local search operator implementing a local search step, denoted by LsStep in the pseudo-code. The local search operator depicted by the box labeled "TSP-2opt-LsStep" representing a 2-opt local search step for the Traveling Salesman Problem (TSP) (cf. Section 2.2) takes over this responsibility. The local search operator thus built in turn is used by yet another local search operator depicted by the box labeled "SimpleILS". As can be seen from the pseudo-code inside this box, this local search operator implements a simple and standard ILS. The ILS needs for its operation three other local search operators implementing a perturbation, a local search procedure and an acceptance criterion which are named Perturbation, LsProcedure, and Accept in the pseudo-code, respectively. Additionally, a start state named s_0 in the pseudo-code has to be provided, too. Again, after the application of the ILS, this parameter will contain the end state. The required local search operators are represented by a box labeled "LsProcedure", by a box labeled "TSP-DoubleBridge", and a box labeled "BetterCost", respectively The latter local search operator implements an acceptance criterion that only accepts a newly found local optimum as new current state, if it has better cost than the last current state. The local search operator represented by box labeled "TSP-DoubleBridge" a local search step in the form of a perturbation which is a 4-opt perturbation for the TSP. Additionally, a termination criterion named TermCrit in the pseudo-code for the ILS is needed. The box labeled "MaxΔIterations" represents a termination criterion which indicates termination as soon as a number of iterations have been executed. Note that a termination criterion is not considered to be a local search operator here.

Analog to the hierarchal composition of local search operators, actions being extended local search operators are arbitrary hierarchical compositions of other actions. Each such hierarchy finally is based on some basic actions that do not use any other actions but change a state directly, either by manipulation of the search state or by changing information of the heuristic state part. The concept of viewing actions as arbitrary hierarchical compositions of other actions is called *hierarchical action structure*. It is one of the core concepts of the GAILS method. Any hierarchy of actions is called *action-hierarchy*. An action-hierarchy basically is a tree. The *individual actions* used within the action-hierarchy coincide with the nodes. The root node of an action-hierarchy is also called *root-action*. An action-hierarchy can be illustrated or rather represented as a tree as is shown in Figure 4.1. The boxes in this figure are the nodes of the tree illustration and correspond to actions, the arrows are edges and indicate affiliation and usage. Such a tree illustration is called *hierarchy-tree* for short. Subtrees of an hierarchy-tree correspond to sub-hierarchies of an action-hierarchy which are action-hierarchies by themselves. In Figure 4.1, the subtree with root node labeled "LsProcedure" represents an action-hierarchy by itself. The same holds true for the two subtrees with root nodes labeled "SimpleILS" in Figure 4.2 which are affiliated to the node labeled "RoundRobin" there. Some subtrees and sub-hierarchies of an action-hierarchy eventually are leaves. Leaves of an action-hierarchy finally must be some basic actions which do not use further actions. They are called *leave actions*. Leave actions might only change the search progress and thereby only change heuristic information without manipulating the search state of a state directly such as an acceptance criteria of an ILS. Other leave actions mainly and directly manipulate the search state part of a state such as actions implementing a local search step. Actions of the latter type are also called *elementary actions*. In figures 4.1 on page 75 and 4.2 on page 76, the node labeled "TSP-DoubleBridge" represents an action implementing a 4-opt perturbation for the TSP that does not need any other action for its operation. It manipulates the search state directly and

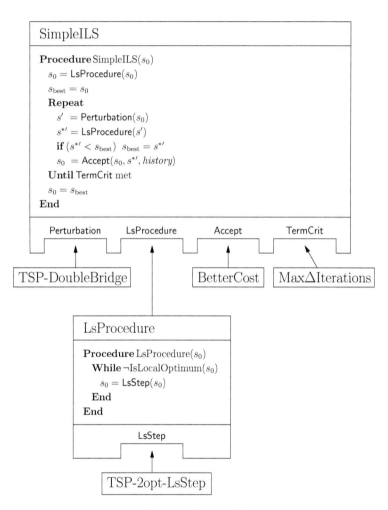

Figure 4.1: Simple ILS

thus is an elementary action. In the same figures, the node labeled "BetterCost" represents an acceptance criterion which is a leave action.

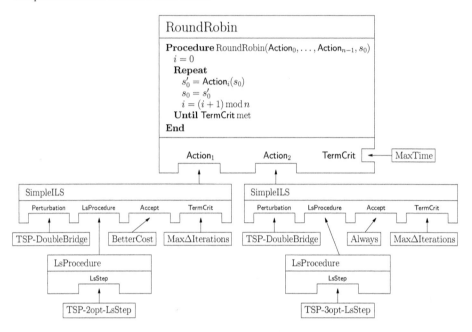

Figure 4.2: Round-Robin of two ILS

Figure 4.2 is a larger illustration of an action-hierarchy. There, the node labeled "RoundRobin" represents an action implementing a round-robin action application scheme. This action takes a number of n ($n \in \mathbb{N}^+$) other actions and executes them in turn as expressed by the pseudo-code inside its representing node. The n actions are executed in turn until the termination criterion named TermCrit in the pseudo-code holds. This, for example, can be a maximum time criterion, depicted by the node labeled "MaxTime" in this example. The start and the end state of the round-robin action application scheme is transferred via parameter s_0 again. The n actions should terminate by themselves, of course. Figure 4.2 illustrates the alternation of two actions that both implement a simple and standard ILS. The two ILS only differ in the local search step their local search procedure component is based on and the acceptance criterion used. The ILS affiliated to Action$_1$ is the same as is illustrated by Figure 4.2. The other action implementing an ILS and affiliated with Action$_2$ employs an acceptance criterion represented by node labeled "Always" which is supposed to always accept a newly found local optimum as new current state. The local search procedure component of this ILS employs a 3-opt local search step for the TSP represented by node labeled "TSP-3opt-LsStep". As long as termination of the alternated actions is ensured, it is all the same for the root-action implementing the round-robin scheme what kind of actions are input. It only needs a number of actions, regardless whether these are leave or elementary actions, or actions consisting of a whole action-hierarchy. The action implementing the round-robin scheme only sees the root-action node of any action-hierarchy it uses anyway. Figure 4.3 illustrates this. There, the two ILS from Figure 4.2 are replaced by the actions implementing their local search procedures which is reflected by replacement of the corresponding nodes.

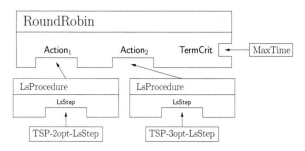

Figure 4.3: Round-Robin of two local search procedures

Many actions use other actions for their operation thus forming action-hierarchies. According to this original view on what actions are the nodes of an action-hierarchy are actions. From a certain perspective, one can also argue that actions *consist* of other actions instead of using other actions. In this new view, whole action-hierarchies are considered to be actions. Both views are reasonable. Each action-hierarchy can be viewed as black box. However, the only node of an action-hierarchy that actually is visible from outside in the black box view is the root node. It can be considered to be the interface of an action-hierarchy, everything else of the action-hierarchy is transparent for any user of the action-hierarchy even if in fact the whole action-hierarchy is supposed to be used. The root node of an action-hierarchy is representative of the whole action-hierarchy. Therefore, essentially each root node can be considered to be an action, hence also called root-action. Recall that sub-hierarchies of an action-hierarchies are action-hierarchies and according to this view then are actions by themselves. Since any node of an action hierarchy is the root node of a sub-hierarchy (and hence a root-action) and each sub-hierarchy can be considered to be an action-hierarchy, each node of an action-hierarchy essentially is an action again. On the other hand, a root-action of an action-hierarchy cannot work without the other actions of the action-hierarchy. The whole action-hierarchy together then essentially is an action. As a result, both views are just two sides of the same coin and are adopted in parallel during this discourse.

The concept of a hierarchical action structure allows to compose action-hierarchies and hence actions according to a building blocks principle and in turn enables massive reuse. This reuse can have several occurrences. On the one hand, by viewing action-hierarchies as black boxes and using them the same way as leave actions, complete action-hierarchies or sub-hierarchies thereof can be reused. The action-hierarchy illustrated in Figure 4.1 is reused as is to act as the action input for Action$_1$ in Figure 4.2. Reuse can also take place by reusing action-hierarchies only partly in that only certain parts are replaced. The sub-tree of the hierarchy-tree from Figure 4.2 that represents the action-hierarchy which is input for Action$_2$ basically is the same as the one input for Action$_1$, only the acceptance criterion and the local search step of the local search procedure are replaced. Figure 4.3 illustrates reuse as well. On the one hand, the action-hierarchies implementing the local search procedures from the two ILS depicted in Figure 4.2 are reused and input for Action$_1$ and Action$_2$ of the action that implements the round-robin scheme. On the other hand, the root-action of the action-hierarchy depicted in Figure 4.2 is reused, only the two ILS action-hierarchies from this figure are replaced by the actions implementing the local search procedure employed by the two ILS.

The replacement kind of reuse basically reuses blueprints for how to build proper action-hierarchies (and hence actions) such that certain local search methods such as an ILS are properly implemented. This kind of reuse reflects the general-purpose nature of metaheuristics where the concrete forming of individual local search methods that are used by a metaheuristic is irrelevant, in particular as regards the problem type to operate on. The individual local search methods and analog the actions used simply must approximately do what they are intended for by the using metaheuristic or action. For example, consider an ILS. It is not important in principle which local search procedure, perturbation, or acceptance criterion is input. Even a simple local search step could be input instead of a local search procedure. Of course, this does not make sense, but, in the context of assembling action-hierarchies, ensuring building proper and useful actions-hierarchies falls in the responsibility of the action-hierarchy builder. As was shown, the concept of an action is rather abstract. In particular, except for elementary actions, any action-hierarchy mostly is independent of a concrete problem type. In the case of the simple and standard ILS illustrated by Figure 4.2, only the perturbation and the local search step action are specific to the TSP type. This is exemplified by the ILS pseudo code: The type of a underlying COP instance to solve does not occur. Exchanging the perturbation and the local search step makes the ILS work for any other problem type immediately. Action-hierarchies as a result can reuse substantial parts across problem types as well.

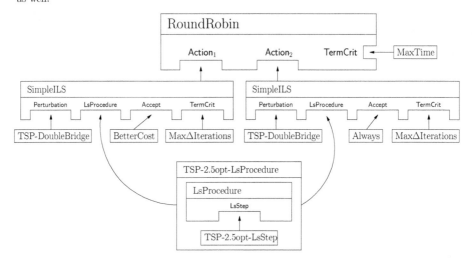

Figure 4.4: Round-Robin of two ILS with a mutually shared action

The kinds of reuse covered so far basically work by copying and therefore rather reuse blueprints. The copied action-hierarchies work as templates. A third type of reuse is that of sharing concrete instantiations of action-hierarchies as is illustrated in Figure 4.4. There, the actions implementing the local search procedures of the two ILS from figure Figure 4.2 are replaced simultaneously by one, now mutually shared action-hierarchy represented by the box labeled "TSP-2.5opt-LsProcedure". This box contains a hierarchy-tree and represents an action or rather action-hierarchy implementing a local search procedure based on the 2.5-opt local search step for the TSP [JM97, GP02], which is represented by node labeled "TSP-2.5opt-LsStep" inside the box. Mutually sharing action-hierarchies and hence actions presupposes that they do not store a state by themselves. This, however, is no handicap, since all information about the state of an LSA can be stored in its state, but sharing actions requires to coordinate the changes made to a state in order to keep it consistently.

Local search operators can implement arbitrary local search methods including metaheuristics. All they have to do is to take a starting search state and induce a state transition to the search state;

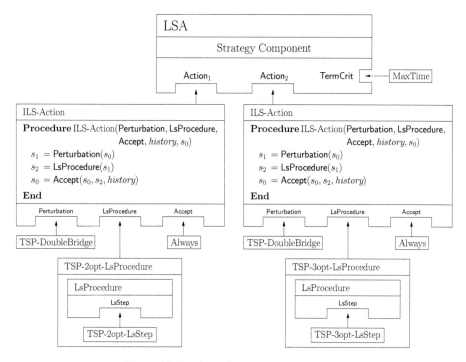

Figure 4.5: Local search agent using two actions

of any granularity. The same holds true for actions since they are extended local search operators and basically also for an LSA. An LSA uses a set of applicable actions as is illustrated in Figure 4.5. There, the node labeled "LSA" represents an LSA that employs two actions denoted Action$_1$ and Action$_2$ in this node. Which one is used in a state is determined by the LSA's strategy component denoted by "Strategy Component" there. This strategy component can as well have learning capability resulting in a learning LSA. The actions or rather action-hierarchies represented by the two hierarchy-trees labeled "ILS-Action" in Figure 4.5 are input for Action$_1$ and Action$_2$. As can be seen from the pseudo-code in the root nodes of the hierarchy-trees, they are ILS-actions consisting of a perturbation, a local search procedure, and an acceptance criterion named Perturbation, LsProcedure, and Accept in the pseudo-code, respectively. The perturbation and the acceptance criteria are the same for both ILS-actions (and the same as depicted in Figure 4.2) and are represented by nodes labeled "TSP-DoubleBridge" and "Always", respectively. The two ILS-actions only vary in the local search procedures they employ represented by the sub-trees of the two hierarchy-trees with root nodes labeled "TSP-2opt-LsProcedure" and "TSP-3opt-LsProcedure", respectively. Parameter *history* of the ILS-action provides the state history for the acceptance criterion. Not only the hierarchy-trees for the ILS-action represent action-hierarchies, but the whole tree of figure Figure 4.5 can be regarded as a hierarchy-tree that represents an action-hierarchy implementing an LSA. Hence, an LSA can be considered to be a action, too. Conversely, LSAs will be implemented by actions. This makes sense, since an LSA also starts in a certain state, makes some state transitions to its state in the form of moves and ends up in an end state. Viewed as a black box, an LSA simply induces a move in the form of a state transition to its state and hence *is* an action.

4.2 Learning Perspective

The previous section has introduced the idea and basic concepts of the GAILS method. This section discusses details related to learning within the GAILS method such as learning opportunities, possible actions, and hazards and other problems which have to be kept in mind and pondered when designing and implementing concrete algorithms later. To start with, this section will present how learning can in principle be organized in the context of learning LSAs in Subsection 4.2.1. Next, possible variations in designing actions for an LSA will be introduced and debated in Subsection 4.2.2. Finally, several conceivable learning variants and scenarios are demonstrated in Subsection 4.2.3.

4.2.1 Learning Opportunities

Several basic design decisions concerning who, what and when to learn can be made when employing learning LSAs. Depending on the design decision, the nature of learning can greatly vary and several problems but also potentials can emerge. The possible design decisions will presented and described next. A discussed will follow thereafter.

- Learning can be confined to one problem instance only (*problem instance specific learning*) or generalization from one or more problem instances to others can be attempted (*problem instance non-specific learning*).

  Problem instance specific learning can and will be started completely anew for each new problem instance. In the beginning, a learning LSA will be mostly concerned with learning and only bit by bit build up knowledge that in turn can be used to improve its behavior. This kind of generalization takes place within one single problem instance and is instance-specific. The learning assumption here is that some regions of the search space exhibit similar regularities. Experience collected from some regions can be used for improved processing of new regions of the search space.

Problem instance non-specific learning tries to detect regularities that are valid across all problem instances of a given class of instance, typically a problem type. The assumption in problem instance non-specific learning is that a class of instances exhibits certain general regularities invariant in each problem instance, most of all independent of the size, that can be exploited. In principle, any class of problem instances can be tackled such as the subclass of all clustered instances of the TSP.

- Learning can be *offline* or *online*.

  Online learning takes place during the search process. A learning LSA's learning component is fed with training examples and the learning component perform computations during the search process in order to update their learned strategy for action selection. In the case of a reinforcement learning LSA, the function approximators representing the strategy functions update their models. Anything learned this way can be used immediately to guide the subsequent search process.

  An identifying trait of offline learning is that learning does not take place within the run of a learning LSA. Instead, training examples are collected and stored during a run in order to subsequently present it to the learning component and trigger learning. Accordingly, offline learning does not affect the run that collects the training examples except for the computational overhead to collect and store them. Several proposals for batch learning collecting examples from one or more runs before learning have been made [ZD95, Zha95, ZD96, ZD97]. A priori search space analysis or parameter tuning – if an algorithm to be tuned is treated as black box – can be regarded as typical examples of offline learning. Several methods for search space analysis and automatic tuning of algorithms have been proposed [ADL02, BSPV02, SS03].

- Learning can be performed by a machine or by a human researcher or a combination of both, in particular in the form of computer supported human learning.

  Typical reinforcement learning techniques do not require interaction by a user. On the other hand, tuning and algorithm improvement can be viewed as a learning process on the part of the algorithm implementor (cf. Section 2.7). Intermediately, a human researcher can use machine learning techniques to guide his or her own learning process. He or she can interpret what was learned by a computer and identify utilizable regularities that result in hard-coded strategies in an improved next version of an algorithm. A learned strategy function or policy in the scope of a reinforcement learning LSA for example might express a general trend suggesting to hard-code the respective policy.

After this presentation of principle design decisions for learning LSAs they will now be discussed. The simplest combination of design decisions certainly is to learn and use what was learned by the learning component online immediately. After some moves, the learning component is confronted with the collected training examples and is updated. What was learned then can be used immediately. Generalization across problem instances is not required, since learning can be confined to the problem instance at hand. On the other hand, no transfer of knowledge between problem instances in the form of what was learned by the learning component of a learning LSA can be undertaken. Everything has to be learned completely anew for each new problem instance. Besides the computational overhead for online learning which might well be prohibitive for many real-world applications, the time until the learning component has learned something useful for guiding the search might be long and another feasibility obstacle. Since time typically is a scarce resource, pure online learning with subsequent forgetting might be too wasteful. Online learning on the side of

humans does not make much sense either, since no metaheuristic is interactive yet and is not likely
to be so in the future.

Offline learning in between runs of a learning LSA seldom makes sense without generalization across
multiple problem instances. It seldom makes sense to first process a problem instance to collect
training examples, next start the learning process and finally process that problem instance again.
Consequently, generalization across multiple problem instances of what was learned in the form of
instance independent learning component seems very much desirable if not mandatory. Not only
will it be possible to reuse what was learned, learning as well can be postponed reasonably and thus
enables offline learning. Generalization enables to learn a priori when time is not critical on a couple
of problem instances. Afterwards, a trained learning component guiding the search can be applied
only in subsequent runs. Whether learning then is done in an on- or offline manner is irrelevant as
long as generalization to new unseen instances works. One variant is to additionally collect training
examples during the time critical runs and update the learning component in between runs, if time
admits. Learning can be done online as well, refining the learning component in each run, if it does
not take too much time.

Reinforcement learning LSAs will produce a policy represented by a strategy function which can be
analyzed by human researchers afterwards and which can lead to new insights. Some learned policies
might not have been predictable by humans yet are very successful. Policies might exploit very
inconspicuous yet important regularities leading to completely new insights that would not have
been made otherwise, this way truly adding to the human learning capability. For example, consider
having a reinforcement learning LSA using several ILS-actions that only vary in the strength of the
perturbation. In other words, the learning LSA is to learn to choose the proper strength for the next
action which basically means to choose the proper balance between exploration and exploitation
in each move. A learned policy can now be analyzed and regularities and underlying laws based
on features that indicate when to do what possibly can be extracted yielding new insights in this
crucial topic. One result could be that several perturbation strengths are not used at all since
they have been identified to be completely useless. Or, only one ILS-actions is used all the time
suggesting that flexible alternating several ILS-actions is not necessary. Instead, a standard ILS
with the winning ILS-action can be used. Such a result then is nothing else than automatic tuning
of ILS parameters by a computer. Finally, it might occur that a learned policy suggests to vary
some actions depending on the time: one action is preferred at the begin of the search, the other
towards the end. Altogether, these comments show that computer supported human learning can
be very fruitful in its own, independent of the actual performance of a reinforcement learning LSA
and also learning LSAs in general. The more a learning LSA thereby can generalize, most of all
across problem instances, the less work is left to a human researcher.

4.2.2   Actions

The GAILS method is intended to solve COPs by training a virtual learning LSA to take the
proper action at a time. Trivially, several actions must be available; without a choice, nothing can
be learned. In general, the more actions are available, the more likely it is that a subset of them
will work fine together in enabling learning to guide a learning LSA. The disadvantage of providing
too many actions is that learning might be more complicated while in fact not all available actions
will be used finally. There might well be a significant trade-off between the number of actions
and learning ability. According to [ZD97], the branching factor for applicable actions must be
small, i.e. not too many actions should be applicable for each move. Otherwise one-step look-ahead
search for example is too costly for a state-value function implementation. Even for an action-value
strategy function implementation, too many function approximators might have to be evaluated.

Nevertheless, the success of a concrete GAILS algorithm – beside predictive features – certainly centrally depends on a useful set of applicable actions.

To exemplify the possibilities in constructing a set of applicable actions for an LSA, consider an ILS-action based LSA. One apparent key issue for the success of local search methods and metaheuristics in particular is to find a proper balance between exploitation and exploration of the search progress (cf. Section 2.4). Consequently, the ILS-actions used should exhibit some flexibility with respect to their strengths. In the case of the ILS metaheuristic, this balance foremost is controlled by the perturbation of an ILS-action, supported by the acceptance criterion (cf. Section 2.5). Therefore, ILS-actions can be built based on a number of increasingly strong perturbations or by providing an explicit strength or strengths equivalent parameterization of one perturbation. A parameterization of a perturbation with respect to varying strengths for example can easily be done by repeating the application of a perturbation. The more perturbations are carried out directly following each other, the stronger the change or "perturbation" to the search state. Other possible variations of ILS-action can stem from different local search procedures and from variations in the acceptance criterion. With ILS-actions varying in the acceptance criterion also, a learning LSA can also learn to choose the proper acceptance criterion at a time. Acceptance criteria can vary from always accepting a new local optimum, even if the cost worsens compared to the currently visited one, to accepting only better local optima for continuing the search. In the latter case, called *better acceptance criterion* any new state accepted will also be an overall best one. Intermediate between these extremes, new local optima can also be accepted after some interval of time, steps, iterations, or moves in general has expired or with some small probability in each decision (the latter is called *ε-better acceptance criterion* where ϵ ($\epsilon \in [0,1]$ denotes the acceptance probability). An acceptance criterion can be based on a history of states, so even more complicated acceptance strategies can be conceived.

Since local search steps and hence local search procedures and perturbations are based on neighborhood structures, different perturbations and local search procedures can be devised by inventing several neighborhood structures for perturbations and local search steps. One can also partition one large perturbation neighborhood into several smaller equivalence classes of neighborhoods thus yielding different kinds of perturbations. Local search steps (and derived procedures) in turn are also based on an exploration scheme over a neighborhood structure, so, secondary, several neighborhood exploration schemes can be designed, too. In general, the space of local optima, S^*, is dependent on the neighborhood structures used. Recall that an ILS-action typically is randomized, for example induced by a randomized perturbation component of the ILS-action. The set of all potential move targets of an ILS-action can be considered to be a neighborhood for a the current state. This holds true for any state. ILS-actions accordingly induce a neighborhood structure. If several neighborhood structures induced by several ILS-actions are used simultaneously, any local optimum for any such ILS-action is a state an LSA can potentially move to. This increase of the search space can be of advantage but it also can be disadvantageous. In general, the larger the search space, the harder it is to handle. On the other hand, introducing new actions and accordingly induced neighborhoods for an LSA changes the cost-surface and new high quality states, typically in the form of local optima might appear that were not reachable or not reachable easily or likely before. Having more neighborhood structures means being able to compose and choose among more actions to make the next move.

Another issue in action design and usage is that of randomization. Randomized local search operators typically work better than non-randomized ones [HS04], so actions as extended local search operators will be randomized also. The randomization of actions should not be too huge. Randomization inherent to the actions basically is noise that makes it harder to predict the outcome and effect of actions. If actions are randomized, more than one application per intended move might be required to get a representative trend of the value and effect of an action. Learning will take longer the more randomization is involved, since the global trends have to be filtered from the random effects. Obviously, randomization impedes learning.

Concluding this subsection, an exemplifying collection of action proposals for the TSP is made and briefly discussed. The typical local search steps available for the TSP are n-opt ($n \in \mathbb{N}^+, n \geq 2$) local search steps, either in first or best improvement fashion. Perturbations for the TSP mostly are based on 4-opt variants (cf. Section 2.3 and Subsection 6.4.1. The most widely used among them is called double-bridge local search step [LK73, MOF91]. In general for the TSP, neighborhoods for 4-opt local search steps or beyond are very large. Exploring such a neighborhood exhaustively, for example in a look-ahead-search manner, is prohibitive. Instead, double-bridge perturbations commonly are randomized in the context of ILS. As an example for obtaining different perturbations by partitioning a neighborhood structure, all combinations of edge exchanges for an n opt perturbation can be enumerated according to a fixed deterministic scheme and the first m ($m \in \mathbb{N}^+$) of such exchanges then implement perturbation number one, the next m exchanges implement perturbation number two, and so on. The size of such equivalence classes can be as low as one exchange while the number of perturbations can be confined arbitrarily, too. The enumeration of edge exchanges can be made dependent on state features and other characteristics. For example, the edge exchanges can be sorted according to the average length of the edges involved. One drawback of such a restricted perturbation neighborhood might be that not all states of the original neighborhood can be reached or become less likely to be reached. Also, the perturbations become more deterministic and less far reaching with the effect that they perhaps do not escape an attraction basin of an employed local search procedure or that they initiate a loop over only a limited number of local optima. To remedy this, perhaps some randomization has to be introduced again.

4.2.3   Learning Scenarios

To provide a more detailed impression of how different learning LSAs can be designed, several learning scenarios are presented in this subsection.

STAGE Learning Scenario

One of the conceptually simplest applications of a strategy function for solving COPs is by using a state-value function that represents the final cost obtained when starting a certain local search procedure in a certain state. This kind of state value function can be used by means of any local search procedure where the original cost function is substituted by the state-value function. A local search descent then is an optimization with respect to find the best starting state for a certain local search procedure and thus can be used to escape a local optimum found. The search process will have two alternating phases: A given local search procedure will descent to a local optimum while a state-value function based local search descent will escape this local optimum – just like a perturbation of ILS – and provide a new start state. This works until both optimization phases have come to a common local optimum. Then, some kind of restart has to be carried out. This approach was exactly taken in the STAGE algorithm [BM98], [BM00, p. 98].

Learning a state-value function for the STAGE learning scenario can be achieved by simple Monte-Carlo (MC) sampling (cf. Subsection 3.2.5). The search trajectory naturally splits into episodes, each comprising one local search descent to a local optimum of the unmodified local search procedure. After each descent, for each state of the descent training examples are built consisting of of the feature vector for the state combined with the cost of the local optimum reached. These training examples then are used to train the function approximator representing the state-value function.

Searching a complete neighborhood using a state-value function substituting the cost function, might be very slow. Computing the value of a state-value function for a neighboring state probably is substantially slower than computing its cost, especially if this is done incrementally during the neighborhood enumeration as is typically the case. Beside the cost of a neighboring state, several other features have to be computed as well and next input to the function approximator that represents the state-value function. To alleviate this impediment, a state-value function need not be asked for any possible neighbor. Instead of exploring a complete neighborhood, only a subset of potentially good neighbors is evaluated. The set of potentially good neighbors can be as simple as a random sampling of a neighborhood and a subsequent filtering according to the state-value function. This approach has been taken in [ZD97] and was called RSGS (random sample greedy search) there.

The STAGE learning scenario can work, but it has to be contemplated whether performance really stems from the guidance of the learned strategy function in the form of the learned state-value function. The strategy function guidance can also be viewed as a perturbation from the ILS metaheuristic that intelligently escapes local optima before starting a new local search descent. The STAGE approach then is nothing else than a disguised ILS. Learning simply produces a good perturbation based on information how to escape local optima attraction basins while maintaining good state quality. The main performance might well be due to the ILS concept itself, though.

Concluding, note that the state-value function from the STAGE learning scenario is not an approximation of the cost function. The value of a state indicates what *final* cost can possibly be reached from this state by means of the specific local search descent. which is the cost of the local optimum of the attraction basin in which the state is located (with respect to the given local search procedure). This in principle has nothing to do with the costs of the states.

Perturbation Probing Learning Scenario

A state-value function from the STAGE learning scenario can be used in a way other than substituting the original cost function. It can be used to estimate the result of an ILS-action. Given a set of different perturbations or one randomized perturbation, several perturbation applications are done sampling some potential start states for a following fixed local search procedure for which the state-value function was learned. The potential start state with the best value according to the state-value function gets selected to actually start a local search descent from. Here, the state-value function serves as an abbreviation of an application of the local search procedure. This scenario then tries to implement a faster version of sampling a neighborhood for the set of local optima S^*. Even if a local search descent only comprises some local search steps, learning to predict the outcome of a local search procedure can be used to speed up the search process. Learning works the same way as in the STAGE learning scenario. The only difference is that the episodes might better be produced during the execution of respective ILS-actions.

The perturbation probing learning scenario can work with several local search procedures in parallel, too. An action-value function, perhaps represented as a family of state-value functions, is used where each action corresponds to a local search procedure. Each potential start state after a perturbation application gets evaluated for such action and the start state with the highest value is used to actually start state the respective local search procedure from. Learning has to involve local search descents from all local search procedures, of course.

As simple and appealing as the STAGE and the perturbation probing learning scenarios are, some problems need to be discussed. The problem with these learning approaches is that splitting a search trajectory into episodes according to local optima comes with some hazards, if each episode effectively only comprises the descent to one single local optimum. Note that typically the episode will contain the local optimum reached. Such an episode division entails that rewards effectively are connected to one single local optimum only. Using rewards for one local search descent only means that the returns, which intrinsically are intended to represent a long-term goal will be computed based on information that stop at the next local optimum found. All return models discussed before (cf. Subsection 3.1.3) then necessarily will be connected only to a next local optimum and thus will represent rather short-term goals independently of how rewards are assigned. It is debatable whether this kind of short-term information can be used to represent the goal in solving COPs which reaches beyond finding a single preferably good local optimum.

To remedy the problem of locality in value function applications, learning can be extended to episodes with several local optima escapes and local search descent or can perhaps be based on continuous episodes in the first place. Then, using a return model that reflects the cost of the best local optimum reachable from a current state rather represents the long-term effects compared to the expected short-term success of only a single local search decent. In the case of the perturbation probing learning scenario, a state-value function will evaluate the long-term expected success of continuing an ILS from a state which was obtained after a perturbation application. This emphasizes learning a policy that directs the whole procedure of repeatedly local visiting optima instead of directing only one single local search descent. Even if used repeatedly, in the original perturbation probing learning scenario no real guidance will be learned, because no real policy that guides the search globally will be learned. What is learned instead is used to increase the number of probed perturbations substantially which in turn might improve the effectiveness of an ILS.

Local Search Probing Steps Learning Scenario

In this learning scenario, a number of actions in the form of different local search steps is provided. The local search steps can vary in the neighborhood structures and enumeration schemes employed. A learning LSA effectively learns via its strategy function to choose the best local search step in each state. It learns to decide about single basic moves. It can happen that no local search step is applicable, since the learning LSA resides in a local optimum common to all local search steps. Hence, local optima escape techniques are required, also. This can be remedied by incorporating large local search steps that perhaps are randomized and that practically never get stuck in a local optimum such as perturbations from the ILS metaheuristic.

As was concluded before (cf. Subsection 4.1.2), visiting local optima once in a while is mandatory. Any LSA consequently must ensure occasional local search descents also. If a learning LSA only uses actions yielding basic moves as in this learning scenario, this is not ensured automatically. In general, when employing actions that can visit suboptimal states, a guiding policy will have to know how to move from these intermediate suboptimal states to local optima at least occasionally. It is not clear how a policy can be learned that achieves this and directs periodical local optima visits while simultaneously aiming at the long-term goal. In fact, it seems at first very unlikely that a policy based on basic moves can be learned that guarantees descents to local optima *and* still keeps the long-term goal in "mind". Another problem with this learning scenario is that the moves used are too fine grained which might prevent the strategy function from generalizing properly for the whole state space, for example because of too many examples are needed for training (cf. Subsection 3.3.3).

In general, all learning scenarios that are based on actions that can yield suboptimal states as move targets as well suffer from two main disadvantages. First, occasional local search descents have to be ensured which requires additional arrangements. Second, the problem of locality of feedback in the form of rewards is given. If feedback does not reflect information about escaping local optima also (as is the case for the STAGE learning scenario for example), it cannot be hoped that escaping local optima can be learned other than as a side effect. Hence, if the aim is to learn a global guidance, it then rather seems advisable to use learning scenarios which will span episodes over several or better many local optima or at least incorporate those moves that escaped previous local optima, too, in order to capture the crucial information about how to escape local optima.

For learning to achieve the goal in solving COPs it is crucial for a learning LSA to learn to escape local optima. If no information about such escapes is provided to a learning LSA, it cannot be hoped that this be learned. Any goal a learning LSA follows then can only be based on local information and hence can only refer to a next local optimum to find. Consequently, one can only hope to improve behavior with respect to doing the best in a single descent. A learning LSA will learn to guide the search greedily and locally to a good starting state for a subsequent local search descent. This is comparable to a reinforcement learning LSA that always can do only one move – here composed of the local optimum escape and subsequent descent – until its return must be computed. Such a learning LSA will, when learning to maximize its return, in effect learn how to maximize its immediate reward and hence learn how to act greedily with respect to only the next action to take. This cannot be regarded as learning to follow a long-term goal.

Combined Local Search Steps and Procedures Learning Scenario

One possibility to remedy the missing guarantee to visit local optima is to ensure periodical local search descents. This can be achieved by providing complete local search procedures as actions also in conjunction with conventional local search steps. Any application of an action representing a local search procedure can be viewed as delegation of control by the policy for visiting a local optimum. The local search steps actions can well be perturbations. The purpose of a policy then conceptually will shift and become to guide the search towards promising start states for local search descents and trigger the descents. The policy is flexible when to trigger a local search descent in contrast to a learning LSA based only on ILS-actions where a local search descent is triggered after each perturbation.

It is not obvious how a policy can be learned that guarantees periodical applications of local search procedures. This, however, is necessary, since these will most probably yield the best states. It can be argued that as a trend, the better the start state for a local search descent is in terms of the cost, the better will be the cost of the resulting local optimum (cf. Section 2.5). Therefore, it seems not necessary for a policy to explicitly learn to find good start states for local search descents. Instead, a policy can be learned that strives towards good states and occasionally does a local optima probing in the form of a local search descent. Such a policy can be trained to have a "global view" with the potential to find promising regions and in terms of the long-term goal. This is in contrast to the problem of locality for learning scenarios trying to find only good start states for a single local search descent and which thus can be considered to have a "local view".

Nevertheless, it remains unclear how and whether it can be learned by a policy to trigger local search descents effectively. This basically means to learn when to apply the local search procedure actions and when to apply the other actions. Of course, local search procedure actions can be initiated policy-externally, e.g. after every n moves, but this initiation then will happen regardless of whether the current state is a good start state for the initiated local search procedure or not.

Local Search Procedures Learning Scenario

The problems of all learning scenarios presented so far in this subsection mainly are due to the fact that they employ actions that can move a learning LSA to suboptimal states also, for example in the form of local search steps. Visiting suboptimal states entails that local optima descents must be ensured somehow by a policy. As a result, it seems reasonable to devise better learning scenarios where a learning LSA only proceeds among local optima. This and the next learning scenario presented exactly do so.

The learning scenario described now only employs actions based on local search procedures. A learning LSA chooses a local search procedure, applies it and winds up in a local optima. Next, a local search procedure different from the one that lead to the current local optimum is chosen. This process can be repeated until local optimum common to all local search procedures has been reached. In this case, or after some number of local search descents in general, a perturbation is applied which escapes the (common) local optimum. This way, only local optima will be encountered. In the local search procedures learning scenario, a learning LSA learns to choose the proper local optimum to move to next based on a set of available local optima corresponding to the available local search procedures. If the available local search procedures are based on local search steps with varying strengths, learning to select the proper strength for the next move can implement a proper balance between exploitation and exploration. Some local search procedure might be designed and used to escape local optima common to many local search procedures available, thus acting as a perturbation.

Standard GAILS Learning Scenario

In the previous learning scenario, a learning LSA basically does only perform local optima moves. Still, escaping local optima is necessary. Having only a finite number of local search procedures as actions, however, it cannot be ensured that there will always be a local search procedure that can act as a local optimum escape. In the worst case, a learning LSA might arrive in a local optimum common for *all* available local search procedures. In order not to rely on luck, some escape mechanism for (common) local optima has to be included. This can be done with the help of a fixed perturbation which always is applied in a common local optimum. Instead of such a fixed escape strategy, the escape mechanism in the form of a perturbation can be incorporated in the learning process, too. This can be done by only allowing for ILS-actions where a perturbation with a subsequent local search procedure and perhaps a following acceptance criterion application is regarded as one coherent action. The learning scenario just described is the standard GAILS or GAILS standard learning scenario for which the GAILS method originally was contrived for.

ILS-actions can easily be built by providing n different perturbations, m different local search procedures, and perhaps k ($k \in \mathbb{N}^+$), different acceptance criteria and combining them in any of $n \times m \times k$ possible compositions, since the three components of an ILS-action are completely independent. By providing varying strengths for the ILS-action components, a learning LSA will have to learn to balance between applying exploring actions and intensifying ones. Any policy learned can be interpreted in this context directly. ILS-actions move a learning LSA from one local optimum to the next and effectively work on the smaller state subset S^*, $S^* \subseteq S$, of local optima. ILS-actions are of a rather large scale. A policy then can be viewed as strongly directing the progress of an ILS towards the long-term goal by repeatedly visiting local optima. Using local optima moves for awarding feedback in the form of rewards excludes some problems encountered when assigning rewards based on basic moves that can lead to suboptimal states: Ensuring regular local search descents and the problem of locality are not an issue, since any episode must contain several states in the form of local optima thereby automatically including some information about local optima escapes. Additionally, the set S^* of local optima must contain any goal state. Altogether, moving an LSA based on this set is more promising (cf. Section 2.5).

Any search trajectories of basic state transitions containing several local optima will have phases of local search descents and phases of escaping local optima, even if these phases might be blurred. Hence, any such search trajectory can somehow be considered to have been produce by a variant of the ILS metaheuristic. As has been argued, any local search method must ensure occasional visits to local optima and accordingly will produce such search trajectories. Guiding adaptively a learning LSA in doing such repeated visits of local optima is the main concept of GAILS. Seen this way, the GAILS method is a very general approach to solve COPs with many conceivable learning scenarios.

4.3 Goal Design

In the learning scenarios just presented, learning LSAs are designed in terms of applicable actions and partly how they learn. Actions in turn imply what kind of moves an LSA makes and accordingly when a learning LSA may receive reinforcements in the form of rewards. Yet, to completely specify a reinforcement learning problem variant for solving a COP and derived a learning LSA, the reward signal and the return model have to be devised also. Recall from Subsection 3.1.2 that the return model computes the performance measure. An actual return then is an actual performance assessment value. The computation of actual returns is based on rewards. The number and magnitude of rewards used for computing an actual return and hence a performance measure value thereby depends on the reward signal used, when and to which kinds of moves rewards are assigned, and how many moves can be made by a learning LSA during an episode used for computing a return. Accordingly, the performance measures only primarily depends on the return model used, but secondary also on the reward signal used and on the reward assignment design. Altogether, these design options induce or represent the performance measure for a learning LSA and hence the goal a learning LSA wants to achieve in learning and using a policy and accordingly are called *goal design option*. It is not always clear what kind of goal a given combination of return model, reward signal, and reward assignment design will actually induce, especially as concerns long-term effects, so the actually induced goal by a given combination of these goal design options for a learning LSA is called *actual goal*. These goal driving components must be designed properly such that the induced actual goal in fact reflects the true goal any learning LSA wants to achieve which is the goal in solving COPs: Find a global optimum or a reasonably good feasible state, secondary as quick as possible.

There are several return models available and several ideas for assigning and using rewards for a learning LSA. These comprise variations such as to which kind of moves rewards are to be assigned and how to compute rewards from the cost of states including the choice whether to include move costs or not. Another goal design option is whether the search process can be split into episodes of various types or whether it is continuous. Altogether, the following goal design options and combinations of concrete variations thereof (also called *goal design combinations* for short) have to be discussed in the context of goal design for an LSA:

- Timing of learning in the form of episode design.

- Timing of reward assignment in the form of the design of moves.

- Design of a reward signal in the form of a computation function computing rewards from cost of states.

- Choice of return model.

It has to be contemplated, which actual choices for the presented four kinds of goal design options are available and what actual goals for a learning LSA will result and whether these actually coincide with the goal in solving COPs.

This section presents several proposals how to classify moves, and how to define reward signals and move costs. All these proposals are discussed in the context of the known return models in terms of the actual goal the individual variations will direct a learning LSA towards. This section starts with a brief classification of moves indicating when rewards can be assigned in the first subsection, Subsection 4.3.1. Next, variations in how and when to split a search trajectory fruitfully into episodes are given and analyzed in subsection Subsection 4.3.2. Subsection 4.3.3 then will propose two reward signal designs and will combine them with other goal design options. The resulting concrete goal design combinations will be discussed regarding the actual goal induced by them. The next subsection, Subsection 4.3.4, deals with move costs and the last subsection, Subsection 4.3.5, finally concludes with a summary and overall conclusions.

4.3.1   Agent Moves

Anything that changes the current state of an LSA is considered to be a action (cf. Section 2.8). Any action an LSA applies results in a move from one state to another. Actions are extended local search operators, so moves are mainly determined by search state transitions. Depending on how actions are built, in the general case a move of an LSA can comprise any number of basic search state transitions. Any action eventually must be translated into local search operators inducing basic search state transitions available for the problem type at hand. Any basic search state transition is induced by a most basic local search operator in the form of a local search step. Local search steps work based on a neighborhood structure and an exploration scheme to select a neighbor to become the new current state (cf. Section 2.3). Any composition of local search steps such as a local search procedure can be virtually merged and regarded as a homogeneous local search step as well. Since there is a one-to-one correspondence of local search steps to neighborhood structures (cf. Section 2.3), this combined local search step can be thought of being based on a corresponding neighborhood structure which somehow is composed from the neighborhood structures of the composing local search steps. This new neighborhood structure for the new combined local search step is called *effective neighborhood structure*. A single neighborhood then is called *effective neighborhood*. Transferred to actions as extended local search operators, abstractly, any moves taken by an LSA involving any number of basic search state transitions can be regarded as being induced by actions that are based on exactly one adequately built local search operator in the form of exactly one adequately built local search step now only inducing exactly one basic search state transition. Special cases are local optima moves that make an LSA do not visit suboptimal intermediate states, for example resulting when using ILS-actions. Without loss of generality and viewed abstractly, for any theoretic contemplations it is assumed from now on that the moves an LSA makes (with the according actions) either:

1. are based on only one basic search state transitions, called basic moves (cf. Section 2.8), in the form of local search steps for an appropriately defined effective neighborhood structure (together with an appropriately defined scheme to select next neighbors, of course) that well can lead to suboptimal states, or

2. are local optima moves (cf. Section 2.8), i.e. the LSA only moves from one local optimum to the next.

If the cost of a state an LSA moves to, i.e. the move target, increases compared to the cost of the initial state of the move, the move is called an *improving move*. If the cost decreases, it is called a *worsening move*. If the cost remains the same, technically, it can be called either name.

4.3.2 Episodes

Trajectory method metaheuristics such as the ILS metaheuristic sample the search space in one continuous search trajectory. Many reinforcement learning techniques such as MC methods or direct policy estimation methods, however, work based on the notion of (finite) episodes. These reinforcement learning techniques can only be applied, if the search trajectory process can be split into episodes. This subsection briefly discusses how continuous search trajectories can in principle be split into episodes and which consequences the split might entail concerning learning.

One first glance possibility to split one continuous search trajectory into several episodes is by restarting an LSA after some time. Each episode begins after a restart and ends with the next restart (overall start and end of the search are "restarts", too). Restarts can be incurred for example when detecting stagnation. Stagnation can be defined as not finding a new overall best state for some period of moves such that it is expected that further improvements of the overall best state is unlikely to happen. Stagnation can be defined and detected for example with the help of runtime distributions according to [HS98, HS99, Hoo99a, SH01]. Also, a maximum length for a period that an LSA is allowed to proceed without improvement of the overall best state found so far can be set a priori. The period length can be expressed in the form of a maximum number of local search steps, move or other iteration counters allowed or a maximum amount of elapsed time, for example. Another possibility to produce episodes is to split search trajectories at each move with some small probability ϵ ($\epsilon \in \mathbb{R}^+$). This can either induce a restart of an LSA or the LSA continues as usual, only its continuous search trajectory is interrupted. The last state of an episode split this way needs not be a local optimum. Therefore, it makes sense to do or wait for ending a final local search descent to give the chance for a final improvement.

Another possibility to split single continuous search trajectories into episodes is to split at local optimum occurrences. This can be done in certain frequent intervals for example indicated by a local optimum counter, maximum episode size, or in terms of counters for some other criterion extracted from the search trajectory such as step or move counters. Such episodes will comprise several local optima. Another possibility is to subdivide single continuous search trajectories into episodes anytime a local optimum has been reached, either starting before or after a local optimum has been escaped. If parts representing the escape of a previously visited local optimum are not included, episodes of this kind of division basically only comprise local search descents. In the case of the GAILS standard learning scenario, each moving in the from of several basic moves from one local optimum to another can be viewed as an episode. The local optimum escaping perturbation can be included or not. In general, if local optima escapes are included and are moves, worsening moves of an LSA will be included as well. Finally, if learning scenarios are used that employ several local search steps and procedures simultaneously, the search trajectory can be split into episodes each time a local optimum common to all or some local search steps and/or procedures has been found.

Summarizing, four types of episodes can occur:

1. Episodes comprise only one descent to a local optimum without worsening moves in the form of so-called *pure local search descent*. This entails that an LSA does basic moves and hence visits suboptimal states, too.

2. Episodes comprise moves from one local optimum to the next local optimum including local optimum escape parts and hence potentially including worsening moves, again entailing that an LSA does basic moves and can visit suboptimal states, too.

3. Episodes comprise several local optima potentially including worsening moves on a basic move or local optima move basis.

4. No splitting is employed and continuous episodes are used comprising the whole continuous search trajectory on a basic move or local optima move basis.[1]

Note that episodes contain many states a learning LSA moves to. Learning can be triggered for all moves at any state of an episode. Note also that for an arbitrary state s_t of an *original episode* $s_0, \ldots, s_T$ ($s_t \in S, t \in \{0, \ldots, T\}, T \in \mathbb{N}$) – when a learning LSA starts in start state s_0 and move via states $s_1, \ldots, s_{T-1}$ to end state s_T – only successor states $s_{t+1}, \ldots, s_T$ can be used to compute the return for this state s_t. Hence, only the successor states of state s_t are relevant for the computation of a return for state s_t; any preceding states can be discarded. For this reason, the *effective episode* (in contrast to the original episode) for a state s_t, $\mathit{eff}(s_t)$ ($\mathit{eff} \colon S \to (\mathbb{N} \to S)$) is the state itself together with the successor states within the episode, i.e. $\mathit{eff}(s_t)(k) := s_{t+k}$ ($k \in \mathbb{N}$). Accordingly, the effective episode for a state is not the whole or original episode, $s_0, \ldots, s_T$, that was obtained after splitting a learning LSA search trajectory.

Recall from Subsection 4.2.3 that if episodes only comprise one local search descent and perhaps an additional single local optimum escape part (episodes of type 1 and 2), only short-term goals can be pursuit by a learning LSA. As has been elaborated before, one crucial design part for local search methods is how to escape local optima and find new promising starts for a new descent to a local optimum. Escaping local optima entails to visit suboptimal or even bad states (in terms of cost). Accordingly, especially these states are important and have to be evaluated when guiding a learning LSA by means of a strategy function and hence especially moves to and from these suboptimal states apparently should be reinforced. Perhaps, reinforcement should rather be given to moves to suboptimal states than to moves to local optima. This supports learning to find good starting points for descents to a local optimum. Otherwise, learning to identify good states is supported, which on the other hand, is not bad either, since good states typically are also good starting points for local search descents [JM97]. The problem of locality will be discussed also later in Subsection 4.3.3 when concrete goal design combinations are discussed.

Generally, if the aim is to learn long-term or rather global guidance, it seems advisable to use as much future information, especially about local optima and local optima escapes, as possible and weigh it relatively high. If an episode does not comprise many local optima, learning of long-term effects is not easy, if possible at all. Consequently, it seems more useful to divide a search trajectory into episodes that contain several or better many local optima or at least incorporate those moves that escape the local optima, too, in order to capture information about how to escape local optima. If episodes contain many local optima, the advantage is that more information about local optimum escape is extractable.

4.3.3   Reward Design

The actual goal a learning LSA pursuits is to maximize the return it receives for any of its coherent interactions with its environment. In learning, it therefore will learn a policy that will select actions yielding rewards that then maximize the return. The policy in particular has to take into account long-term effects. The actual goal a learning LSA pursuits accordingly mainly is determined by the design of the reward signal and the return model. Since only a few sensible return models are available (cf. Subsection 3.1.3), the goal driving momentum mainly has to be adjusted via the

---

[1] Strictly speaking, in the context of reinforcement learning this is not an episode. The finite-horizon return model cannot be applied for this kind of episode. For ease of notation, it is assumed to be a special case of episodes nevertheless (cf. Subsection 3.1.2).

design of the reward signal. It has to be contemplated, how reward signals for each return model can be designed such that the combination makes a learning LSA to learn a policy that drives the search process quickly towards good local optima and, secondary, do this quickly. Clearly, any design must ensure that local optima are still very attractive. As a matter of fact, any reward a learning LSA receives has to be connected to the cost function defined by the underlying COP type since this is the only available source of (reinforcement) feedback (cf. Subsection 4.1.1). For this reason, the choices for designing reward signals mainly are centered around the question how to incorporate the cost of a state usefully. Secondary effects in inducing the actual goal a learning LSA pursuits in learning and applying a policy stem from action and episode design. The latter two issues in goal design have been discussed already, so this subsection will be concerned with different ways to define reward signals and putting these definition together with the return models available, and the action and episode designs with the aim to derive and discuss resulting actual goals. The discussion in this subsection will pick up on the discussion of the previous subsections and will treat all possible goal design combinations. To recall, four goal design options with the following concrete variations have to be combined to derive a complete actual goal:

- Move definition (or, in other words, timing of reward assignment): An LSA does basic moves (basic move) or local optima moves only (LO move).

- Type of search trajectory splitting: An episode comprises only one local search descent to a local optimum with (1 local optimum) or without (1 LS descent) the escape part from a previous local optimum (including worsening moves in the former case, excluding any worsening moves in the latter one). Or, episodes comprise several local optima including worsening moves (1+ local optima), or continuous episodes comprising several local optima including worsening moves are used (continuous).

- Return model used: Finite-horizon (finite), or infinite-horizon discounted or average-reward return model (infinite/average).

- Design of the reward signal function: Delta cost reward vs. inverse cost reward signal (as will be discussed next).

Before discussing the various combinations, it has to be noted that the choice of the return model is restricted by the first two kinds of goal design options. Trivially, if moves are local optima moves, an episode must contain several local optima, either as one continuous episode or as a long episode containing several local optima. Otherwise, the return necessarily has to be computed after each reward assignment and hence is equal to the reward. Since maximizing the return then means maximizing the reward, the aim of such a learning LSA will be to learn to act greedily with respect to the next action, in this case to act greedily in reaching only the next local optimum. This only by chance will reflect the desired long-term goal (cf. Subsection 4.2.3, "Local Search Probing Steps Learning Scenario"). In almost the same manner, if a finite-horizon return model is used, non-continuous episodes must be employed.

Treating the infinite-horizon discounted and the average-reward return model as one case, altogether 10 combinations of design choices independent from any reward signal design are conceivable and have to be discussed. Each will be discussed in terms of two reward designs. The resulting 2×10 sensible goal design combinations are listed in Table 4.1 using the abbreviations for the goal design option variations just introduced in the previous listing. These abbreviation will also be used to label the paragraphs in the next subsection that discuss the respective goal design combinations.

Except for episodes containing only search trajectory parts in between two local optima (episode types 1 and 2), the argumentation concerning resulting action goal that will follow in this subsection

LSA move	Episode Split	Return Model
basic move	1 LS descent	finite
basic move	1 LS descent	infinite/average
basic move	1 local optimum	finite
basic move	1+ local optima	finite
basic move	1 local optimum	infinite/average
basic move	1+ local optima	infinite/average
basic move	continuous	infinite/average
LO move	continuous	infinite/average
LO move	1+ local optima	infinite/average
LO move	1+ local optima	finite

Table 4.1: Feasible combinations of design options for actual goal design

are valid both for effective and original episodes (cf. Subsection 4.3.2), so no distinctions are made for these unless explicitly mentioned. Any episode comprising only one local search descent (episode type 1) is also a subset of an episode including the escape of the previous local optimum (type 2 episodes) also. When discussing effects for episodes of type 2 in a certain context later, the effects for subset episodes of type 1 will already have been discussed in this same context and can be transferred to the local search descent part (also denoted *descent part* for short) of the type 2 episodes. If such a transfer is not admissible, the effects will be discussed separately. Note also that for episodes containing both local optima escape and descent parts, the effects for states of the descent part are not as pronounced as they would be, if the episode did not contain the escape part also. This is because function approximation over a feature space over a state space will generalize over *all* states from escape and descent parts, diminishing individual learning effects. For example, update effects for states from a descent part will partly be made up for by update effects for states from the escape part.

Delta Cost Reward

The first reward signal design assigns rewards in dependence of the change in cost for each move made by a learning LSA:
$$r_t := \Delta(c(s_{t-1}), c(s_t))$$
where r_t denotes the reward obtained by a learning LSA moving from state s_{t-1} to s_t (s_{t-1}, $s_t \in S$) involving times $t-1$ and t ($t-1 \in \mathbb{N}, t \in \mathbb{N}^+$) where c, $c: S \to \mathbb{R}$, is the cost function from the underlying COP type. This kind of reward signal is denoted *delta cost reward signal*. Actual reward value are called *delta cost reward*. The delta function $\Delta: \mathbb{R} \times \mathbb{R} \to \mathbb{R}$ discussed here simply is a difference: $\Delta(x,y) := x - y$. The reward signal with the simple difference delta function is denoted *simple delta cost reward signal*, an actual reward is called *simple delta cost reward*. Other delta functions are conceivable, too, and can be more elaborate. Normalization of the cost function c is necessary in order to make the approach work across different problem instances (cf. Subsection 3.3.1). Two variations of a difference-based delta reward signal together with a finite-horizon return model have been used in [MBPS97, MBPS98].

In the case the search process can be split into finite episodes with states $s_0, \ldots, s_T$ the finite-horizon return model can be applied. The accumulated rewards and hence the return according to the finite-horizon return model (cf. Subsection 3.1.3) will become the net gain in cost from the start to the end of an episode:

$$R_t = \sum_{t=1}^{T} r_t = c(s_1) - c(s_0) + c(s_2) - c(s_1) + \ldots + c(s_T) - c(s_{T-1}) = c(s_T) - c(s_0)$$

The end state of an episode, s_T, needs not be a local optimum. However, since it does not make much effort to do a last local search descent, the last state of an episode can well be considered to be a local optimum. If a learning LSA can do worsening moves during an episode, instead of using the local optimum reachable from the last state of an episode, the best state in cost encountered during the episode can be used also as the last state s_T by stopping the episode there. Using the simple delta cost reward signal and the finite-horizon return model, the return of a state will reflect the net gain in cost a learning LSA is capable to obtain (anytime) during an episode following the policy that produced the episode.

Since the net gain is relative to the cost of the starting state s_0 of an episode, care has to be taken: if starting in a very bad (high cost) state finding only an average cost local optimum during an episode possibly will yield a higher return than starting in a very good (low cost) state yielding only a small or no improvement at all. It can even be the case that the starting state of episode of the second kind have a better cost than the best local optimum found in episodes of the first kind. It seems advisable to almost always start an episode in good state, since in principle it is true that having the biggest improvement relative to the start state of an episode will maximize the return and hence is desired when using the simple difference reward signal and the finite-horizon return model. But it is also true that the best state found must be reasonably good in absolute terms according to the goal in solving COPs.

basic move	1 LS descent	finite

Assigning rewards to basic moves if episodes comprise only one local search descent to a local optimum suffer from the locality problems mentioned before (cf. subsections 4.2.3 and 4.3.2). The finite-horizon return model based on simple delta cost rewards then will reflect how good the local optimum reached from a state will be in relative terms expressed by the net gain in cost. This only holds true, of course, if a learning LSA in fact is performing a local search descent which is not easy to guarantee for a learned policy (cf. Subsection 4.2.3). The finite-horizon return model together with the simple delta cost reward signal will represent the actual goal to do as good as possible in one local search descent. This, however, is not exactly the goal in solving COPs which is long-term in nature. If a learning LSA can and will do worsening moves, this goal design combination cannot be used to represent the long-term outcome of a learning LSA's search process anymore, since it only represents the special case of local search descent parts of a learning LSA's whole search trajectory. Consequently, it cannot be used to learn to predict the quality of a state in terms of being a good start state for continuing a learning LSA's search from there in the long run, thereby visiting several local optima.

basic move	1 LS descent	infinite/average

The infinite-horizon discounted return model computes a weighted average over an infinite sequence of rewards. Insofar, this return model is similar to the average-reward return model which averages each reward of a potentially infinite sequence with equal weight. The weight for later rewards decreases geometrically in the case of the infinite-horizon discounted return model, i.e. usually very fast. If using the infinite-horizon return model in the case of simple delta cost rewards assigned to basic moves and episodes comprising only pure local search descents, returns no longer reflect the net gain during the episode, but rather emphasizes the gains made at the beginning of the local search descent. The return computation then emphasizes the delta cost improvements obtained by moves to states visited directly afterwards over moves to states visited in the more distant future. That's why this goal design combination will foster steep local search descents in maximizing returns. The emphasis can be smoothed by parameter γ. The higher this parameter is, the higher rewards acquired later in a local search descent are weighted. For $\gamma = 1$ ($\gamma \in [0,1]$) the finite-horizon return model results. Employing the average-reward return model and simple delta cost rewards together with episodes consisting only of one pure local search descent has a similar effect. The actual goal induced by this goal design combination will be to maximize the delta cost per each move. This actual goal does not necessarily need to be correlated to finding better local optima, but rather reflects the goal of finding steepest local search descents to local optima, i.e. finding improvements quickly. Note that finding the steepest local search descent does not necessarily mean that better local optima will be found as has be shown when comparing best vs. first improvement strategies in [PM99]. Here again, the problem of locality of reward assignments emerges which prevents the resulting actual goal to incorporate long-term effects.

If assigning simple delta cost rewards to basic moves, episodes comprising several local optima and therefore several local optimum escape parts will contain rewards for worsening moves and hence negative rewards. In general, when assigning simple delta cost rewards to basic moves, the actual goal induced by the various combinations of return model, reward signal, and episode division most likely is greatly influenced by the effect that worsening moves and hence negative rewards have. If local optimum escape sequences are included into an episode, the resulting return regardless of the used return model will be of a less local nature and not only be related to the next local optimum reachable (cf. Subsection 4.3.2). By including worsening moves, a return can reflect the success of local optimum escape strategies as well. In the case of using the simple delta cost reward signal, a return will indicate how good a state is with respect to a learning LSA continuing the search from there in terms of the net gain that can be expected over one or more local optima visits, i.e. in the long run.

| basic move | 1/1+ local optimum | finite |

If episodes comprise only one escape from a local optimum with a subsequent local search descent based on basic moves, the finite-horizon return model together with these other goal design option variations will represent the actual goal of escaping a local optimum and subsequently do as good as possible in one local search descent. The return will be in relative terms reflecting the net gain in cost. An actual goal coinciding to the goal in solving COPs is not represented directly when maximizing the return by this kind of return model in this goal design combination because of the short-term nature of the rewards used for the return computation. On the other hand, the longer the episodes for computing returns are, in particular the more local optima and accordingly the more local optimum escape sequences they contain, the more representative for the outcome of longer learning LSA search processes the return becomes in principle. The return for the finite-horizon return model will converge with increasing episode length to the actual long-term net gain of a learning LSA search process starting from a certain state and thus will more closely reflect what a learning LSA can do in the long run. Still, it must be ensured that a learning LSA does frequent local search descents. Otherwise, the return will not be representative anymore (cf. Subsection 4.1.2).

| basic move | 1/1+ local optimum | infinite/average |

Using the infinite-horizon discounted return model, simple delta cost rewards, and basic moves together with episodes comprising one or more local optimum escapes (and local search descents) also will yield that the return for a state will mostly depend on the advances made right after visiting this state. Depending on the discount factor γ, long-term effects might not be influential, even if the episodes are long and contain many local optima including their escapes. This does not reflect the long-term in nature goal in solving COPs as directly as the finite-horizon return model but is still closely related, the more the higher parameter γ is.[2] Since worsening moves are incorporated as well, using the average-reward return model in this setting will promote a preferably high and steady improvement or a least decrease in cost for each move made by a learning LSA. The longer an episode, the more this steady improvement reflects long-term effects analog to the argumentation for the finite-horizon return model. This return model together with the other goal design decisions effectively will encourage to take greedy actions with some long-term influence, but the resulting actual goal does not directly reflect the goal in solving COPs. It rather is only somehow correlated to it.

| basic move | continuous | infinite/average |

The last goal design combinations including basic moves to be investigated is applying the infinite-horizon discounted or the average-reward return model to continuous episodes (implying that the search trajectory need not to be split into episodes). The effects with respect to which actual goal will result then are basically the same as if using very long episodes comprising several local optima as has just been discussed. This holds even more true for the infinite-horizon discounted return model where the discounting factor make episodes exceeding a certain length equal with continuous episodes, since rewards are weighted geometrically decreasing which is a very rapid decrease in computing returns. Concerning the average-reward return model, the rewards are used to compute an average, but beyond a certain number of rewards, the average has stabilized to the exact value anyway. In contrast, not too long episodes might support emerging necessary residuals in the case of an average-reward return model. This return model relies on temporary differences (residuals) in estimated returns for different states (cf. Subsection 3.2.8). If averaging over all possible state sequences, in the limit they all will be equal preventing from learning anything.

So far, only effects of goal designs combinations in the context of a learning LSA doing basic moves have been investigated. The following discussion for the delta cost reward signal will be concerned with a learning LSA doing local optima moves. Visits to suboptimal states are transparent to a learning LSA then; any state encountered by a learning LSA will be a local optimum.

| LO move | continuous | infinite/average |

Using the simple delta cost reward signal for local optima moves enables easy use of the infinite-horizon discounted return model. This combination together with continuous episodes will yield an actual goal to maximize the decrease in cost of the states (in the form of local optima) moved to during the search process. The return for a state will indicate how good one of the states, i.e. local optima, moved in the (near) future will be when starting the search process in this state. The infinite-horizon return model together with local optima moves and continuous episodes emphasizes the delta cost obtained by moves to local optima visited directly afterwards over those for moves to local optima visited in the more distant future. The discount factor can control how important it is to find good local optima quickly fro maximizing the return: the lower the discount factor is, the more important becomes it to find a good local optimum quickly and the more looses the cost quality criterion its importance. High quality local optima in the more distant future become less interesting. Since rewards in this scenario can well be negative, the infinite-horizon discounted return model also favors a preferably high and steady improvement which is the same as a preferably high and steady improvement in cost of the local optima found. This emphasis on a steady improvement is more at the beginning of a search starting in a certain state and vanishes gradually according to the discount factor as the search continues. This supports the desired capability of an learning LSA to find goal states quickly. Later, the moves made by a learning LSA can become worse and worse, allowing for more exploration when the local search process typically

---

[2]Clearly, in the end, if $\gamma = 1$, the infinite-horizon discounted return model is equal to the finite-horizon return model.

stagnates anyway.

The actual goal when using the average-reward return model in conjunction with local optima moves and the simple delta cost reward signal in a continuous episode setting will be similar to the effect of using the infinite-horizon discounted return model. Those moves will be preferable that have the highest reward because they give the best improvement. The average-reward return model in the just mentioned goal design combination induces an actual goal of establishing a trend of successively improving the quality of the local optima encountered, because this will maximize the return. This does at least strongly correlate to the goal in solving COPs. Note that this correlation here centrally is due to the fact that rewards are assigned to search state transitions from one local optimum to the next only, anything else being transparent for a learning LSA and hence for its learning ability. Learning then is in terms of finding good chains of local optima in the form of positive trends in terms of cost, directly, skipping any impeding details such as the need to visit and subsequently escape local optima.

| LO move | 1/1+ local optimum | infinite/average |

The effects of using long episodes including many local optima or a continuous setting with respect to computing returns are almost the same for the infinite-horizon discounted return model, since future weights of rewards decrease geometrically and therefore very fast. The return for the average-reward return model are averages which will only slightly differ between very long and continuous episodes. So the results of the respective discussion apply here too. Only if the episodes are too short, a return might not reflect long-term consequences sufficiently for representing the long-term nature of the goal in solving COPs. The same holds true if the discount factor γ of the infinite-horizon discounted return model is too small. The best long-term representation certainly can be obtained with the finite-horizon return model using long episodes. The resulting returns indicate how good a learning LSA can possibly do with a horizon that is only constrained by the length of the episodes used to compute it.

Inverse Cost Reward

The second proposal for a reward signal design made here is to assign a reward proportional to the cost of the target of a move instead of to the delta in cost of a move. The lower the cost of a state[3], the better is this state, so the reward signal must be anti-proportional in dependence of the cost. On the one hand, this can be achieved by using the inverted cost of a move target state directly as the basis to assign rewards to that move:

$$r_t := -c(s_t)$$

for a move from s_{t-1} to s_t and a cost function c. As an alternative, in order to always have positive rewards instead of always negative rewards and in order to bound the rewards, the distance to an upper bound UB ($UB \in \mathbb{R}$) can be used instead, which in the case of normalization to the interval $[0, 1]$ simply is 1:

$$r_{t-1} := UB - c(s_t) = 1 - c(s_t)$$

This kind of reward signal is called *inverse cost reward signal* and does not exhibit the problem of relativity of improvements made by a learning LSA that comes with the (simple) delta cost reward signal design. Actual reward values are called *inverse cost reward*.

When assigning an always positive or always negative reward to each move of a learning LSA, the return for the finite-horizon return model will keep accumulating rewards and the longer the

---

[3]Note that optimization without loss of generality is considered to be of a minimizing nature during this document (cf. Section 2.2).

episode the greater will be the return. Maximizing a return will be possible for a learning LSA by making as many move as possibly which certainly is not desired. To remedy this drawback, delayed rewards or better pure delayed rewards can be used instead (cf. Subsection 3.1.3). Here, only the last move of an episode will yield a reward at all or at least a substantial reward. The last move will be assigned the reward derived from the last move target:

$$\forall t \in \{1, \ldots, T-1\} \, . \, r_t = 0 \land r_T = -c(s_T)$$

where T denotes the end time point of the episode comprising states $s_0, \ldots, s_T$. If an episode stops in a local optimum this is equivalent to assigning a value proportional to the cost of the local optimum that was reached from a state in an episode. If an episode does comprise several local optima, the single delayed reward assignment for an episode can also be made with respect to the best state found during the episode (T then denotes the time point when the best state of an episode was encountered). The inverse cost reward signal variant that does assign only one reward during an episode is called *pure delayed inverse cost reward signal*. Actual reward values are called *pure delayed inverse cost reward*.

It is arguable whether it makes sense to use the infinite-horizon discounted or the average-reward return model together with pure delayed rewards, since the single delayed reward will be discounted or averaged with lots of zero rewards and the resulting return will typically become infinitesimally small if the effective episode exceeds a certain length. Episodes having a certain length, however, must be ensured to happen often enough to incorporate long-term effects. Of course, if effective episodes are short, even delayed rewards will yield a substantial return in the case of infinite-horizon discounted and average-reward return models, but these short effective episodes will not be able to reflect long-term information by a return. The infinite-horizon discounted and the average-reward return model can be used together with inverse cost rewards directly regardless of the type of episodes used, since they do not accumulate the individual rewards to an potentially unbounded sum as the finite-horizon return model but rather compute weight and unweight averages over an infinite sequence of rewards, respectively. The return in the case of the infinite-horizon discounted return model is bounded independent of the length of an episode if the rewards do not increase faster than the geometric discount, i.e. by γ^n, which is not to be expected in practice and impossible if rewards are based on normalized cost. Accordingly, using an infinite-horizon discounted return model with episodes of either finite or continuous type together with the original, non-delayed, inverse cost reward signal will yield bounded returns.

Several considerations have still to be made before plunging into the details of discussing resulting actual goals for the different goal design combinations in the context of a inverse cost reward signal. Combining the pure delayed inverse cost reward signal with a finite-horizon return model based on finite episodes will reflect an actual goal of finding good states in absolute terms directly: The return for a state will indicate the quality of the best state that can be reached from it. The return computed by the infinite-horizon discounted return model in this case will be the higher the better the visited states are, according to the discount factor γ preferably at the beginning of an episode. Although an actual return according to the infinite-horizon discounted return model, depending on the level of the discount factor γ, is dependent on the length of an episode, an infinite-horizon return model has to be used cautiously. If the effective episodes are not long enough to ensure that their length is at least T such that the final geometric weight γ^T practically is zero, this will still induce returns that are dependent on the length of an effective episode. But in order to capture long-term effects, γ must be able to be set relatively high. It seems to be preferable, to use continuous episodes in the first place to circumvent this problem. The average-reward return model will compute increasing returns the better the visited states are in terms of cost in general. From now on, the finite-horizon return models are only considered in combination with the pure delayed inverse cost reward signal, while the infinite-horizon discounted and average-reward return

model is only analyzed in combination with the inverse cost reward signal. These restrictions will
not always be mentioned explicitly in the following discussions.

basic move	1 LS descent	finite
basic move	1 LS descent	infinite/average

Consider the goal design combinations of using basic moves together with the infinite-horizon
discount or average-reward return model and the inverse cost reward, or using basic moves together
with the pure delayed inverse cost reward signal and the the finite-horizon return model. For these
combinations, if episodes do only comprise one local search descent, the same arguments apply
as were given respectively in the discussion of the simple delta cost reward design for these goal
design combinations. Returns computed by the finite-horizon return model then only represent
a short-term actual goal of finding a preferably good next local optimum regardless of any long-
term effects and will only be representative with respect to local search descent parts of a learning
LSA search trajectory. Since single local search descent episodes are relatively short, the returns
according to the infinite-horizon discounted return model might not be applicable either, since
episodes are not long enough for the discounts to decay and the return effectively is partly or even
mostly determined by the length of an episode. To remedy this, delayed rewards can be used also,
but again, making the returns strongly dependent on the episode length in contrast to the finite-
horizon return model. The returns computed by an average-reward return model will maximal if
the average cost of states found during a local search descent is minimal. On the one hand, this
certainly is somehow connected to preferring finding low cost states at the end of a local search
descent near the local optimum, but on the other hand is certainly not related to this goal directly
and it can be doubted whether this goal design combination will yield a useful actual goal.

basic move	1 local optimum	finite
basic move	1 local optimum	infinite/average

To remedy the short-term character of returns, episodes can be prolonged. The first step is to
prepend a preceding local optimum escape part to an episode only consisting of a local search
descent. Still, rewards can only be assigned to basic moves. The finite-horizon return model with
pure delayed rewards will basically yield the same actual goal as before, now simply taking into
account local optimum escapes in return computation which make returns more representative of
for reflecting long-term effects. The infinite-horizon return model will produce returns that are the
higher the faster a local optimum can be escaped and the sooner the next local search descent can
be initiated, since the less worsening moves occur during such an episode, the less inferior states
are visited at the begin of an episode and the less lower level rewards have to be accumulated
with relatively high weights. Again, effective episodes might be too short so any return might
become dependent too much on episode length. The average-reward return model will weigh all
rewards equally and therefore will induce a preference for fast and humble local optimum escapes
to maximize the return, too, but less developed. The escape and descent rate at the beginning of a
local search descent will be less influential in computing returns when applying the average-reward
return model. Together, the better states during the escape phase are, the better the resulting
return will be and hence the preference of a learning LSA acting according to the average-reward
return model in the goal design combination discussed now will be to produce such episodes. The
actual goals reflected by the infinite-horizon discounted and average-reward return models will then
be to quickly escape a local optimum and to quickly find a local search descent to the next local
optimum. This actual goal might not really correlate to the goal in solving COPs but instead might
foster to find many, low quality local optima, since fast escapes and descents can rather be achieved
for inferior cost local optima, i.e. ones which are not too "deep" as concerns the cost surface of a
COP.

basic move	1+ local optimum	finite
basic move	1+ local optimum	infinite/average
basic move	continuous	infinite/average

As was just mentioned, the relatively short length of episodes can be a problem. To remedy this in the context of (pure delayed) inverse cost reward signals, longer episodes spanning over many local optima or continuous episodes can be used instead. Assuming rewards are assigned to basic moves, the finite-horizon return model with pure delayed rewards using finite or continuous episodes, again, will reflect an actual goal of finding the best cost state possible, independent of when this happens. This way, long-term effects are considered. The infinite-horizon discounted return model will reflect the same actual goal as for using episodes that comprise one local search escape and one local search descent now including more than one escape and descent part. Yet, because of the geometrically decreasing weight only some local optimum escapes and local search descents will be influential and hence will be included in return computation. The average-reward return model is not as susceptible to be constrained to the nearer future and can capture more long-term effects when longer finite or continuous episodes are employed, since it computes an unweighted average. Consequently, the shorter with less delta in cost local optima escapes turn out to be during an episode and the steeper at the beginning and possibly the longer and gentler in the end local search descents are during this episode, the higher will be the return. Or, in other words, the faster the search dwells in the proximity of a local optimum again, i.e. the faster each previous local optimum can be escaped and the faster the next local optimum can be approached, the higher will be the return. This will yield an actual goal of finding as many preferably equal cost state as possible or perhaps plateaus of equal but relatively low cost states. A resulting risk is that this will drive a learning LSA to get stuck in overly exploiting a region of local optima with the same cost. Since the average-reward return model aims at minimizing an equally weighted average, it will foster finding many reasonably good local optima but not necessarily finding at least sometimes high quality local optima. This holds true especially if the rewards are assigned to basic moves as is the case here. Then, the average cost of all states visited during an episode, not only of local optima, is to be minimized and this is most likely achieved by keeping the average cost level of states other than local optima as low as possible. Circling around in a reasonable low cost region will achieve this, without guarantee that this region contains a goal state other than by chance.

LO move	1+ local optimum	finite

After the discussion of basic moves in the context of (pure delayed) inverse cost rewards, only local optima moves are examined next. For local optima moves, the finite-horizon return model in conjunction with episodes including several local optima (and a pure delayed inverse cost reward signal) will represent an actual goal of finding a local optimum with preferably good cost, regardless when. The better any found local optimum was during an episode (assuming an episode stops at the best local optimum found), the better the return will be. This does in fact reflect the primary goal in solving COPs but neglects the secondary one (cf. Section 2.2). This, however, can be remedied by including move costs as will be discussed soon (cf. Subsection 4.3.4).

LO move	1+ local optimum	infinite/average
LO move	continuous	infinite/average

If using the infinite-horizon discounted return model and the inverse cost reward signal, either with very long finite or continuous episodes, anyway containing many local optima, the actual goal will become to find a chain of local optima that decreases steadily, foremost as quickly as possible, since this will maximize the return. This goal design combination again favors finding good local optima quickly, now in absolute terms. This certainly correlates to the goal in solving COPs, however not directly, since finding many reasonably good local optima in the beginning (weighted relatively high) might yield a higher return than finding poor cost local optima at the beginning, but a very high quality local optimum later during the search (weighted relatively low, unfortunately). This is because the infinite-horizon discounted return model in conjunction with the inverse cost reward signal design will yield the actual goal to minimize the weighted sum of the cost of local optima moved to during the learning LSA search process. As has been argued for the delta cost reward signal design already, this return model emphasizes the cost of states visited directly afterwards over the states visited in the more distant future (states now are only local optima). Again the potentially short-term nature of the infinite-horizon discounted return model induced by rapidly decreasing weights γ^n of the future rewards might be a problem, which is compensable by adjusting the discount factor γ, though.

The average-reward return model is less susceptible to concentrate on the nearer future compared to the infinite-horizon discounted return model. The return of an episode will be the average cost of local optima found during the episode. The problem with this kind of averaging is the same as for assigning rewards to basic moves: Maximizing an average of inverse costs of visited states can more easily be accomplished by visiting many reasonably good quality states than by visiting a lot of really bad ones and a couple of really good ones. In practice, nevertheless, the latter scenario is preferred over the first one. This way, the average-reward return model might not induce a proper actual goal.

Other reward signal designs might be possible, too, but are not discussed here.

4.3.4   Move Costs

The primary goal in solving COPs is to find a goal state in the form of a global optimum or a reasonably good local optimum. The secondary goal is to accomplish this as fast as possible (cf. Section 2.2). If using an infinite-horizon discounted return model, the discount parameter γ can be used to control the importance of future rewards. If γ is close 1, all future rewards of an episode will almost equally influence the final return and hence the actual goal. As γ becomes smaller, the emphasis shifts to rewards obtained at the beginning of an episode. After some time, any reward obtained has virtually no influence anymore because of the geometric discounting. Parameter γ therefore can be used to adjust how important is it to find a goal state quickly: The lower γ is, the more important it is to find goal states soon. This kind of control for the secondary optimality criterion in solving COPs is not available directly when using a finite-horizon or average-reward return model. There, all rewards are weighted equally and hence will influence the return and accordingly the actual goal pursuit equally. Finding goal states quickly is not preferred over finding them late. This can well be desired, but it need not be this way necessarily.

One possibility to adjust the importance of the rate of reaching a goal state is to use *move costs*.[4] Move costs typically are positive values m_t ($m_t \in \mathbb{R}$) computed by a function m ($m\colon \mathbb{N} \to \mathbb{R}$) with $m_t := m(t)$ in addition to each reward r_t that are subtracted from the original reward r_t obtained in each move (also called *pure reward*) before the resulting difference (also called *actual reward*) enters the return computation as $r_t - m_t$. Move costs work as penalties or perhaps as additional rewards. The higher the costs for making a move are, the less desirable is making moves for a learning LSA. Since move costs are subtracted from a reward, the more move costs are collected, e.g. the more moves are done during a search in case of constant move costs, the lower the final

---

[4]The notion step cost, although used in the literature, would be a little bit misleading if used here, since a learning LSA does not "step" but "move" to a next state and hence costs are added to rewards that are obtained by moves. The notion of steps is reserved in this document for local search steps and moves of a learning LSA are not necessarily local search steps.

return will be as computed by any return model. Making fewer moves receiving only small rewards might pay off compared to making a lot of moves not receiving substantially better rewards but incurring much more moves costs such that the difference in reward quality between these two options cannot compensate for the move costs. Assigning a penalty for each move in the form of move costs will prefer short search trajectories to reach a goal state over long ones, since a long search will impose a high accumulated move cost penalty diminishing returns.

In general, penalizing each move via move costs can be used to control balancing between state quality and search trajectory length and hence runtime. This is a very general and important issue in combinatorial optimization since both the state quality and the runtime are to be optimized in any optimization problem. In fact COPs are really multi-objective optimization problems (cf. Section 2.2). Longer runtime is only worth it if substantially better states are found. On the other hand, move costs can well be negative, too, giving further special rewards, for example when reaching a new globally best local optimum during a search. In general, move costs can be used to add special reward aspects in the form of one-time or occasional rewards, and so on. Move costs need not be regular reinforcements adding to rewards for each move. Instead, they especially are suited to define special reinforcements that do not incur as frequently as rewards.

There are several possibilities how to design and assign move costs. Some will be discussed next. Fixed move costs can be used that add a fixed value $m_t = m(t) = m'$ ($m' \in \mathbb{R}$) for any move from a state s_{t-1} to a state s_t. Another possibility is to assign dynamic move costs. These dynamic move costs can be made dependent on various aspects or combinations of them such as:

- the state or its representing features: $m \colon \mathbb{N} \times S \to \mathbb{R}$ or $m \colon \mathbb{N} \times \vec{S} \to \mathbb{R}$,

- the time: $m \colon \mathbb{N} \to \mathbb{R}$, e.g. increasing or decreasing slowly as time goes by such as $m(t) := \frac{1}{t}$ or $m(t) := \frac{t}{1000}$,

- the actually taken action, possibly also dependent on the state or its feature representation: $m \colon \mathbb{N} \times \mathcal{A} \times S \to \mathbb{R}$ or $m \colon \mathbb{N} \times \mathcal{A} \times \vec{S} \to \mathbb{R}$ (e.g. a cost proportional to the length or the computational complexity of a move such as an ILS-action in order to penalize long or computation costly moves, or even vice versa, to encourage long search moves),

- special, one-time occasional negative move "costs" called *one-time rewards* that will increase a reward such as for finding a new overall best state so far (in order to direct learning towards finding new best states), or

- a value dependent on the cost improvement of a move: $m \colon \mathbb{N} \times S \times S \to \mathbb{R}$ with $m(t, s_1, s_2) := f(t, \Delta(c(s_1), c(s_2)))$ for some functional dependency f ($f \colon \mathbb{N} \times \mathbb{R} \to \mathbb{R}$).

Making move costs dependent on state properties in the form of state features potentially introduces the possibility to learn, perhaps a priori, move cost designs, too. Using state feature dependent move costs can also be used to integrate a priori knowledge, for example, by penalizing certain states that are known in advance to be very bad with respect to continuing the search from there. Sometimes, local search methods do not exclusively proceed in the space of feasible states but consider other candidate states that have to be repaired to yield a feasible states. Visiting infeasible states might be crucial in order to guarantee to reach any feasible states of the search space. Yet, a search must not stop at an infeasible state. Accordingly moving to an infeasible state can be allowed but has to be penalized to discourage staying at infeasible states for long. This can be done by assigning special move costs when moving to an infeasible states and perhaps by increasing the reward when moving from an infeasible state to a feasible one again. These kinds of move costs will reinforce short traveling times among infeasible states and simultaneously will foster returning to feasible ones.

By increasing move costs with time, very strong emphasis on short search trajectories can be given. Special occasional rewards such as each time when finding a new overall best state can be used to direct a learning LSA to learn policies that frequently lead to new overall optimal states and that might encourage exploration. Using move costs proportional to the delta in cost of a move made might foster making big improvements in the case of the (pure delayed) inverse cost reward design where no delta in cost per move is considered per se. Move costs can also be used to penalize taking certain actions that are not to be excluded in advance but should only be used with care and rather occasionally, for example because they need a long computation time. Nevertheless, it might be necessary to apply such actions from time to time in order to maintain feasibility or to ensure that all states are reachable, or something else. For example, one problem mentioned when discussing the combined local search steps and procedures learning scenario in Subsection 4.2.3 was to ensure local search descents. This can be remedied by rewarding taking local search procedure based actions with the help of move costs which are assigned to these actions and perhaps penalizing taking perturbation actions too often.

Finally, accumulated penalties can be used to detect stagnation in the case of a continuous episodes using an infinite-horizon discounted return model and the simple delta cost reward signal. If a search does not find or improve to a good local optimum within a reasonable amount of time, then the recent move costs will superimpose the original rewards which might indicate search stagnation and a necessary exploration. If the accumulated rewards in the form of the return (including the negatively entering move costs) for a not too recent state fall below a certain level, recent moves were rather unsuccessful, hence the search perhaps must explore.

4.3.5   Summary and Conclusion

The rather theoretic contemplations about designing actual goals for a learning LSA have to be verified in practice. This subsection is intended to summarize the most important insights or rather hypotheses and to draw some conclusions in terms of which goal design combinations seem most promising. These should be tested in experiments first. Note that it can well be the case that some of the hypotheses about goal design combinations turn out to be wrong, that some seemingly inferior ones will in fact produce better results than expected, or that other goal design combinations because of new detected prosperous side-effects emerge. In almost the same manner, predicted effects of goal design combinations might not occur, while other effects might appear that have not been anticipated and hence discussed here.

Summarizing, devising actual goals using basic moves suffers from certain inherent problems. First of all, it is not straightforward to ensure frequent local search descents to local optima and which has been identified to be almost mandatory in practice (cf. subsections 4.1.2 and 4.2.3). In this context, the actual goal has to be designed in such a way as to drive a learning LSA towards learning to escape local optima also. The drawback might be that it thereby misses to keep the goal in solving COPs in mind which is long-term in nature. Using basic moves, the granularity of reward assignments might be too fine and might produce too many too similar training examples which might impede learning when using function approximation (cf. Subsection 3.3.3). Note also that an actual goal trying to establishing a positive trend in encountering local optima is something different from establishing a positive trend while doing basic moves. The latter actual goal will make a learning LSA to seek steep local search descents to local optima, independent from the actual quality of the resulting local optimum.

In general, if assigning rewards to basic moves, the actual goal induced by the various goal design combinations is greatly influenced by the effect that worsening moves (and hence either negative rewards or lower value rewards) have. The issue of escaping local optima, which, of course, is important but rather of a local nature, might handicap better focus on the long-term aspects of the goal in solving COPs which is more than just escaping local optima. Instead, an actual goal should rather make a learning LSA to try to balance exploitation and exploration on the level of local optima visits in order to better reflect the goal in solving COPs.

All these problems with basic moves just mentioned do not occur in the context of local optima moves. Additionally, problems with necessary periodical local search descents to local optima will not occur either. Instead, any actual goals induced by any respective goal design combinations basically will become to find a chain of local optima that decreases, foremost as quickly as possible. This better coincides with the goal in solving COPs. Visiting only local optima seems reasonable for a learning LSA to prevent from producing too many hardly different training examples as well. In a sense, only important training examples are produced, since the set S^* of local optima must contain a goal state. No learning effort has to be waisted for learning how to escape local optima. Altogether, it seems more promising to employ goal design combinations and learning scenarios such that moves are based on local optima moves.

After having decided about what kind of moves are best to employ, another important goal design option concerns which return model to use. All three available return models proposed (cf. Subsection 3.1.3) suffer from certain problems. The infinite-horizon discounted return model is susceptible to take into account only short-term rewards in the nearer future in dependence on discount factor γ. The short-term sensitivity of the infinite-horizon discounted return model is in contrast to the rather long-term nature of the goal in solving COPs. On the other hand, by changing the discount factor, the balance between long-term and short-term quality of an induced actual goal can be adjusted, perhaps even dynamically. Additionally, a short-term oriented view leaves room for later exploration, again aiming at balancing between exploration and exploitation. The average-reward return model is more long-term oriented since it weighs sooner and later rewards equally. It is not clear, though, whether the average-reward discounted return model can in fact reflect the goal in solving COPs. Maximizing an average over inverse costs of visited states, for example, can more easily be accomplished by visiting many average quality states than a lot of really bad ones and a couple of really good ones. Maximizing the average of rewards when using the delta cost reward signal design on the other hand will only induce a steady improvement in general, which not necessarily needs to include the best local optima. Both the infinite-horizon discounted and average-reward return models do not directly reflect the goal in solving COPs but rather are more or less only strongly correlated to it.

Using long effective episodes containing many local optima and the finite-horizon return model with pure delayed rewards perhaps does reflect the long-term goal best. But it is difficult to obtain long effective episodes for many states and hence comparably few training examples might be available which impedes learning. This is because typical trajectory method metaheuristics proceed in one continuous sequence of states and splitting into episodes is rather artificial. It cannot be predicted in advance whether the episode splits are reasonable. Without splitting, however, learning can only take place once and offline after a learning LSA has been run.

An alleged problem for the standard GAILS and other learning scenarios when using the simple delta cost reward signal is that moves can also have negative rewards assigned. In principle, this is perfectly fine. The reinforcement learning techniques presented before can handle negative rewards as well (cf. Section 3.2). For these techniques to work properly episodes must comprise more than one local optimum, though. In fact they should comprise a lot of local optima, in principle, the more the better. If episodes become too long, MC-based reinforcement learning techniques are hardly applicable because, again, learning can only take place once and offline after a learning LSA has been run. But all other reinforcement learning techniques are based on continuous tasks anyway, so no split into episodes is needed. This leads to another issue to be contemplated: Which

reinforcement learning techniques supporting certain return models are available? Entailing, how
elaborate and numerous are these techniques and which kind of episodes do they typically support?
Among the two models suitable to be applied to a continuous episode setting the infinite-horizon
discounted return model has the most reinforcement learning methods developed for it in the form
of Q-learning.

A problem mentioned when introducing the simple delta cost reward signal is that any rewards
obtained and hence any return is relative to the cost of the state from which a learning LSA starts.
This problem is relevant for any kind of move definition and in particular for local optima moves as
occurring in the standard GAILS learning scenario, too. In this scenario, however, the acceptance
criterion can easily control the quality of the next local optimum to continue the subsequent search
trajectory from. By always starting in a good local optimum and only ending in a comparably
good local optimum, the absolute values of the reward assignments according to the simple delta
cost reward signal can be controlled and bounded. Not all new local optima can be rejected by
an acceptance criterion, otherwise no learning examples will be obtained. A typical acceptance
criterion for the ILS metaheuristic is to only accept local optima if they are better than the current
one. This effectively means to only visit local optima that are the overall best ones found so far. The
number of such moves is rather restricted in practice, so one learning LSA run will not yield many
training examples. Hence, it seems advisable to occasionally move on by occasionally accepting
worsening moves also, either with a small probability ϵ or when no new overall best local optimum
was visited for some runtime, measured in local search steps, iterations, time, or whatever. The
probability of accepting a worse new local optimum can be made dependent on the decrease in cost
similar to the procedure of the Simulated Annealing (SA) metaheuristics. By varying the reward
values via the acceptance criterion, a counterbalance to discounted reward accumulations can be
obtained. The decreasing weighting of later rewards can be used to smooth greater and greater
absolute values for rewards when the search progresses. This allows for more exploration when the
search advances.

Summarizing, using local optima moves together with the infinite-horizon discounted return model
in a continuous episode setting and either reward signal design seems to be most promising and
thus should be tested experimentally first (cf. Chapter 6).

4.4   Related Work

The GAILS method tries to incorporate machine learning techniques to further improve metaheuristics. The resulting GAILS algorithms have a central learning and steering component and can be
regarded as being learning ones (cf. Section 2.7). Several local search methods, metaheuristics, and
variants thereof have been coined to be adaptive or learning, too. This section presents work that is
related to and partly inspiring the GAILS method. The related work presented in this section will
comprise metaheuristics and other local search methods or methods related to metaheuristics that
can be regarded as being adaptive or learning. The metaheuristics presented here will be reviewed
with respect to their learning ability and will be related to the GAILS method. Also, work related
to the hierarchical action structure concept (cf. Subsection 4.1.3). is reviewed.

Traditional metaheuristic that exhibit more pronounced adaptivity or even learning ability are introduced in the first subsection, Subsection 4.4.1. Genetic Algorithms (GA) have been considered
to be a learning method in itself. Therefore, GAs and variants attempting to incorporate an explicit
learning component, partly even in the direction of reinforcement learning solution techniques, are
described in the subsequent subsection, Subsection 4.4.2. Existing applications of reinforcement
learning methods for combinatorial optimization are covered in Subsection 4.4.3. This subsection

is concerned with reviewing those optimization methods that are most closely based on reinforcement learning directly; the approaches presented there are most relevant to the GAILS method. Subsection 4.4.4 covers any other applications of machine learning techniques to either directly solve combinatorial optimization problems or that augment existing methods. The last but one subsection, Subsection 4.4.5 reviews work for hierarchical reinforcement learning which is related to the hierarchical action structure concept. The last subsection, Subsection 4.4.6 finally compares the GAILS method to the related work presented and discusses similarities and differences.

4.4.1 Adaptive Metaheuristics

Several metaheuristics can be regarded as being adaptive. This subsection presents all those metaheuristics that more or less have some adaptive component included. The individual metaheuristics will be briefly sketched and any potential adaptation and learning components will be pointed out.

One of the oldest metaheuristics that can be said to incorporate some kind of data structure that adapts itself to the specifics of a problem instance is *Tabu Search* (TS) [GTdW93, GL97]. TS remembers aspects of the history of the search and tries to avoid to visit search states that have either been visited before or which have some common property with previously visited search states. Previously visited search states are deemed to indicate suboptimal search states in order to prevent from redundancy in the search effort. The aspects of the search history are stored in a so-called *tabu list*. A tabu list can store complete search states or it can store only aspects or distinctive characteristics of search states. The name of the metaheuristic and the search history stems from the fact that search states that are contained in the tabu list or that have common property with those stored in the tabu list are tabu and cannot be visited until the tabu inducing elements are removed from the tabu list. The behavior of TS changes during the search and can be viewed to weakly adapt itself to the specific problem instance at hand by means of the tabu list that stores which aspects of search states indicate already searched and hence "suboptimal" regions of the search space. Because TS algorithms remember information, they are also called *memetic algorithms* [MF99, Mos99]. By remembering already visited parts or aspects of the search space, the method can somehow be regarded as being adaptive. The search, on the other hand, is not directly guided by means of the tabu list but is rather constrained. The tabu list is not a central learning component. Transfer of the tabu list across problem instances seems to be possible, if it stores general size and instance independent search state attributes in the form of features, but has not been attempted yet. Another central learning element is not present, so TS cannot be considered to be learning.

One version of TS that dynamically determines the behavior influential length of the tabu list is called *Reactive Tabu Search* (RTS) or simply *Reactive Search* (RS) [BT94, Bat96, BP97, BP99, BP01]. The length adaptation is restricted to a predefined range of lengths. Within this range, the current tabu list length is computed anew in each step or iteration based on the frequency of revisited search states or search state characteristics. Coarsely, the length increases when repetitions of search states or search state characteristics occur and decreases when such repetitions do not occur for a sufficiently long period of steps or iterations. Since the length of the tabu list intrinsically is an algorithm parameter, this TS search variant rather is parameter tuning. Since the tuning happens during the search based on the search experience, it can be said to adapt itself. Learning does not take place, though, since no learned knowledge is remembered or transfered.

Whereas TS implements adaptive behavior by setting search states tabu, *Guided Local Search* (GLS) [VT95, Vou97, MT00, VT02, MTF03] manipulates the cost function of the COP dynamically for its optimizing purpose. The cost function is changed gradually with the aim to smooth out the current local optimum and thus to help the underlying local search procedure to escape it. This is done by adding changing penalties to the cost function for certain search state properties or rather features. If a feature is present in a current search state, e.g. a local optimum, it adds a varying penalty to the cost function, thus making the current search state increasingly less desirable. The idea of GLS is to smooth out and hence eliminate plateaus and to enlarge attraction basin of good or even globally optimal search states. GLS aims at alleviating a lot of the ruggedness of the cost surface of COP search spaces and thus effectively simplifying the search space. GLS adapts the penalties for features dynamically dependent on the current state of affairs of the search. Accordingly, it can be viewed as being adaptive and temporarily learning to penalize features indicating suboptimal search states. A specific component that encapsulates the learning part and that controls the progress cannot be identified, though. Learning experience is not stored persistently and has to be collected anew for each new region of the search space. Transfer of the learned across several problem instances therefore is not possible either, since the learning experience in the form of penalties is only relative to the current stage of a search. By normalizing features, it is conceivable in principle, though. Altogether, GLS can be regarded as adaptive, but not learning.

Another metaheuristic that even contains the word adaptive in its name is *Greedy Randomized Adaptive Search Procedure* (GRASP) [FR95, FR01, RR02]. This metaheuristic combines construction heuristics and local search procedures. It repeatedly constructs a new start search state for applying a local search procedure by using a given construction heuristic. The adaptive behavior of GRASP stems from its construction part. A complete search state is constructed by adding stepwise one solution element after another to a partial search state, i.e. by repeatedly assigning solution elements to solution components (cf. Section 2.1), until a complete search state has been built. Note that being a complete search state implies that it is feasible as well (cf. Section 2.1). The sequence of solution components that get assigned a solution element remains fixed. For each solution component get an assignment, a list of m ($m \in \mathbb{N}^+$) possible solution elements, i.e. possible assignments to the solution component, is ranked by means of some heuristic value. Among the first $n \leq m$ ($n \in \mathbb{N}^+$) such solution elements, one is picked at random. At each step, the heuristic function that induces the ranking among possible solution elements to be assigned is updated as to increase or decrease the desirability of some solution elements to represent the respective solution component in a complete search state. Because of this changing ranking that is supposed to reflect the usefulness of a solution element with respect to the optimization goal, GRASP can be viewed as being adaptive. It can be said that a GRASP learns to identify good solution elements which in turn can direct the construction method. The construction heuristic can benevolently even be interpreted as a controlling and learning component. Transferring the heuristic value is not attempted, because they typically change with the instance size but it could.

One variant of GRASP is called *Reactive GRASP* [PR00]. This variant tries to dynamically adapt a range parameter that controls which solution elements are possibly placed in the candidate list. Instead of using only one fixed value for this parameter, a set of such values is provided and for each one a probability is maintained. In each step of the search state construction, a parameter value is selected randomly according to the probabilities maintained and used. Thus, the candidate list changes dynamically and accordingly the search state construction process itself changes dynamically. The probabilities for the available parameter values are updated periodically. This is done according to the following scheme: the better the search states found after construction and subsequent local search procedure application with the help of a certain parameter value, the more its probability gets increased and vice versa. That is, those values that tend to lead to better search states in total will be used more frequently in the future. Again, adaption only comes through parameter tuning during the search process. This can be denoted adaptive behavior and but certainly is not learning, since no transfer of what is learned is undertaken.

One early variant of ILS is the adaptive multi-start algorithm from [BKM94, Boe96], called *Iterated Global Optimization* (IGO). The execution of IGO is divided into two phases. In the first phase,

n local optima are sampled randomly by generating n random start search states and starting a local search procedure from them. In the second phase, a set of $m < n$ best local optima found so far is maintained and a new start search state is constructed from this set of best local optima by combining good solution elements from the m best local optima. Then, the local search procedure is started from this newly constructed start search state. If the new local optimum is better than the worst of the current m best local optima maintained, it substitutes the worst local optimum. The steps of the second phase are repeated until a termination criterion holds. The rational behind this approach is that empirical experiments for several problems [MGSK88, Müh91, BKM94, Boe96] have revealed a tendency of good local optima to cluster in relatively close distance around the global optimum or other good local optima. Distance typically is measured in terms of local search steps needed to transform one search state into another. Effectively, distance indicates how many solution elements are common or similar, since local search steps work by incurring little changes of solutions elements. Accordingly, by combining aspects of the best local optima found so far, new, improved start search states for local search descents yielding even better local optima after applying a local search procedure are hoped for to be found. This resembles the procedure and rational of Genetic Algorithms which are described later in (cf. Subsection 4.4.2). In a sense, IGO is adaptive, but it certainly is not learning.

WalkSAT [SKC94, MSK97] is quite a successful local search procedure for the SAT problem [GPFW97, HS00b, HS00a]. Empirical investigations [MSK97] have revealed some relationship between the optimal performance of WalkSAT on the one side and the ratio of a randomization parameter of the WalkSAT algorithm and some search space property on the other side. The relationship is exploited in [PK01] to dynamically adjust the randomization parameter during the search process. This is done by repeatedly sampling search states and computing the search space properties involved. The randomization parameter is adjusted such that a certain ratio indicating best performance according to the empirical investigations is retained. A similar version of adaptive WalkSAT is described in [Hoo02]. The randomization parameter is adjusted according to how long no improvement has been found. Both versions of WalkSAT are not true learning approaches but can be called adaptive, since the search space property is computed from experience found so far and hence the algorithms somehow adapt themselves to the search progress and problem instance at hand. True learning has taken place only on the side of the investigator finding out about the invariant for WalkSAT only.

The *Ant Colony Optimization* (ACO) metaheuristic [DD99, DDG99, DS02, DS04] is a population-based construction metaheuristic. It is derived from the behavior of ants. A number of artifical ants construct a search state by adding new solution elements step by step in such a way to a partial search state that only *feasible solution elements* are added. A solution element is feasible, if the partial search state after adding the solution element can be finished to a complete and accordingly feasible search state and accordingly does not represent a dead end in the construction process. The ants are performing a randomized walk on the *construction graph*. The construction graph's vertices are the solution elements and the edges reflect adding the adjacent solution elements subsequently during a search state construction. It is completely connected. Each vertex and edge get a dynamically changing *pheromone trail* or *value* (or pheromone for short) that represents information gathered during the search and a heuristic information that represents a prior information such as the (increase in) cost of adding a solution element is assigned. Any artificial ant now traverses the construction graph building a complete search state. Some variants of ACO further employ a subsequent local search procedure to additionally improve the search states found by a single ant [Che97]. The transitions from one vertex to another, i.e. from one partial search state to the next, are randomized. The probability distribution that governs the selection of the next solution element to add, i.e. that governs the selection of the next vertex to visit in the construction graph, is governed by the pheromone and heuristic values. The probability of selecting an edge and an adjacent vertex is monotonically increasing with increasing pheromone and heuristic value. After some ant has constructed a complete search state, the pheromone values are updated such that the pheromone intensity for vertices and edges for solution elements of good search states found are increased and are decreased for bad search states. By means of marking edges, too, it can identify even combinations of single solution elements. An ACO metaheuristic adapts itself to a problem instance at hand by using information gathered during the search which means it is adaptive. It can be said that an ACO metaheuristic learns to identify good solution elements and combinations thereof. The pheromone trails are used to direct the search state construction process towards more promising search states. In this way, ACO is related to reinforcement learning. In each step, an ant has to extent its current partial search state. Each possible extension can be viewed as an applicable action. The value of an action is computed from the heuristic value and its pheromone trails for the potential solution element to add and the edge between this and the previously added solution element. The pheromone trails now can be interpreted as Q-values and the pheromone update rule can be regarded as an update rule similar to those from Q-learning methods including γ and α parameters (cf. Section 3.2). When all ants have finished their search state construction process, final delayed reinforcements are given to the Q-values, i.e. pheromone trails, depending on the quality of the search state found. A special implementation of an ACO for the Traveling Salesman Problem (TSP) that adopts updates and actions, i.e. next solution element selection rules, to the standard rules of reinforcement learning methods was presented in [GD95] and is called Ant-Q. The ACO metaheuristics cannot easily, if possible at all, transfer the learned pheromone values across different problem instances, in particular if they have different sizes. Since pheromone trails are attached to solution elements and combinations thereof and the number of solution elements typically changes between two problem instances and practically always between problem instances with different sizes, the learned experience is not easily or not at all transferable between different problem instances. The heuristic values and pheromone trails certainly influence the search progress and can even be said to guide it. Seen this way, the ACO metaheuristic certainly adapts to the problem instance at hand more than simple local search methods do. A central learning component that additionally controls the search process is not discernible directly. However, the pheromone trails associated to the components of the construction graph can be interpreted as action-values where actions consist of adding a new solutions element. Pheromone trails thus induce a policy which then can be regarded as a central and controlling learning component.

Rollout algorithms [BTW97, BC99] are a combinatorial optimization scheme that has connections to reinforcement learning and to Dynamic Programming (DP) in particular. Rollout algorithms are based on a construction heuristic that sequentially adds one solution element after another to finally form a complete search state. The construction heuristic can be stochastic but must ensure that each partial search state along the way can be extended to a complete search state. The construction heuristic is augmented to a rollout version following the policy iteration idea. At each construction step, the rollout augmented construction heuristic picks some solution component in the current partial search state with no solution element assigned to yet. Only those solution elements are added that do not make further extensions to a complete search state impossible, i.e. it only adds feasible solution elements. Consequently, n feasible solution elements to add will yield n new extended provisional partial search states. To each such n extended provisional partial search states, the original construction heuristic is applied m times yielding m complete search states per each of the n extended provisional partial search states. For each extended provisional partial search state, the average over the m complete search state resulting from is computed. This average is considered the approximate cost of the extended provisional partial search states. Finally, the extended provisional partial search state which has lowest approximate cost thus computed is used to repeat this process in the next step. This procedure continues until a complete search state has been found. The whole process can be viewed as a sequential decision problem and under some conditions [BTW97], as an MDP: For each step, the set of solution elements that can be added can be regarded as the set of applicable actions. By characterizing the search state via features independent of the size of a partial search state and using the final cost of a complete search state as a final (and delayed) reward, an action-value function can be learned, e.g. using TD(0), that in turn can induce a policy. Reward signals other than just final rewards are conceivable, too. This method was implemented in a slight variation in [MM98, MMB02] and applied to scheduling code as part of a compiler. There, an action-value function was learned over pairs of actions indicating whether to prefer one action over the other. The resulting partial ordering then was used as a preference structure to select an action according to the ϵ-greedy policy. Rollout algorithms in their basic version do not store any approximate costs computed which could be used to learn value functions, neither do they employ a central and controlling learning component. Whether the rollout approximations can be considered to induce adaptive behavior is controversial. The version of [MM98, MMB02] on the other hand does employ an explicit value function which is learned and used. This method certainly can be regarded as learning.

A value function application to construction local search methods is presented in [TS01, TS02]. The method described there maintains a training set of complete search states together with the respective costs. In each iteration of the main loop of the method, the training set is used to learn a value function which is based on features that can be computed for both partial and complete search states. The value function is intended to predict the cost of a completed search state. Then, a new search state is constructed step-wise by adding new solution elements using the value function previously learned to perform a one-step look-ahead search over feasible search states elements that can be added. Finally, at the end of the main loop, the training set is updated in consideration of the newly constructed search state and the loop is repeated. The value function is represented by a function approximator. The function approximator used in [TS01, TS02] is a weighted local learning function approximator and kernel adjustment is done using cross-validation. The method used described in [TS01, TS02] basically is reinforcement learning by Monte-Carlo (MC) sampling. Since the decision which solution element to add next is based on the value function and since this value function of course represents the learned information and transfer of the value function can be accomplished under the same restriction that apply for the other reinforcement learning based approaches discussed before (cf. Section 3.2), this approach can be said to be learning. The method strongly resembles the reinforcement learning approach using Rollouts presented before [MM98, MMB02].

4.4.2 Genetic Algorithms and Reinforcement Learning

The idea of *Genetic Algorithms* (GA) [KP94, Bäc96, Mit96, BFM97, RR03] is to mimic the evolutionary process of nature for optimization. GAs are population-based. They work by iteratively applying in each iteration a number of operators to a population of search states, called *individuals*, to generate a new population of individuals. Any new population of individuals resulting after each iteration is called *generation*. The first operation that typically is applied to a generation is called *recombination* or *crossover*. This operator works on two or more individuals called *parents*. It usually mixes and recombines solution elements or even more basic elements of solution encodings called *genes* to yield a number of new individuals, called *children* or *offspring*. Next, a *mutation* operator is applied to the children thus generated which changes some solution elements or genes in a randomized fashion. The (mutated) offspring population produced in each iteration typically is larger than the parental population. Each offspring can be further improved by applying a local search procedure. In order to maintain feasibility, search states might have to be repaired after recombination and/or mutation operator application. Each potentially improved offspring individual finally is evaluated with respect to the optimization goal. The evaluation normally is done by the computing its cost. A final *selection* operator then selects children that will form the next generation. Typically, those with highest quality expressed by least cost get selected. Other selection criteria are conceivable, too. The best children regarding the selection criterion used are called the *fittest*. For the same reasons recited for the GRASP and the ACO metaheuristic, it can be said that a GA learns to identify good solution elements or solution element combinations by concentrating samples of good search states in the respective current population. This is done via the crossover and the selection operator. The more favorable a combination of specific solution elements is, the more often will it occur in the individuals of a next generation. These recurring combinations of solution elements can be called building blocks and can be regarded as partial search states [MP99b].

The drawback of recombination operators can be that they too often split good coherent building blocks. One extension to GAs has been ventured with respect to further improving the recombination operator by strengthening the reuse of good solution elements and even bigger building blocks. Methods extending conventional GAs with probabilistic modeling are called *Estimation of Distribution Algorithms* (EDA) [MP96, PGL99, LELP00]. They try to establish a probabilistic distribution over the search space which assigns higher probabilities to better search states. The probability distribution then is used to bias a sampling of the search space hoping to find better search states. The sampling mechanism typically substitutes the mutation and recombination operators and can be combined with any local search procedure independent of any GA context to achieve further improvements of sampled search states. Feasibility of search states must be ensured, of course. The original rational is that GAs can be regarded as implicitly creating a probabilistic model of what good search states are by sampling search states during the generations with the help of recombination and mutation operators and filtering out bad search states by means of a selection operator. The population retained is a role model for good search states. In the case of EDA algorithms, for each new generation probabilities are updated according to some update rule in order to increase the probability of good solution elements or building blocks and to weaken the probability of bad solution elements and build blocks. Typically, the best search state among a new generation is used as a prototype that guides the update of probabilities. In one subclass called *Population-Based Incremental Learning* (PBIL) algorithms, the probabilistic models used are represented by assigning probabilities to individual solution elements or even to single genes or bits of the search state solution encoding indicating that they are deemed differently desirable for occurring in a search state [Bal94, BC95, dBJV97, BD97, PGCP99]. EDA methods explicitly employ a learning component which is the sampling operator. The learned experience is stored in the probability distribution. Such an algorithm directly learns to identify good building blocks for high quality search states. One problem is the maintenance of feasibility. Sampling according to probabilities assigned to solution elements or combinations thereof cannot ensure feasibility per se. Instead, repair or other methods have to be employed to yield feasible search states again. Another problem with this approach is that the experience learned is not easily or not at all transferable across different problem instances with or without different sizes, since solution elements are related to probabilities and the number of solution elements typically at least changes with respect to problem instance size. Another possibility to assign probabilities is by assigning probabilities to arbitrary combinations of solution elements forming arbitrary n-tuples of single solution elements. This approach represents mutual information dependencies and was implemented in algorithms called MIMIC [dBJV97]. This algorithms tries to model even more complex interdependencies by quite complex probability distributions representation methods. [LELP00] gives a quick overview over them.

One possibility to interpret probabilities associated to 2-tuples of solution elements or bigger building blocks is to interpret them as action-values that indicate the utility of adding the second solution element just after having added the first solution element while constructing a complete search state. This is similar to the Ant-Q view on ACO as was discussed before where the pheromone are used to compute such 2-tuple probabilities. The approach presented in [MP99a, MP99b] combines GAs with reinforcement learning techniques in such a way that new children are generated by first applying a recombination or a mutation operator to one or more parents such that they only will produce partial search states as children. These then are completed using the 2-tuple probabilities as action-values effectively yielding a policy for constructing complete search states. The update of the probabilities or rather action-values is done according to reinforcement learning update techniques after each new generation has been constructed. The reward needed to apply reinforcement learning techniques is computed from the difference in cost from the average fitness of the parent generation and the fitness of the new children after application of the selection operator normalized by the average fitness of the parent generation. No transfer of the learned over different problem instances has been undertaken in [MP99a, MP99b], The approach of deriving probability distributions for stochastically generating search states was formally embedded in the reinforcement learning framework in [Ber00]. The attempt there is to interpret probability distributions as a policy for stochastically and sequentially generating new complete search states from partial ones as has just been exemplified for 2-tuple probability assignments. In [Ber00], gradient descent update rules are proposed to improve the construction policy.

All EDA-based methods surely are adaptive. The solution finding process crucially depends on the sampling process or simply only consists of it. The process is guided by the probability values. Regarding the learning ability discussion for the ACO metaheuristic, the same arguments apply here when viewing the sampling process not in parallel but as a sequential construction process which necessarily is the case in practice. Perhaps, in the case of EDA-based methods, a learning component is more clearly discernible.

The COMIT algorithm [BD98] extents the EDA approach to not only work with GAs but with any local search procedure that can be iterated in independent runs. The COMIT algorithm maintains a set of n best search states and derives a probabilistic distribution over solution elements or combinations thereof as is done in other PBIL algorithms such that the probabilities represent the peculiarities found in the n best search states. Here, the probability distribution is computed potentially anew each time. The derived probability distribution is used to stochastically generate or rather samples k ($k \in \mathbb{N}$) complete search states. These n search states are intended to be the start search states for a local search procedure. These k potential starting search states are evaluated and the best $m < k$ such search states are actually used to start a local search procedure from. The best resulting local optima then can be inserted into the set of n best search states and used to derive a new probabilistic distribution for start search state generation. Instead of maintaining complete search states to derive a probability distribution, building blocks can be maintained, too. The COMIT algorithm does not provide for transferal of anything learned across different problem instances but certainly is adaptive. Its learning ability and adaptiveness basically is the same as was just discussed for EDA.

In a sense, a GA which employs an additional local search procedure in order to improve offspring before selecting the next generation can be viewed as a population-based version of ILS. The recombination and mutation operators together or individually can be interpreted as a perturbation. The subsequent local search procedure coincides with the one from the ILS while the selection operator of a GA is the population-based equivalent of the acceptance criterion of an ILS. The recombination operator is special in so far as it operates on two or more individuals instead of only one as a typical perturbation for ILS does. On the other hand, if an ILS is implemented with a number of simultaneous search trajectories interpreted as individuals, a perturbation that is applied on a number of individuals is a straightforward extension. Viewing GAs this way, methods for adaptive operator selection in GAs as reviewed in [SF97] can be regarded as ancestors of the GAILS method. The data structures and update rules used in [Jul95] to adapt operator application probabilities are comparable to utility values for operators and are similar to those used in standard reinforcement learning techniques. The RL-GA system [PE02, PE05] most closely resembles the GAILS approach. The RL-GA system explicitly employs a reinforcement learning agent that can select between several recombination and mutation operators in each iteration. The operators lead from one generation to the next. Each operator will produce one or more children which immediately replace the worst individuals of the current generation to form the next generation. The reinforcement learning method employed is Watkins' $Q(\lambda)$-learning [Wat89]. The action-value function is represented as a table. The search state information consists of features such as the duration of the search so far, the average fitness and the entropy of fitness of the current generation which have been cast into discrete intervals. This allows for a table-based implementation, but continuous features and function approximator application can be used in principle as well. Each reward needed for updating action-values is computed as the improvement of the best children generated over the best parent, normalized by the fitness of this best parent. The RL-GA system does not employ local search descent to further improve the children arising from operator application. One outstanding characteristic of this approach, however, is that it seems possible to transfer the action-value function across different problem instances even with different sizes. The experiments conducted in [PE02, PE05] with the RL-GA are done on quite small TSP instances of size 40 and 150, respectively. The instances for training and for application of the learned actions-value function were generated randomly. The RL-GA algorithm does in fact improve its behavior with respect to a GA of the same make but without the learning component. According to [PE02, PE05], the learned action-values indeed indicate some preferences for some actions and hence exhibit that something was learned. The learning version performed slightly better than the basic and standard GA employed. The RL-GA certainly employs a central learning component that learns to direct the search by learning to choose the proper operators in each iteration.

4.4.3 Reinforcement Learning Based Combinatorial Optimization

The first attempts to apply reinforcement learning techniques to a combinatorial optimization problem stem from [ZD95, Zha95, ZD96, ZD97]. The problem tackled there is a resource constrained scheduling problem from NASA. In resource constrained scheduling problems, the succession of tasks is not only constrained by dependencies among the tasks but also because of resource constraints. The reinforcement learning approach builds upon a local search approach for solving the problem [ZDD94]. A schedule in the reinforcement learning variant is built by first producing a preliminary potentially infeasible schedule with some constraints relaxed and then applying repair-based local search steps that reschedule tasks until a feasible and preferably good schedule is found. The reinforcement learning approach to this problem is to learn a value function that can predict which potential successor search states from a number of neighborhoods to move to next. Rewards consist of a little step cost for each step while the resulting schedule is still infeasible and a final delayed reward in the form of a normalized measure of the schedule length that is independent of the problem instance, and its size in particular. The step cost is intended to encourage to find feasible search states quickly. The value function is approximated by a feed forward neural network with one hidden layer and standard sigmoid activation functions. Input for the neural network is a schedule representation in the form of problem instance independent normalized features. Training works by solving a problem instance using the current value function approximation implementing an ϵ-greedy one-step look-ahead search over the set of possible successor search states. This one-step look-ahead is applied using the value function approximation until a feasible and reasonably good schedule is found forming a sequence of schedules. Each such sequence of schedules is used as an episode. For each schedule visited during such an episode tuples of features representing an infeasible schedule and the reward obtained when arriving at the respective complete schedule are built. These tuples then are used to compute in batches of episodes the updates according to TD(λ) for the neural network. For technical reasons, the updates are computed going through the training examples of an episode in reverse order. Additionally, the examples of the best schedules found so far are remembered to retrain the neural network for these schedules in order to prevent the neural network to forget about these good schedules. Since the number of possible successor search states can be quite huge for each repair operator that induces a large neighborhood, a sampling procedure called random sample greedy search (RSGS) is proposed as an alternative for best improvement search. This step variant only samples a representative portion of a complete neighborhood. The learned value function is instance independent and can be transferred across multiple problem instances even of different sizes. The value function centrally also guides the optimization process and can be seen as a central learning component. Accordingly, this approach can be regarded as a learning method for solving COPs.

Another early attempt to improve local search methods with reinforcement learning ideas is the STAGE algorithm (cf. Subsection 4.2.3, [BM98, BM00]). This algorithm learns a value function that evaluates how good a search state is with respect to the quality of the local optimum that can be reached from it when starting some fixed local search procedure there. This value function then can be used to find new promising starting search states for local search procedures when a local search procedure got stuck in a local optimum. The algorithm works by interleaving two local search procedures based on a common neighborhood structure. One local search procedure optimizes with respect to the cost function. When this procedure found a local optimum, a local search descent with respect to the learned value function is performed. This optimization effort can be interpreted as searching for a new promising start search state for the original local search procedure. When both local search procedures converge to a common local optimum, the whole procedure is restarted to a randomly generated new start search state. The implementation of STAGE as described in [BM98, BM00] uses search state features that represent condensed search state properties as input for a linear function approximator that is used to learn the value function. The samples for learning the value function are obtained by extracting for each search state encountered during the descent of the local search procedure that optimizes with respect to the original cost function some search state features and relating them to the value of the cost of the subsequently found local optimum. These examples are used to train the function approximator in supervised learning mode. This is a kind of MC-method for value function evaluation. It can be replaced by other standard reinforcement learning value function evaluation methods.

The algorithm called XSTAGE is variant of STAGE. In XSTAGE, value functions obtained for several problem instances can be transferred to unseen problem instances. One way to make value function transfer work is to normalize the features and the cost of a search state used to train the function approximator with respect to the instance size. Normalizing the features and the cost of a search state enables direct reuse of a learned value function for a new problem instance. Since only features are normalized for STAGE but not the cost, this approach is not taken in XSTAGE directly. Instead, a special property is used to combine previously learned value functions from n problem instances unchanged and letting them vote on the search state transition decisions to be made for the new problem instance. This works since the behavior induced by value functions are scale- and translation invariant. The absolute values of predicted costs become irrelevant and voting ensures equal weights for each value functions integrated regardless of the actual magnitude of the predicted cost. Averaging the cost predictions to choose the next neighbor for example would not be scale independent.

STAGE and XSTAGE can certainly be regarded as learning local search methods since they employ a central learning and decision making component as well as a means to represent and transfer the learned. There is a close connection between STAGE and the Rollout strategy. If one interprets a Rollout as a local search descent counterpart of a construction local search method, a single rollout is an estimation of the cost of the local optimum which is reachable from a certain search state (in this case a partial search state). If reinforcement learning is added to Rollouts recording and approximating such estimations, a STAGE variant for a construction local search method is obtained.

A variant of the STAGE approach to combinatorial optimization is due to [MBPS97, MBPS98]. This variant tries to learn a problem instance independent value function in a learning phase on a number of training problem instances. It then uses the value function in order to improve optimization performance on previously unseen problem instances in a performance phase. The problem tackled in [MBPS97, MBPS98] is the Dial a Ride Problem (DARP), a variant of the TSP [Ste78]. The value function is approximated by a linear function approximator employing four instance independent features. The cost of a tour, i.e. the tour length, is normalized using a theoretically proven lower bound estimate due to [Ste78]. In the training phase a 2-opt variant for the DARP is used in a number of episodes, each run on a randomly generated problem instance of random size (in a certain range). Even in the training phase, the 2-opt neighborhood is used together with the current approximation of the value function as guiding cost function. A step is only done, if it produces an improvement of at least ϵ ($\epsilon \in \mathbb{R}^+$) (introduced as transition costs) over the value of the value function approximation of the current search state. If no more such steps can be done, a final unmodified 2-opt local search procedure is applied with respect to the original cost function. This final local search descent with respect to the original cost function, is not recorded, though; only the cost found is stored. The search trajectories are saved as episodes by extracting features for each search state of the search trajectory and a batch version of undiscounted TD(λ) is applied to learn a value function. Two kinds of reward signals are used. One is the drop in the cost function plus a little step cost ϵ. A final reward is given at the end of a stored episode by the drop in cost to the cost found by the finishing 2-opt local search procedure. This reward signal is called Z transition cost. The other reward signal is composed of a step cost ϵ and a final reward in the form of the cost found by the closing 2-opt local search procedure. This reward signal is called M transition cost (cf. Subsection 4.3.3). The function approximator trained with Z transition costs is supposed to model the expected tour cost obtained when doing a 2-opt local search descent with respect to the value function (and a little transition threshold ϵ) and next doing a final unmodified 2-opt local search descent (with respect to the original cost function). In the M transition cost case, the function approximator estimates the 2-opt local search procedure for the original cost function results directly. By employing a value function that guides the search in the performance phase and by being able to transfer the learned value function, the STAGE variant for DARP from [MBPS97, MBPS98] truly can be said to be a learning local search method.

An approach to combine local search methods with some form of reinforcement learning is described in [Nar03]. This approach tries to learn to choose the appropriate next local search step for moving to a new search state in each current search state. Whenever a new search state has to be proceeded to, an algorithm according to this approach assumes to have some choice alternatives in the form of a set of available local search steps. For each such alternative, the algorithm maintains a weight that works as a utility value indicating the utility of the respective local search step to be applied next. The weights are used to induce a softmax (cf. Section 3.2) or randomly tie-breaking greedy policy for choosing the next local search step to apply to the current search state. After an advance to a new search state has been done, the weights are updated according to several possible update schemes; none of these, however, is a standard reinforcement learning update scheme. The updates generally work as follows: If the cost improved during a step, the weight for the local search step that was applied is increased, otherwise it is decreased. Since the next local search step to be applied is chosen in dependence of the weights which are updated dynamically dependent on the success of such a step, the method can be regarded as adaptive. The weights are problem instance independent in principle, since they are completely independent on any search state properties. The magnitude only depends on the search progress itself. Transferring them across multiple problem instances is not excluded per se. The main difference to the other approaches presented so far in this subsection is that the value function learned is completely independent from any search state features. Whether this method can be regarded as learning is debatable. The weights which are computed are non-stationary, i.e. they change dynamically during the search process and hence rather are derived statistics indicating how often a certain local search step application yielded an improvement.

4.4.4 Other Adaptive Approaches to Combinatorial Optimization

Other not yet presented extensions to local search methods that can be regarded as adaptive have been proposed, too. Additionally, reinforcement learning techniques have been applied to problems that are of combinatorial nature in principle, but are rather aimed at learning controllers or schedulers. Several methods for learning online controllers or schedulers for several problems using reinforcement learning have been proposed such as for switch packet arbitration [Bro01], production scheduling [SBM98], and other scheduling and allocation problems [GC96, CB96, SB96]. These problems are of combinatorial nature and are subject to optimization. Yet, the applications aim in a different direction. Instead of solving a specific problem instance once, a changing problem setting has to be solved dynamically in an online fashion. Reinforcement learning in these cases is used to learn a policy once which is suitable to act as controller or scheduler for all changes in the problem setting. One approach that stands out a little bit in this context, however, will briefly be mentioned now.

One extension to local search incorporates changing fuzzy sets for neighbor selection in a local search step. This approach, called FANS [PBV02a, PBV02b], employs a typical neighborhood structure and a number of fuzzy sets to generate in each step a so-called semantic neighborhood to choose the next search state from. The semantic neighborhood contains all the neighbors that are both neighbors with respect to the original neighborhood of the current search state and whose fuzzy set evaluation exceeds a certain level. The selection of the next neighbor from the semantic neighborhood then can be based on any combination of cost and fuzzy set evaluations, e.g. selecting the neighbor with the highest value for some or more evaluations. The fuzzy sets can reflect properties such as "acceptability" trying to mimic the idea that search states with improved cost will be more likely to be accepted. Whenever a new successor search state has been found, the fuzzy sets are updated using the according properties of the new current search state as representative. Local optima are escaped by using a VND approach, i.e. by using the next of a number of different neighborhood structures as soon as the currently used neighborhood structure together with the fuzzy sets has lead to a local optimum. The selection of the next neighborhood structure can change dynamically, too. Since the fuzzy sets are adapted using the search trajectory, FANS can be counted among the adaptive local search methods. Since the fuzzy sets are not centrally controlling the search and since they are not intended to be transferred across multiple problem instances, this cannot readily be denoted a learning method.

As has been pointed out in Section 2.7, and as is emphasized in [Bir04, p.4]

> [...] tuning should be considered as integral part of the research in metaheuristics.

Automatic tuning of parameters has been already mentioned to be something that could be done by GAILS algorithms (cf. Subsection 4.2.1). The GAILS method is not the first to put the tuning problem in the context of machine learning, though. According to [Bir04], the problem of tuning of metaheuristics has the characteristics of a machine learning problem:

- Training examples are given by results of running an algorithm in a number of candidate parameter settings called *configurations* on a set of tuning problem instances.

- The objective is to find the configurations that fit best, of course with respect to the tuning problem instances seen. The best configurations or rather their common traits then are the learned hypothesis or model.

- There is a need to generalize from behavior or performance observed on training examples in the form of results obtained for tuning problem instances to general behavior or performance on arbitrary problem instances. In the context of automated tuning, the problem of properly and correctly generalizing from training examples arises also. This entails the danger of over-fitting a learned model to the training examples seen for tuning metaheuristics, then better called *over-tuning*.

Based on these observations, a formal and theoretical analysis of the problem of generalization in the context of metaheuristic tuning and from a machine learning point of view is conducted in [Bir04]. The results of this analysis lead to a class of so-called *racing* or *race* algorithms for automatically tuning metaheuristics. Race algorithms have been proposed in the machine learning community for selecting models or rather hypotheses [MM94, MM97, DV99, Mon00]. Tuning metaheuristics the aim is to estimate the performance of candidate configurations on a set of tuning problem instances in an incremental manner. As soon as enough evidence is collected proving the inferiority of a candidate configuration in terms of performance (with respect to the tuning problem instances seen so far, of course), a candidate configuration is discarded and not considered for the following tuning problem instances anymore. This way, no effort is wasted for seemingly poorly performing candidate configurations; effort rather is concentrated on good candidate configurations. The race algorithms proposed in [Bir04] more precisely work as follows: Given a so-called *stream* of tuning problem instances and a set of candidate configurations for a given algorithm, the algorithms is run for each of the candidate configurations on each new test problem instance from the stream. Statistical tests are applied after obtaining new results for one or more new tuning problem instances from the stream, however based on the results of *all* tuning problem instances seen from the stream so far. After each such testing all candidate configurations which are significantly worse than the best candidate configuration are discarded. These two previous steps are repeated until all tuning problem instances of a stream have been processed. In the end, a set of candidate configurations which are not significantly different in performance with respect to the tuning problem instances seen is left. In [BSPV02, Bir04], a special race algorithm based on the results of the formal analysis of the metaheuristic tuning problem is proposed and implemented adopting a Friedman two-way analysis of variance by ranks as statistical test used.

4.4.5   Hierarchical Reinforcement Learning

Building actions hierarchically as done in the GAILS method by the concept of hierarchical action structures is not new. Hierarchical reinforcement learning is also concerned with hierarchies of actions. The aim is to abstract from details of a reinforcement learning problem in order to improve an agent's learning speed and performance. Abstraction is introduced by subgoals that are to be solved such as opening and closing doors for robots. Afterwards, individual high-level or *abstract actions* (abstract actions are also called *macros* or *activities*) [BM03]) solving such subgoals can be used. Abstract actions are built in hierarchies from so-called *primitive (one-step) actions* of the originally given set of applicable actions. Abstract actions can use other abstract actions and primitive actions and the original action set typically is augmented as the union of primitive and abstract actions.

Hierarchical reinforcement learning methods are able to solve two kinds of problems:

1. A *partial* or *sub-policy* (since only defined for a subset of the original state space) for achieving a given subgoal is to be learned. Sub-policies then define abstract actions by executing the policy and the included actions autonomously and preferably transparently as a whole. Hierarchies of actions occur by allowing sub-policies to call other sub-policies in turn. Besides learning to achieve subgoals, achieving the original goal has to be learned also by means of the augmented action set. In a more difficult version, proper subgoals are to be identified autonomously by an agent, otherwise they are given a priori.

2. Abstract actions are given and replace the original action set. An agent is supposed to learn a policy for the reduced state space of all states that are reachable for the new action set.

The second kind of problem in principle is the same as the original reinforcement learning problem, if the action abstraction remains transparent, because the abstract actions are used as blackboxes. The first kind of problem is more complicated. For solving it, reinforcement learning methods must be augmented to deal with the temporal abstraction induced by abstract actions essentially executing as a sequence of primitive actions. This requires new methods for proper value function estimation such as adopted Q-learning algorithms. The same way, hierarchically built abstract actions imply the need to learn a hierarchical construction of an overall policy. Typical reinforcement learning problems can usually be described as MDP. If several primitive actions form a coherent abstract action as is the case for hierarchical reinforcement learning, temporal abstraction must be included for an augmented action set yielding a so-called *semi-Markov decision process* (SMDP) [How71, BM03]: In SMDPs, the time between two action selection decisions can vary and is modeled as a random variable. With the help of SMDPs, some work has been done to solve the two kinds of problems just presented:

So-called *options* are proposed in [SPS98, SPS99] to tackle the first kind of problem. An option consists of a sub-policy, a set of states eligible for starting the sub-policy in, and a stochastic termination criterion. Options can call other options and primitive actions. The aim is to treat options as much as possible as primitive actions in adopting existing reinforcement learning methods. The adopted reinforcement learning methods are supposed to learn sub-policies for options in intra-option learning as well as a policy for the augmented action set to achieve the original goal [SPS98, Pre00]. As a result, gain of speed-up due to option usage can be attained without excluding usage and learning for primitive actions at a finer level, if necessary. Subgoals and options are provided a priori in this approach.

The MAXQ framework described in [Die00] decomposes an original MDP into a set of subgoals, called *subtasks*. The actions for solving the original goal are primitive actions or sub-policies for solving other subtasks. These sub-policies hence work as abstract actions and can be built hierarchically by allowing sub-policies to call other sub-policies. Each subtask has defined a pseudo-reward upon termination. With the help of pseudo-rewards and based on hierarchical subtasks and policies, value functions are decomposed hierarchically also. This decomposition is called *MAXQ hierarchical value function decomposition*. The MAXQ value function decomposition is used to derive a reinforcement learning algorithm for learning hierarchical policies. The work in [Die00] mainly is concerned with a proper decomposition of the action-value function for given hierarchies of abstract actions and deduces and augments Q-learning algorithms.

The concept of multi-step actions is introduce in [SR02]. Abstract actions in [SR02] are formed as so-called *multi-step actions* by repeating n times the same primitive action. The objective is not to learn abstract actions that are built hierarchically but to learn abstract actions automatically in terms of which action to repeat and how often. The method implicitly identify subgoals. Other work such as [Dig96, Dig98, HMK+98, MB01, Hen02, GM03b, SB04, MMHK04] also tries to identify useful subgoals and construct abstract actions for solving these subgoals.

The second kind of problem in hierarchical reinforcement learning is addressed in [PR97, Par98]. There, so-called *hierarchies of abstract machines* (HAM) are introduced which can be viewed as a priori given abstract actions. The intention is to reduce the original state and hence policy space. An expansion of the original action set and hence primitive actions are not used, only abstract actions are allowed to choose from. The abstract actions are seen as programs that execute autonomously and automatically, but nevertheless based on primitive actions and therefore the original states space. The use of reinforcement learning methods for the reduced state space is possible by computing rewards for the abstract actions by accumulating the original rewards obtained during the execution of the primitive actions building an abstract action.

The concept of hierarchical action structures in GAILS is similar to the approaches just presented in some ways. Hierarchies of actions finally are based on primitive actions which means that the same structural, since hierarchical, building principle for abstract actions and hence abstraction is employed. Also, action abstraction remains transparent for callers because abstract actions are used as black-boxes. In case of the GAILS method, only a reduced state space S^* is visible to an LSA when employing ILS-actions. This resembles the HAM approach. Therefore, the same justification apply reinforcement learning methods to the reduced state space directly applies to the GAILS method as long as rewards for abstract actions are accumulated sensibly (cf. Subsection 4.3.3). The differences of the GAILS method to existing hierarchical reinforcement learning methods mainly are due to what is learned. GAILS algorithms work with a reduced state space and applies reinforcement learning methods directly like the HAM approach, but do not bother to learn policies for abstract actions or even form abstract actions; GAILS does not identify subgoals automatically. Actions in GAILS algorithms are given a priori and execute transparently as a whole and automatically, so to say already have a fixed policy provided. The hierarchical action structure of the GAILS method as of now still is merely a structural concept. In contrast, work in hierarchical reinforcement learning mostly is concerned how to learn sub-policies and based on these an overall policies and to partly identify subgoals automatically.

4.4.6 Comparison

Lots of local search methods and metaheuristics have been invented and have been characterized as adaptive and even learning (cf. Section 2.7) In most cases, adaptiveness can be granted, but this is almost trivially true for any metaheuristic. In a sense, they all incorporate models in the form of memory structures that are updated according to the search progress so far and that somehow influence the choice of the next search step to take. But basically none of the traditional metaheuristics presented in Subsection 4.4.1 can readily be regarded as performing learning. Most do not have an explicitly identifiable learning component at all or it is not used centrally. Also, for most of the approaches presented in this section any strategy component remains the same in principle; it only varies along modes of operation that are stored in a model. Such behavior rather has to be regarded to be some kind of automated tuning which in its layout essentially is problem instance specific. Online tuning yielding adaptiveness is a nice feature to have, but learning as discussed in Section 2.7 requires more. It requires an explicit learning component that centrally controls and directs the search process. Finally, only some of the methods presented in this section attempt to transfer learned knowledge across multiple problem instances.

First attempts to really incorporate learning and transfer of knowledge across problem instances have been undertaken with the help of reinforcement learning methods. These establish value functions which can be transferred. The first endeavor to use reinforcement learning to solve COPs is due to [ZD95, Zha95, ZD96, ZD97] and indeed employs a central learning component. This component is realized via a value function and learns to choose actions in the form of construction operators. Learned value function approximations are transferred also by putting essential effort in inventing and proper normalization of features and costs. However, the approach described in [ZD95, Zha95, ZD96, ZD97] is restricted to construction heuristics. This entails that this method only applies for an episodic setting. Also, the neural networks used to approximate the value function there do not seem the first choice for value function approximation since neural networks require retraining (cf. Subsection 3.3.4 and [ZD97]) and hence episodes. This somehow restricts the applicability of the method. Also, no means such as action-hierarchies to speed-up learning by reducing the search space are considered.

Another early method using reinforcement learning to solve COPs is the STAGE algorithm and its variants [MBPS97, MBPS98, BM98, BM00]. STAGE can be interpreted as an instance of ILS metaheuristic. The local search procedure component of an ILS is represented by the local search procedure of STAGE optimizing with respect to the original cost function. The local search procedure that optimizes with respect to the learned value function in STAGE can be identified as the perturbation component of an ILS, since it is used to escape local optima. The acceptance criterion is not present in STAGE which amounts to an acceptance criterion which always accepts the next local optimum to continue the search from. As such, STAGE learns a good perturbation rather than a component that explicitly controls the search as is the aim of the GAILS method. The XSTAGE variant of STAGE additionally tries to transfer learned value functions and as such a learned perturbation scheme.

The RL-GA algorithm [PE02, PE05] most closely resembles the GAILS method The RL-GA reinforcement learning approach to solve COPs also can be viewed to learn to choose among actions in the form of operators for offspring creation. However, it is based on GAs and not on trajectory methods and hence does not directly fit in the notion of a single virtual agent in the form of an LSA, although the RL-GA approach also explicitly speaks of an agent and not only of value function approximation as in the context of the STAGE algorithm. The idea of abstracting actions for agents in reinforcement methods and hence a reduction of the search space is not employed in RL-GA, though, and hardly applicable, anyway. The RL-GA method has not been tested on larger instances yet, so their performance remains unclear as of now. Also, the value function is only learned in the form of a table over discrete feature values. This impedes transfer of learned functions and it is still to show that it scales for large instances without using function approximation.

Summarizing, the GAILS method is a consequent advancement of the attempts to solve COPs with the help of machine learning techniques presented in this section and tries to overcome some of the drawbacks or weaknesses of these methods such as easy transfer of learned knowledge or conceptual simplicity and clearness.

4.5 Summary and Discussion

This section summarizes briefly the GAILS method (cf. Subsection 4.5.1). Next, potential hazards for the practical application of GAILS methods are discussed in cf. Subsection 4.5.2 and based on these and preceding contemplations, the most promising choices for conducting first experiments to empirically verify the GAILS method are proposed in (cf. Subsection 4.5.3).

4.5.1 Summary

Local search methods and metaheuristics in particular must be extended with machine learning techniques to further improve them. The goal is to find and exploit helpful search space regularities by means of machine learning in the seeming "chaos" that huge and complicated search spaces such as those for COPs exhibit. The problem is that direct supervised learning is not possible, since nobody knows the global or reasonably good local optima. The only feedback available is of evaluative nature based on the cost functions of the COPs to be solved. Fortunately, in this case, reinforcement learning techniques can be applied. This is done by means of the notion of a learning LSA. A local search method can be viewed as a virtual local search agent (LSA) that moves from state to state by applying actions. It observes its environment in the form of the current search state and perhaps some information about the search hitherto. The search state encodes a solution to the COP to solve and together with additional information becomes an agent state. The applicable actions are extended local search operators which in turn come in the form of local search steps, procedures or any combination thereof that are available for the COP to be solved. The local search operator combinations are carried over to actions such that actions can be arbitrary hierarchical compositions of other actions forming so-called action-hierarchies. The problem of a learning LSA solving a COP then in effect is an MDP which in turn is a special case of a reinforcement learning problem. The GAILS method at its core is the concept of such a learning LSA applying actions in the form of arbitrary action-hierarchies that learns via reinforcement learning techniques to fulfill its long-term goal of finding a global or a reasonable good local optimum.

The bigger the moves induced by actions are, the more likely is the learning success. If actions available to a learning LSA are too fine grained, learning might be impeded. Moving in the space of local optima S^* is especially desirable, since this subspace must contain a goal state in the form of a global or a reasonable good local optimum. Seen abstractly, ILS is moving exactly in this space S^* so it is only natural to use it as the basis for designing a learning LSA. The actions then will be compositions of a perturbation, a subsequent local search procedure and possibly an acceptance criterion, together forming a so-called ILS-action. By learning to choose the right ILS-action from a set of different possibilities for each move, the seemingly key issue of balancing exploration and exploitation for success of metaheuristics can be addressed and learned directly. The conceptual simplicity and clearness of ILS-actions can foster human understanding of what was learned and why and how to use it for improving local search methods in general. This ILS inspired application and learning scenario is called the standard GAILS learning scenario. Since visiting local optima at least occasionally intuitively and empirically is superior over not doing so, all local search methods in the end repeatedly have to visit local optima. These can consequently be viewed as doing an iterated local search also. If this iterated local search furthermore is guided as in the case of a learning LSA, this observation also leads to the name and concept of GAILS, underlining the universality of the guided adaptive iterated local search (GAILS) method and centrally for the learning LSA concept. In this spirit, the GAILS method certainly is inspired by the ILS metaheuristic but is far more general along the lines that ILS is far more general as a schema than a special metaheuristic instantiation. Accordingly, ILS-actions are the standard and most promising learning and application scenario of GAILS, but other scenarios resulting from other action designs are well within the concept of learning LSA and hence of the GAILS method.

4.5.2 Hazards

Some remarks concerning foreseeable practical limitations or hazards when applying the GAILS method will be given in this subsection. Learning does impose some computational overhead. The question always has to be whether this overhead does in fact pay off. Perhaps learning is possible and does have some improvement effects, but in practice it might still be better to use a simple but fast local search method that simply visits more states. Often, the pure speed and the resulting quite extensive coverage of the search space of local search methods is one of the keys for success for these methods [LMS02]. This trait should not be given up easily. In order not to impose a computational overhead that rules out any efficient usage, some other requirements with respect to an efficient application of what was learned must be regarded. Trivially, the computation of the features must be efficient. Preferably, they have to be computed incrementally. The same holds true for the efficient computation of strategy function values through function approximators and to a smaller extent for the learning rate of the function approximators employed. On the other hand, there are as well use cases where runtime is of secondary priority. Finding better search states in just a reasonable amount of runtime is an improvement as well, so runtime should not be overemphasized either [BGK+95]. Also, there are several COPs that employ a cost function which is very computation-expensive to compute [EEE01, VBKG03, Kno04, KH05], for example when training neural network by means of local search methods, where the computation of cost function values basically requires to train and evaluate a neural network sufficiently [AC04]. The computational overhead introduced by learning and applying a strategy function might be negligible compared to the effort to compute cost function values in these cases. In contrast, it might pay off especially to exploit search space regularities to speed-up the search and reduce the number of cost function evaluations.

The GAILS method is intended to improve performance by finding and exploiting search space regularities automatically. This endeavor can fail. Besides a principle weaknesses of the GAILS method, performance problems might well be caused by and accounted to the underlying function approximators. In a learning LSA approach to solving COPs randomized actions will occur. The problems associated with noise that is introduced by the great variety of possible outcomes of action applications, in particular if actions are randomized, are the same for all strategy functions. It is hoped that the noise inherent in local search actions can be handled successfully by the function approximators used. If function approximators fail to stabilize learning in the face of noise this might well happen because of the yet too limited power of machine learning techniques and need not indicate the futility of applying machine learning techniques to capture search space regularities in combinatorial optimization. Even if learning of regularities does not work at first, this not necessarily means that it does not work at all, but might well have happened because the machine learning techniques are not matured and powerful enough yet to handle the huge amount of data in the form of training examples that will accumulate when taking on COPs. Existing regularities in this case are simply too hard to detect for existing function approximation techniques.

The contemplations regarding reward and move cost design were mainly aimed at analyzing long-term effects: The value function indicates, how good it is to move to a certain LSA state or to use a certain action in an LSA state when following the policy implemented by the value function thereafter. The acceptance criterion for ILS-actions influences the policy being followed also. For example, consider a better acceptance criterion (cf. Subsection 4.2.2) which only accepts improving moves. Since it can be expected that improving moves do not occur too often, the policy actually being followed mainly is determined by the decisions of the better acceptance criterion. This influence can be diminished by softening the better acceptance criterion to an ϵ-better acceptance criterion which accepts improving moves or any move with a probability of ϵ ($\epsilon \in [0,1]$). If ϵ is 1.0, any move will be accepted. Experience for several COP, however, show that the better acceptance criterion typically is performing best [MO95, JM97, JM02], so the influence of the acceptance criterion choice has to be investigated carefully, too.

Recall that the results from [BKM94, Boe96] suggest that good local optima cluster together and perhaps around global optima. Using a better acceptance criterion then exactly might be useful to establish a trend of improving moves among LSA states of these clusters. Instead of learning to follow a policy that is based on arbitrary LSA moves, one could combine value functions with the better acceptance criterion and learn for improving moves only. The change simply is that search trajectories will have monotonically improving costs, but they nevertheless are search trajectories. As a consequence, it will have to be learned to choose the proper action at a time that will most probably yield the next improving move and, as long-term effect, is more likely to yield similar improving moves thereafter, if following the learned policy further. No balancing for temporary escapes from search space areas by worsening moves probably will be learned then but perhaps this is not necessary. One problem with this approach will be that very little training examples will be available, since there are not too many improving moves during a search. This is true in particular for the later stages of a search when the best solutions are found and hence probably the most important improving moves occur.

Finally, care has to be taken when applying anything what was learned: Is what was learned used for what it was trained? Certainly, a character recognition learning algorithm is not fruitfully used to recognize and distinguish between human faces. The same holds true for the learning scenarios of this section. They have in common that, depending on how the strategy function was trained, different policies will result. Trivially, the actions used for training and for which the strategy function will be tailored for, must be the same during application. A strategy function can also be trained to evaluate the immediate result of an action. This does not reflect any long-term effects, though, which are to be included since the goal in solving COPs is long-term in nature. As an example consider the STAGE learning scenario (cf. Subsection 4.2.3). Consequently, care has to be taken when using learning scenarios and when designing other goal design options in order to make the reinforcement learning components together with the learning scenario reflect the true long-term goal one wants to achieve. In general, when analyzing the performance of GAILS algorithms and hypothesizing about the reasons for success or failure, the hazard of unintentionally misusing what was learned has to be considered.

4.5.3   Proposals for First Experiments

As has been summarized at the end of Subsection 4.3.5, using local optima moves for a learning LSA together with an infinite-horizon discounted return model in a continuous episode setting seems most promising. With respect to available learning scenarios, only the local search procedures probing and the standard GAILS learning scenario support local optima moves for a learning LSA. Considering the discussion about the learning scenarios in Subsection 4.2.3, the standard GAILS learning scenario seems more promising and should be tested first. This will keep the number of training examples lower than for basic moves, more likely preventing to overstrain function approximators and costing less computational overhead for training, since runtime issues have to be regarded to some extent, too. Accordingly, actions will be ILS-actions. To keep things simple, the acceptance criterion should be kept fixed among the ILS-action for first tests but should be varied among better, ϵ-better, and always accepting ones to check for these influences, too.

Using a state-value function implies performing a look-ahead search which certainly will take too much time taking into account the action design. Instead, an action-value function and accordingly a Q-learning approach to reinforcement learning should be used. Several Q-learning methods using the infinite-horizon discounted return model are available and can be tested using both reward signal designs from Subsection 4.3.3 in a continuous episode setting easily. In order to tune the balance between short-term and long-term bias of the resulting actual goals, parameter γ can be adjusted. For subsequent experiments, move costs are an interesting addendum for controlling the actual goal and especially the importance of the second optimization criterion in combinatorial optimization which is runtime. One-time rewards that increase regular rewards when finding a new overall best local optimum might be interesting to play with as well. Also, penalizing computational costly actions might be useful when runtime tuning is on the agenda.

Clearly, the size of the search spaces of typical COPs prohibits using tabular representations of strategy functions. Consequently, feature-based strategy functions have to be used. The number of features should not be too high in order not to overstrain function approximators and not to induce a lot of computational overhead on their side and when computing the features for each state encountered during search. A strategy function can be learned online according to the Q-learning method, but transfer across multiple problem instance is strongly recommended, inherently because of the computational overhead needed for learning: Reuse of what was learned is strongly advised. Using strategy functions this way based on features with the aim to transfer them across different instances of a problem type entails the need to normalize all features including costs in order to make them problem instance independent, in particular independent of the size of an instance. The method proposed before in Subsection 3.3.1 using lower and upper bounds seems appropriate and sufficient for the first tests. Action-value functions are best represented as one function approximator per action. As function approximator, for the first experiments, a selection of powerful, yet easy to implement or easy available function approximators should be used. Secondly, any used function approximators should be able to learn incrementally or should easily be adaptable. Several implementations for SVMs are available which can be reused and which can do regression, sometimes even on an incremental learning basis [SLS99a, Rüp01, Rüp02, EMM02, TPM03, Ban04]. Since SVMs are very fast and powerful, these are the first choice. Besides this, regression trees [BFOS93, Tor99, Gim05] offer themselves to be integrated and used for the first tests.

Altogether, first experiments are intended to test several Q-learning methods using several types of function approximators. Of course, this should be done for several problem types in order to examine the dependence on problem type specific properties, too. It can very well be that the GAILS method will work better for some problem types than for others. As a consequence, many combinations of learning strategies, function approximators, and problem types have to be tested extensively and exhaustively. Since the components learning strategy, function approximator, and problem type are basically independent from each other concerning any concrete implementation, they consequently can be combined arbitrarily, proper interfaces assumed, thereby reusing the implementations for each component. In order to enable such a reuse, which will speed up experimentation substantially and will enable extensive coverage of all important combinations of components, building a framework first that exactly reflects and enables this independence will pay off soon. Therefore, complementary to the abstract GAILS method, a concrete framework for the implementation of GAILS method based algorithms has been built within the scope of this thesis. The next chapter is concerned with developing the respective requirements and presents a concrete framework design reaching as far as describing important aspects of its concrete implementation.

Chapter 5

The GAILS Implementation Framework

Any new generic method proposed such as the Guided Adaptive Iterated Local Search (GAILS) method has to be evaluated. In order to estimate its quality, the principle workings have to be examined. This requires comparing many different algorithm variations originating from the method to find out which ones are best and under which conditions. The number of variations to be analyzed can be huge. The GAILS method presented in the previous chapter, Chapter 4, allows for many possible alterations yielding many different variations of concrete algorithms. Since the resulting algorithms are too complex to be analyzed analytically and since they are inherently randomized, investigating the GAILS method approach must necessarily be empirical in nature. Theoretical contemplations as done in the previous chapter can be made to exclude some variations beforehand while suggesting others, but in the end, research must necessarily continue in the form of conducting experiments by running algorithms. Aiming at a comparison of the principle workings, the different algorithm variations thereby not necessarily have to be efficiency optimized in the first place.

To support an empirical investigation of many different algorithm variations, it crucially must be able to instantiate or rather implement them rapidly in the form of executable programs. Since any new method has to compete with existing ones, in the case of GAILS with existing local search methods such as metaheuristics, rapid implementation of algorithms should be available for potential competitors also. In order to exclude as many method-external performance influences as possible, influences such as coding skills and tuning mostly should be eliminated, ideally leaving only the method underlying principle workings as source for varying performance. Summarizing, a tool for rapid prototyping or rather implementation of GAILS algorithms and other local search methods is searched for.

Rapid prototyping and implementation requires enabling reuse of code. A rapid prototyping tool can come in the form of an object-oriented *application framework* which exactly is intended to enable reuse [FS87, Fay99, FSJ99]. In order to emphasize the algorithm implementation aspect, it will be spoken of an *implementation framework* or simply *framework* indexframework!implementation for short. A framework for the GAILS method, called *GAILS implementation framework* (*GAILS framework* for short) must enable rapid implementation of trajectory-based LSAs (Local Search Agents) utilizing action-hierarchies and must provide for incorporation of reinforcement learning mechanisms in the case of learning LSAs (cf. Section 2.8). Framework and concepts of code reuse will be discussed in more detail later in (cf. Subsection 5.4.1). when reviewing related work in the form of other frameworks in the context of local search methods.

Such a code reusing, rapid prototyping framework for the GAILS method will be described in this chapter. The first section, Section 5.1, will develop concrete requirements for the GAILS framework. The following section, Section 5.2, will derive an object-oriented design that can be implemented with the C++ programming language [Str91]. The third section, Section 5.3, covers some implementation-specific issues, while related work is surveyed in the last section, Section 5.4. This section also reviews concepts of reuse and the GAILS framework in the context of related work and the reuse concepts.

5.1   Requirements

What are the requirements for the design of a framework for rapid prototyping LSAs in general and learning LSAs more specifically according to the GAILS method? Three top-level requirements, besides the need to reuse code, stem from the design of (learning) LSAs as developed in the previous chapters. These are:

- Model the state of an LSA.

- Model actions of an LSA.

- Model the learning extensions of an LSA to become a learning LSA according to the reinforcement learning approach.

These top-level requirements will now further be refined in this section. The refinements will result in general concepts that have to be represented in the design of the GAILS framework as components. The concepts names will be introduced set in italics. Note that learning LSAs are an extension of original LSAs and accordingly LSAs are a subset of learning LSAs – they simply do not learn. From now on, only learning LSAs are considered including the special case of non-learning. All such LSAs will be called LSA for short.

First of all, the concept of the state of an LSA must be represented, i.e. the concept of an *agent state* must be represented with the GAILS framework. This representation must capture any actual state an LSA can be in at any time during execution. Any (concrete) GAILS algorithm (variation) basically consists of an LSA (cf. Section 2.8). Such an LSA can be viewed as an action that is executed (cf. Subsection 4.1.3). As a consequence, an LSA state is the overall state of an GAILS algorithm. Note that in contrast to previous chapters, notion LSA state is not abbreviated to state (cf. Section 2.8) in this chapter. An LSA state certainly comprises the current solution encoding for the underlying combinatorial optimization problem (COP) instance to solve. Recall from Section 2.3 that such a solution encoding is given by the concept of a *search state*. An LSA state supposed to model the overall state for an LSA and hence a GAILS algorithm comprises more, though. When running an LSA, the problem instance specification must be stored. The concept of a problem instances representation (e.g. a distance matrix for the Traveling Salesman Problem (TSP)) simply is called *problem instance* and can somehow be thought of being attached to an LSA state, too. Furthermore, the best solution (encoding) found so far should also be memorized. This concept is denoted by *best search state*.

In general, any information about the progress of the search is of interest and certainly is information that determines an LSA state, too. Looking at metaheuristics such as GAILS, Iterated Local Search (ILS), Tabu Search (TS), and Simulated Annealing (SA), information about the search progress can consist of step and iteration counters, changes in cost or changes in counters since the previous action application, a search state history, a tabu list, a temperature, strategy function representations, termination criteria, and so on. Recall from Section 2.8 that this information is called heuristic information and is stored in the heuristic state part of an LSA state. The search of an LSA progresses by applying actions. Actions in the context of an LSA are extended local search operators (cf. Section 2.3 and Subsection 4.1.3). Any information changed by actions describing aspects of the search progress apart from the current search state is heuristic information. It is important to notice that heuristic information can be used to alter action behavior. In a sense, heuristic information represents states of actions. Heuristic information as the state representation for actions can be encapsulated yielding the concept of so-called *heuristic states*. Each action potentially can have its own heuristic state or perhaps several actions can also share heuristic states. Since heuristic information certainly belongs to the heuristic state part of an LSA state. In principle, the heuristic state part of an LSA state consists of several heuristic states. However, any heuristic information that is to be shared by *any* node of *any* action-hierarchy or action employed such as the best search state or the problem instance is not attached to a specific heuristic state but to the heuristic state part in general.

In practice, an LSA state is also determined by other global information from several resources besides search states and heuristic information. Such resources for example comprise timers and random number generators. They define an LSA state as well. Altogether, the components of an LSA state just mentioned are illustrated in Figure 5.2. They are labeled following the concept names as "pSearchState", "pBestSearchState", "pProblemInstance", "heuristicStates", "pTimer", and "pRandomNumbeGenerator". Note that Figure 5.2 and most of the other figures are drawn according to the unified modeling language (UML) [BRJ97, RJB98]. Each class is represented by a rectangle or box with up to three parts. This box is called *UML box*. The class name is written at the top, member variables are listed in the middle, while methods are listed at the bottom (cf. Figure 5.4). Member variables and method names are put in typewriter font. Class rectangles may also only contain a class name. Inheritance is indicated by a triangle, aggregation by a diamond shape. In aggregation figures such as Figure 5.2, member variable names are listed above the class rectangles that represent member variables. Names beginning with a small 'p' indicate pointers. Numbers attached to constituents and arrows indicate how many of such elements are contained or connected, respectively. A one is omitted by default. Class names long, double, and bool refer to the corresponding built-in types of C++ [Str91].

Besides the representation of LSA states, actions must be modeled yielding the concept of an *action*. Recall that an action besides being an extended local search operator is *anything* that changes the whole LSA state – not only the search state (cf. Section 2.8). Actions can use other actions and can build arbitrary action-hierarchies which again can be seen and used as a coherent action according to a black box view (cf. Subsection 4.1.3). An action-hierarchy thereby is represented by its root-action. In the context of the GAILS framework, it should be possible to build action-hierarchies according to the building blocks principle. Hence, action-hierarchies should be reusable. Recall that any action-hierarchy must have some interface actions specialized to the underlying COP type in the form of leave actions. Those leave-actions that are problem type specific are called elementary actions (cf. Subsection 4.1.3). Note that not all leave actions must be problem type specific as is the case for acceptance criteria for example. Elementary actions are used to implement for example local search steps and local search procedures. Elementary actions consequently are the interface relating any problem type independent action-hierarchy parts and hence any LSA itself to its necessarily problem type specific parts. The concept of an action in principle is rather abstract and problem type independent except for elementary actions which, however, can easily be exchanged (cf. Subsection 4.1.3).

Lots of local search methods such as an ILS metaheuristic do not terminate by themselves, so termination must be modeled explicitly. This is true for actions being extended local search methods also. As depicted in the pseudo code for the ILS metaheuristic (cf. Algorithm 2.1 and Figure 4.1), this can be done by means of the concept of a *termination criterion* (abbreviated to TermCrit in this figure) which unifies and encapsulates many conceivable means of determining when to terminate. The working of termination criteria can dependent on several information sources; in general any information from an LSA state such as time, aspects of the search state, and in particular heuristic information such as counters can be utilized. Since termination criteria certainly alter action behavior, they can be viewed as heuristic information as well (cf. Figure 5.2, represented by box labeled "terminationCriteria" there). Examples of termination criteria make an action stop after some maximum number of local search steps done or when some other iteration counter has reached a given maximum, as soon as a certain feature value is exceeded, or as soon as a maximum time has elapsed. Note that an action in principle can employ several termination criteria and that termination criteria can be shared.

In order to also implement learning LSAs according to the GAILS method with the GAILS framework, learning ability according to the reinforcement learning paradigm must be enabled. Recall that reinforcement learning methods use function approximators to realize policies. The special type of a function approximator is irrelevant in principle. Due to the size of the search space, function approximation must be based on features (cf. Subsection 3.2.2). Accordingly, the two concepts of a *function approximator* and a feature vector must be modeled (recall that the feature vector itself is also called features and the concept is also called *features*). Function approximators can work as black boxes and are independent from any action that uses them as well as from any actual problem type. All they have to do is learn from training examples consisting of a feature vector and a real-valued target and to compute a real value for a given feature vector. As such, features are the interface to function approximators. Since function approximators can be action-specific, e.g. for an action implementing a Q-learning with several other actions it contains, and intrinsically also contribute to the state of an action since they certainly alter action behavior, they are heuristic information as well and have to be contained in heuristic states (cf. Figure 5.2, represented by the UML labeled "functionApproximators" there). Other ingredients of reinforcement learning methods are *policies*, *rewards* and *move costs*. These concepts can come in different variations and hence need to be modeled flexibly, too.

All necessary GAILS framework components in the form of the concepts as described so far in this section can be arranged into three relatively independent parts as depicted in Figure 5.1:

- Any search state specific and hence problem type dependent components which only affect the search states of an LSA state are assigned to the *problem-specific part* or *problem part* (illustrated by the leftmost area in Figure 5.1 labeled this way). The components of this part comprise foremost solution encodings or rather search states and the implementation of local search operators such as local search steps and procedures. A search state is shown as box denoted by "Search State" in Figure 5.1 The smaller boxes attached and labeled according to local search step names for the TSP represent the local search operators in this figure. The attached box labeled "Features" represents the feature vector of all search state features. The problem instance also belongs to the problem-specific components and is represented by the box labeled "Problem Instance".

- Any actions (or action-hierarchies) as well as other GAILS framework components not specific to a certain problem type such as policies, rewards, or termination criteria belong to the *action-hierarchy part* (or *action part* for short). This part is visualized in the middle area of Figure 5.1. Two action-hierarchy sketches for a standard ILS and a learning variant according to Q-learning named "ILS" and "Q-ILS", respectively, are shown. Solid arrows indicate containment relation (the UML box named "Accept" represents an acceptance criterion). Those boxes for leaves actions of the action-hierarchies that represent elementary actions are separated by a dashed line in the action part to indicate their problem type dependency and

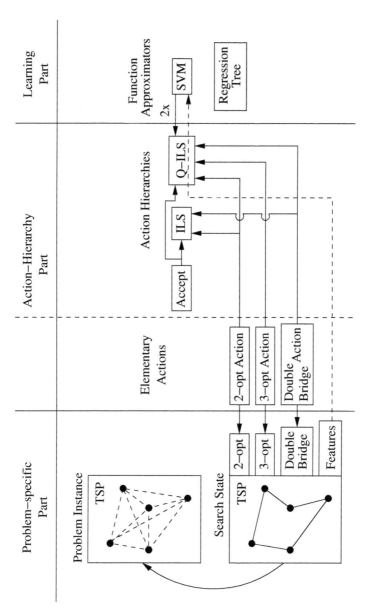

Figure 5.1: GAILS components

interface character with respect to the problem-specific part. The solid arrows leading to the search state's local search operator representation reflect the fact that elementary actions are extended local search operators and must eventually manipulate the search state with their help. Local search operators can be thought of being realized via corresponding search state methods.

- All function approximator representations finally are allocated in the *learning part* since they are the true learning components. The right hand side of Figure 5.1 exemplifies two function approximators, namely Support Vector Machines (SVM) and regression trees represented by the UML boxes labeled "SVM" and " RegressionTree", respectively.

The separation of the GAILS framework components into three independent parts reflects the three in principle independent types of components of any GAILS algorithm. Each GAILS algorithm eventually is composed of some solution encoding and solution manipulation means, some components that provide function approximation, and some components that coordinate the former two kinds of components to actually form a complete algorithm implementing a (learning) LSA. By making the three parts of the GAILS framework independent from each other, they can easily be interchanged independently in any concrete GAILS algorithm. Once concrete components such as search states with their solution manipulation methods realizing local search operators, LSAs in the form of action-hierarchies, and function approximators have been integrated, they can be reused and allow for arbitrarily recombining them. The prerequisite are proper and flexible interfaces between the three parts. These, however, have already been identified in the form of elementary actions and features. The desired objective of the GAILS framework, rapid prototyping and code reuse, can be achieved quite naturally by a design that models the concepts derived in this section and that separates them into three independent parts as just described.

5.2   Design

The GAILS framework is written in the C++ object-oriented programming language [Str91]. Object-oriented design can be viewed as a process of identifying concepts, their functionality, and next relating the identified concepts, mainly in terms of functionality. Any concept must be represented somehow in the GAILS framework. In any GAILS algorithm instantiation as actual executable program that can be run (also called *GAILS algorithm program instantiation* or *GAILS algorithm instantiation* for short) the concepts will be represented by *objects*. The general functionality and interactions of the various objects representing concepts are manifested in hierarchies of individual *classes*, called *class hierarchies*. Class hierarchies in the simplest case are build via trees with individual classes as nodes. In case multiple inheritance is supported, two class hierarchies can also be interleaved yielding together an acyclic directed graph. The children in the class hierarchy of a class are called *subclasses*. Parents of subclasses are called *superclasses*. Subclasses are said to *inherit* from their superclasses indicating that they reuse or adopt from them. A class hierarchy describes the functionality and interactions of a concept as a whole including the interactions with other concepts and hence represents the concept as a whole. A concept can have several occurrences or subtypes. Each individual class describes the functionality and interaction of exactly one "class" of objects representing exactly one subtype of a concept. So, several individual classes are needed to describe and represent a concept as a whole. They are organized in a class hierarchy due to their strong coherence in together describing a concept. Also, a class hierarchy has the side-effect that it saves code by reusing it. This way, all individual subtypes of a concept can be described and represented more efficiently. The nomenclature is that concepts are *represented* or *realized* by objects, classes, or class hierarchies, while, important for the GAILS framework, local search methods, local search operators, or actions are *implemented* by objects or classes. Classes *instantiate* objects and objects are *instantiated from* classes, while class hierarchies *realize* objects. The class instantiating an object is also called an *object's class*, *its class*, or *class of an object*. *Object of class* X is a short cut for saying that an object's class is X. The nodes from figures 4.1 on page 75, 4.2 on page 76, and 4.3 on page 77 named "SimpleILS", "TSP-2opt-LsStep", and "LsProcedure" can be seen as examples for objects realizing actions and which are instantiated from classes. The respective classes realizing actions and instantiating respective actions are "SimpleIls" from 5.12 on page 162, "Tsp2OptLsStep" from 5.11 on page 158, and "Tsp2OptLsProcedure" from 5.11 on page 158.

A GAILS algorithm instantiation executes as a chain of methods calls. The functionality of objects finally is defined in so-called *methods* which can be thought of being functions or procedures attached to objects and which centrally manipulate an object. For each object these attached methods build a set of methods which can be called for it. The set of methods callable for an object is determined by its class. All methods that are *declared* by an object's class in its *class declaration* can be called for the object. A class declaration simply states the methods that can be called for an object of the class together with their signature and other properties. Additionally, object internal data structures in the form of variables, called *member variables* are declared. All methods that are declared by any of the superclasses of an object's class, i.e. all methods of classes an object's class transitively inherits from, can be called for an object, too. The set of all methods that can be called for an object is its *interface*. The set of all methods that can be called for an object of a certain class analog is called the *class interface*. An object or class with a certain interface is said to *support* the methods of its interface in that these methods are callable for the object or for an object instantiated from the class, respectively. The methods of the interface of object x of a class X are also denoted by *methods of/for/from object* x or *methods of/for/from class* X, respectively, or *methods provided by class* X. Declaration of methods does not include provision of actual code. Therefore, methods must also be *defined* by providing actual code that can be executed and which describes or rather implements the method's functionality. Instead of saying that methods are defined, one can also say that they are *implemented*. The same is true for classes: Classes are *defined* or *implemented* also by providing appropriate code from methods, i.e. by defining or implementing methods.

The interaction of objects with other objects formally is only restricted by their interface. In principle, whenever an object with a certain interface, i.e. a certain set of callable methods, is used, another object that also supports these methods and hence has the same interface can be used instead. This, for example, happens if an object x of class X is replaced by an object y of a class Y which is a subclass of class X. Class Y will only *additionally* declare methods and hence extend the interface of its superclass X. It will support *any* method of its superclass also. In the case of the C++ programming language, substitution of an object x of class X by another object y of class Y is only allowed, if and only if class Y is a subclass of class X (and the substitution is realized via pointers or references; which will always be the case for the GAILS framework). An object's interface can be seen as determining an object's so-called *type*. Each subclass also complies to the interface and hence has the type of any of its superclasses. This is important in the context of multiple inheritance as supported by C++ and used for the GAILS framework design. Types (and type denominators) coincide with classes (and class names), since interfaces are accomplished by class declarations. In principle, an object *has the type* or *is of type* of any classes that declare an interface which is a subset of the object's interface. In case of C++, this rule for typing objects is constrained to inheritance relationships induced by class hierarchies, so classes of different class hierarchies are not deemed to have the same type, even if their interfaces are identical The type of an object x is its class X *and* any of its superclasses, so objects can have several types. The nomenclature is to speak of an *object's type*, the *type of an object*, or *its type*. *Object of type* X is a short cut to say that an object's type is X, i.e. that an object's class is X or any of its subclasses.

Defining a method can be done in the class that declared the method or it can be postponed to subclasses. C++ supports polymorphism. A polymorphic method must be declared as "virtual" and is therefore also called *virtual method*. A virtual method can be defined several times in several subclasses of the class that originally declared it. The process of defining a method anew in a subclass is called *overwriting* and the respective method is also labeled *overwritten*. Generally, for a method called for an object the method definition that is actually used stems from the "nearest" superclass of the object's class (assuming that each class also is its superclass). The definition of virtual methods can be delegated to subclasses yielding *pure virtual methods* with no definition at the point of their declaration in some class. Classes that have not defined all pure virtual methods they inherited from *any* of their superclasses (including themselves) cannot instantiate objects. Such classes are called *abstract classes*. All classes that can instantiate objects are called *concrete classes*. All methods which not only are declared but are also defined are called *concrete methods* and can be *applied* or *executed*.   In general, methods put in parentheses in UML boxes in figures illustrating classes according to the UML notion, usually in the root class of a class hierarchy, indicate that these methods are pure virtual methods which must be overwritten to yield a concrete subclass. UML box representations of concrete subclasses in figures then have in common that they also list the names of all pure virtual methods they inherit indicating that they in fact define them. An object's class, i.e. the class that instantiates the object, can only be a concrete class. An abstract class can only be an object's type.

Recall that each concept just identified in this section and in Section 5.1 will be represented by an object in a GAILS algorithm instantiation. For a concept named "XYZ" such as action, heuristic state, search state, and so on, the representing objects are called "XYZ" *objects* such as action objects, heuristic state objects, search state objects, and so on. These objects are realized by class hierarchies which will be presented in the following subsections. The class hierarchy realizing "XYZ" objects, i.e. for objects representing concept "XYZ", is called "XYZ" *class hierarchy* such as action class hierarchy, heuristic state class hierarchy, search state class hierarchy, and so on. Any class of this class hierarchy is called "XYZ" *class* or *type* such as action class or type, heuristic state class or type, search state class or type, and so on. An objects instantiation of a "XYZ" class is called "XYZ" object. The root class of a "XYZ" class hierarchy is called *root* "XYZ" *class* such as root action class, root heuristic state class, root search state class, and so on. Root classes of class hierarchies are also called *base* classes. They typically are also abstract.

The identification process of concepts for the GAILS framework has already been undertaken in the previous section, Section 5.1. This section will be concerned with designing proper class hierarchies for the concepts found in the previous section. All design decisions thereby are centrally guided by the ambition to reuse code and to enable rapid prototyping by ensuring independence and interchangeability of the GAILS framework parts. To recapitulate, the concepts that will become components of the GAILS framework in the form of class hierarchies are listed next. The respective root class names are given in parentheses behind the concept names (all class names are always written in typewriter font; root class names are self-declarative and coincide with the concepts they represent):

- LSA or agent state (AgentState),

- search state (SearchState),

- problem instance (ProblemInstance),

- heuristic state (HeuristicState),

- action or action-hierarchy (Action),

- termination criterion (`TerminationCriterion`),

- function approximator (`FunctionApproximator`),

- features (`Features`),

- policy (`Policy`),

- reward (`Reward`),

- move cost (`MoveCost`),

- timer (`Timer`),

- random number generator (`RandomNumberGenerator`).

These class hierarchies will now guide the further design process and will be addressed in succession in this section. Each of class hierarchies with root classes `AgentState` until `Policy` will be presented in one subsection (subsections 5.2.1 until 5.2.10). The class hierarchies for classes `Reward` and `Move-Cost` will be presented in Subsection 5.2.11, classes `Timer`, `RandomNumberGenerator` and other utility classes will be discussed in Subsection 5.2.12.

5.2.1 Agent State

Objects of class `AgentState` realize the global and overall state of an LSA during execution. For trajectory-based LSAs there will be exactly one agent state object. This is called *global agent state object*. Anything that is needed for computation or does affect the overall LSA state has to be contained in the global agent state object. The global agent state object is the *only* means of passing around information. As such, it must be accessible from everywhere. When changing one constituent, the whole global agent state object and hence state is changed implicitly, too. Its constituents are presented in Figure 5.2. Besides the already known constituents such as search and heuristic state objects, an agent state object also contains a features history (represented by UML box labeled "featuresHistory") plus corresponding eligibilities (represented by UML box labeled "eligibilities"), and an associated action history (represented by UML box labeled "actions"). Together, these are needed to model eligibility traces for some Q-learning algorithms (cf. Subsection 3.2.6). The features history has to store the feature vectors for the preceding n ($n \in \mathbb{N}^+$) LSA states visited and is organized as a ring buffer as well as the action history which stores the actions taken in the n previously visited states. The applied actions are stored in the form of indices representing them. The features and action histories are stored directly in agent state objects and not in heuristic state objects since the features history is with respect to an LSA state as a whole and is the same for any action and hence heuristic state of an LSA.

Further implementation-specific utility objects that foreseeably are required to be globally accessible are also stored in an agent state object. These are (command line) program parameters and means to support algorithm output in a specific format. The output support is encapsulated in objects of class `StandardOutputFormat`, while the set of all program parameters is realized by class `ProgramParameters`. Objects of this class in turn contain objects of class `Parameter` which realize single program parameters and objects of class `PerformanceMeasure` which represent single performance measures (cf. Subsection 5.2.12). Objects of classes `StandardOutputFormat` and `ProgramParameters` are accessible via pointers stored in member variables named "pStandard-OutputFormat" and "pProgramParameters", respectively, as can be seen from Figure 5.2. The individual objects representing program parameters as realized by class `Parameter` are stored as a list of pointers to them which is stored in member variable named "registeredParameters" and labeled this way in Figure 5.2. The objects representing performance measures are also stored as a list of pointers to them in a member variable named `registeredPerformanceMeasures` and labeled this way in this figure.

5.2.2   Search State

One of the objectives of class `SearchState` is to realize current, best and any other search states in the form of search state objects. As well as the global agent state object, there will be exactly one current search state and exactly one globally best search state for trajectory-based LSAs. Hence, there will be exactly one of each of such search state objects be stored in the global agent state (cf. Subsection 5.2.1) for any GAILS algorithm instantiation. These unique current and globally best search state objects are called *global current search state object* and *global best search state object*, respectively.   A search state object essentially must contain data structures that encode a solution, i.e. a solution encoding, also called *solution encoding data structures*.   Additionally, local search operators in the form of methods for class `SearchState` that manipulate the solution encoding (data structures) are needed. In order to encapsulate as many details as possible, these methods should be as high-level as possible and as fine-grained as necessary. They will typically implement basic local search operators such as local search steps and procedures. Yet, they can implement also more fine-grained local search operators such as neighborhood exploration schemes, or, for efficiency reasons, more high-level local search operators such as complex metaheuristics.

Any methods of class `SearchState` for manipulating solution encodings implementing local search operators must also compute and provide features. The features that stem from the search state of an LSA state, i.e. that are computed based on information stored in the search state, are called *search state features*.   Search state features mostly are problem type dependent, but there are some standard ones that arise for *any* problem type and any search state such as cost and changes in cost induced by local search operators. Figure 5.3 presents the aggregation view for class `SearchState`. Cost and also changes in cost can be normalized according to the procedure described in Subsection 3.3.1 using lower and upper bounds. The bounds can be theoretically computed or they can additionally be estimated empirically. The respective bounds for the costs are stored in member variables `empiricalLowerBoundOnUnnormalizedCost`, `empiricalUpperBound-` `OnUnnormalizedCost`, `empiricalLowerBoundOnTheoreticallyNormalizedCost`, and `empirical-` `UpperBoundOnTheoreticallyNormalizedCost`, respectively. Each time a cost or changes in cost are set, these will be theoretically and perhaps empirically normalized and stored in member variables `unnormalizedCost`, `theoreticallyNormalizedCost`, and `empiricallyNormalizedCost`. If member variable named `doNormalizeCost` in Figure 5.4 representing a flag is turned on, the cost will be normalized. Member variable named `normalizeCostType` there indicates which kind of normalization is to be applied. These member variables are represented by the UML boxes labeled "doNormalizeCost" and "normalizeCostType" in Figure 5.3.   Method `getCost` then will return the already computed normalized or unnormalized costs by calling the appropriate method among `getUnnormalizedCost`, `getTheoreticallyNormalizedCost`, or `getEmpiricallyNormalizedCost`. The same procedure is applied to changes in cost, too. The respective methods and member variables are named analog replacing cost by `deltaCost`. In figures 5.3 and 5.4 the methods just mentioned are not shown due to lack of space. Also, the member variables for cost, changes in cost, and their bounds are also only represented by pseudo member variables represented by UML boxes labeled "cost" and "deltaCost" there. These pseudo member variables are also depicted in Figure 5.4 named `cost` and `deltaCost` there.

Other standard search state features realized as member variables in class `SearchState` are a local

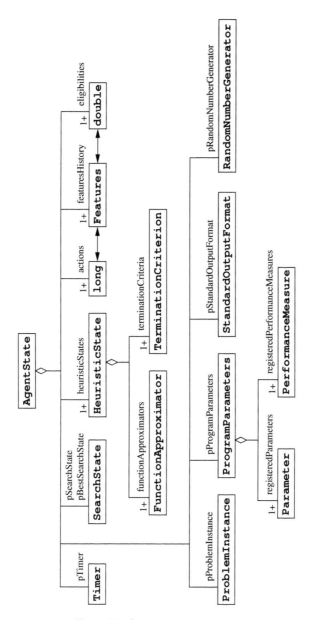

Figure 5.2: Agent state aggregation

optimum indicator (named localOptimum in Figure 5.4, represented by UML box labeled "localOptimum" in Figure 5.3), and step and iteration counters (named stepsDone and iterationsDone,
respectively, in Figure 5.4, represented by boxes labeled and "stepsDone" "iterationsDone", respectively in Figure 5.3).

Solution encodings and local search operators differ from problem type to problem type. Therefore,
class SearchState is abstract and each new problem type will result in a new subclass of root search
state class SearchState as can be seen exemplified in Figure 5.4. This figure presents the class
hierarchy for search state objects. There, two subclasses represented by UML boxes named Tsp-
SearchState and FspSearchState for the TSP and the Flow Shop Problem (FSP), respectively,
have been created. All member variables needed by *any* search state object are located in the
root search state class. These foremost are standard search state features as just mentioned. All
non-standard and potentially problem type specific search state features a search state object might
additionally provide are stored in a feature vector (pointed to by pointer stored in member variable
named pFeatures in Figure 5.4 and represented by UML box labeled "pFeatures" in Figure 5.3).
The member variable storing the pointer to the feature vector is declared in the root search state
class SearchState, but it is filled with concrete features (or their values) by the method defined
in the concrete problem type specific subclasses. Also, any problem type specific member variables
for solution encoding data structures or methods implementing local search operators are added in
the problem type specific concrete subclasses only. In the case of the TSP, for example, a solution
is a tour and is represented as a permutation of nodes and stored in member variable named tour
(see UML box named TspSearchState in Figure 5.4). Figure 5.1 illustrates a tour encoding as
a Hamiltonian circle in the box labeled "SearchState" there. The nodes inside the box represent
cities. Local search operators are depicted as attached boxes labeled according to local search
steps and procedures there such as the 2-opt and 3-opt local search procedures and the double-
bridge perturbation in this figure. These local search operators are implemented by methods of
the concrete subclass TspSearchState as depicted in Figure 5.4. They are named first2Opt-
Step, best2OptStep, first3OptStep, first3OptStep, and perturbationDoubleBridge there.
The TSP tour is stored in member variable named tour, its length in member variable named
tourLength as is shown in this figure. In case of the FSP, local search operators are implemented
by methods named firstExchangeStep, bestExchangeStep, firstInsertStep, bestInsertStep,
and exchangePerturbation in Figure 5.4. The solution encoding data structures for the FSP are
accessible via member variables named permutation and numberJobs there.

Initializing and deconstructing a search state object is general to search state objects for any problem type, but can only be implemented problem type specifically. Therefore, methods initialize
and initializeRandomly defined in root search state class SearchState are used to deterministically or randomly build solution encoding data structures completely anew which, however, use pure
virtual methods computeDataStructures and computeDataStructuresRandomly, respectively, to
do the actual work. These *must* be overwritten in any concrete subclass as indicated in Figure 5.4 by
putting their names in parentheses in the UML box representing the root search state class Search-
State and by repeating their names without parentheses in the UML boxes representing concrete
subclasses TspSearchState and FspSearchState. Methods restart and restartRandomly are
used to deterministically or randomly reset the solution encoding data structures. These use pure
virtual methods resetDataStructures and resetDataStructuresRandomly, respectively, which
must be overwritten in one concrete subclass, too. Virtual method deleteDataStructures finally
is used to deconstruct any data structures.

For the methods of class SearchState and subclasses to work, they need resources in addition
to the solution encoding data structures. Typically, they need to access the problem instance
specification, might have included some element of chance, or need to output a solution (encoding).
Therefore, pointers to such resource and utility objects are also stored in a search state object

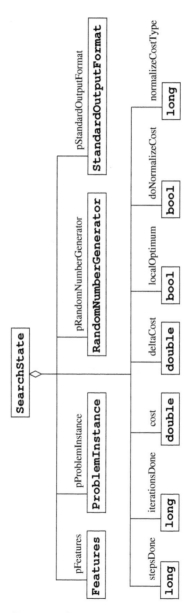

Figure 5.3: Search state aggregation

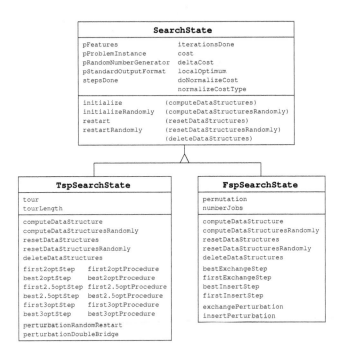

Figure 5.4: Search state class hierarchy

redundantly to their storage in the global agent state object as illustrated in figures 5.3 on page 139 and 5.4 on page 140). They are named pProblemInstance, pRandomNumberGenerator, and pStandardOutputFormat, respectively, in Figure 5.4, and are represented by UML boxes labeled "pProblemInstance", "pRandomNumberGenerator", and "pStandardOutputFormat", respectively in Figure 5.3. Besides ease of use, the reason to use this redundant storage is that search state objects and their methods generally do not have access to the global agent state object to retrieve these needed resource and utility objects from there.

5.2.3   Problem Instance

Objects of class ProblemInstance encapsulates problem instance representations. They can read in problem instance specifications from files and build an internal representation of the problem instance. Additionally, auxiliary information helpful for local search operators can be derived and stored, in order to be accessed by search states objects. A special kind of auxiliary information are features. The features that are computed from problem instance representation are called *problem instance features* and mostly are problem type specific. A standard problem instance feature that is problem type independent for example is the instance size which is stored in its own member variable. Problem instance features do not change during the search of an LSA. Since problem instance representations are problem type dependent, root problem instance class, ProblemInstance, is inherited for each new problem type integrated analog to the class hierarchy for search state objects (cf. Figure 5.4). Again, common member variables storing the instance size, input filename, and so on, are stored in the root problem instance class. The problem type specific problem instance features are provided as feature vector in the root problem instance class, which, however, is filled by methods defined by the concrete problem type specific subclasses. The root problem instance class also provides common methods. Some of them must be overwritten such as methods for reading in files and building and unbuilding internal data structures, since these inherently are problem type specific. For example, the problem instance object for the TSP of class TspProblemInstance stores a distance matrix as depicted by the graph in the UML box labeled "Problem Instance" on the left hand side of Figure 5.1. A problem instance object for the TSP also computes and stores auxiliary information such as nearest neighbors lists.

5.2.4   Heuristic State

Heuristic states contain heuristic information. The heuristic state class hierarchy with root class HeuristicState is intended to model heuristic states via heuristic state objects. Different actions will produce different kinds of heuristic information. For example, actions implementing local search steps produce a change in cost. This change in cost can be stored in a member variable named deltaCost as can be seen in Figure 5.5 in the UML box named LsStepHeuristicState. After doing a local search step in the form of a basic move, an LSA can be in a local optimum or not. A member variable named localOptimum in Figure 5.5 implementing a respective flag indicates this. Many other kinds of heuristic information arises upon action application which can be stored in member variables of heuristic state objects (their names are showing up in the UML boxes in Figure 5.5): Actions implementing local search procedures perform a local search descent to a local optimum via several local search steps and hence additionally can count the steps needed for the last local search descent (deltaSteps). Or, if called repeatedly, such actions can count how many steps they have done altogether (steps). Also, the altogether step count at the time when a new globally best solution was found last can be stored (currentGlobalBest-Iteration). The same holds true for iteration counts in the case of an action implementing an ILS metaheuristic (iterations, deltaIterations, and currentGlobalBestIteration, respectively).

An ILS further maintains a search state history (`searchStateHistory`), the current length of the search state history (`historyLength`) and a best search state found so far locally to the ILS (`bestSearchState`). An action implementing an SA maintains a temperature (`temperature`) and a probability (`probability`) value in addition to step and iteration information. The heuristic information just listed – the current cost is accessible via the global current search state object as stored in the global agent state object and hence is not reproduced in any heuristic state object – can be viewed as being added incrementally. Heuristic information aggregations therefore can be regarded as building a hierarchy of increasingly comprehensive aggregations. These can be mapped directly into a class hierarchy. New subclasses are added when new actions need to add new heuristic information. For example, actions implementing local search procedures need to have several kinds of step counters in addition to the heuristic information associated to local search steps. Figure 5.5 shows the resulting class hierarchy for heuristic state objects (prefix `Ls` is an abbreviation for "Local Search", prefix `Ils` is an abbreviation for "Iterated Local Search", prefix `Sa` is an abbreviation for "Simulated Annealing" there and in all other figures).

As Figure 5.5 also shows, multiple inheritance is used to invoke a building blocks principle in establishing new heuristic state classes: When designing a new heuristic state class, some member variables for storing heuristic information the new class is supposed to store also can be inherited from classes that already provide for it. Anything still left has to be added anew, either encapsulated in a separate heuristic state class to inherit from as well or as add-on member variables in the new heuristic state class directly. For example, class `IlsHeuristicState` completely inherits its heuristic information storing member variables from its superclasses as can be seen in Figure 5.5. As another example, class `QIlsHeuristicState` adds a reward heuristic information (stored in member variable named `reward` in Figure 5.5 in the UML box named `QIlsHeuristicState`). Any heuristic information stored in any heuristic state object then stems from member variables declared in different superclasses as is depicted in Figure 5.6. Heuristic state objects of class `QIlsHeuristic-State` are intended to be used by actions that implement Q-learning algorithms on the basis of ILS-actions in Figure 5.6. All member variables that are accessible in an heuristic state objects of class `QIlsHeuristicState` are grouped according to the heuristic state class they are inherited from as listed on the right hand side of the figure. Dashed lines and curly braces indicate which member variables belong to which class. Each member variables are represented by a UML box containing the type of the member variable and labeled with the name (cf. the notes at the begin of this section). The names are the same as mentioned earlier in this subsection.

By adding heuristic information storage in the form of member variables incrementally to heuristic state classes, heuristic state objects individually adjusted to the needs of different types of actions can be used to store states of action. This prevents from wasting memory in contrast to using only one monolithic, overly complex heuristic state object for any kind of action that stores any heuristic information ever contrived. Since termination criteria are needed by almost all actions and since the GAILS framework pivotally is to incorporate learning aspects, objects realizing function approximators and termination criteria are contained in any heuristic state object and hence are stored in member variables defined in the root heuristic state class `HeuristicState`. These member variables are represented by UML boxes labeled "functionApproximators" and "terminationCriteria" in Figure 5.6, respectively. In Figure 5.5, they are named `functionApproximators` and `terminationCriteria`, respectively. They represent vectors – indicated by label "1+" in Figure 5.6 – storing pointers to respective objects.

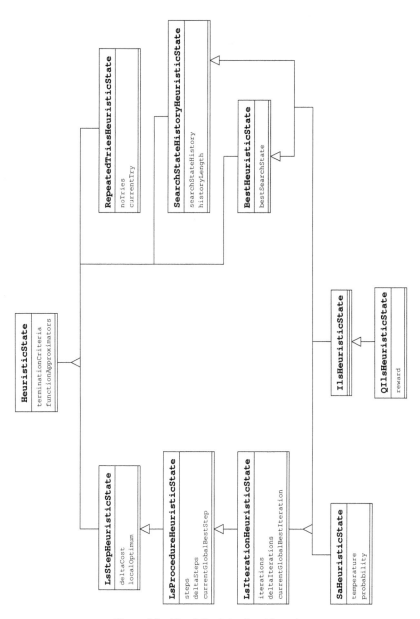

Figure 5.5: Heuristic state class hierarchy

5.2.5   Action

Individual actions in the form of nodes of action-hierarchies (cf. Subsection 4.1.3) are represented by action objects in the GAILS framework. Depending on the kind of action the action objects represent they are also called *root-action object*, *individual action object*, *leave action object*, or *elementary action object*

(cf. Subsection 4.1.3). The action class hierarchy has to be designed such that it enables building action-hierarchies flexibly. Action-hierarchies are built by connecting individual actions in a containment hierarchy. Objects can hold on to other objects also, for example via pointers. Given some set of action objects, containment hierarchies of action objects realizing action-hierarchies therefore can easily be built according to the building blocks principle by connecting them via pointers. The resulting containment hierarchies of action objects are called *object action-hierarchies* since they exactly realize action-hierarchies. Action-hierarchies and hence object action-hierarchies should be as independent as possible from any problem type in order to foster building generic (object) action-hierarchies in the form of reusable blueprints for how to built (object) action-hierarchies.

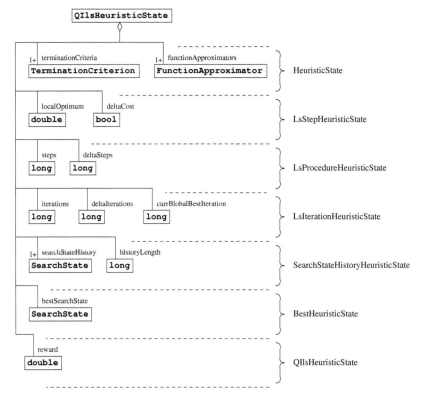

Figure 5.6: QIlsHeuristicState Aggregation

Recall from Section 5.1 that the state of action objects is sourced out to heuristic states. This allows to model actions as action objects representing pure functionality. The action objects themselves then do not have a state which is stored in member variables. In contrast, each action object will hold on to its so-called *associated heuristic state object* by means of a "pointer" to this object. An action object's associated heuristic state object stores the state for it and represents the heuristic state the modeled action operates on. The "pointers" come in the form of indices with respect to the list of heuristic state as stored in the global agent state object. These indices can be used to access heuristic state objects via the global agent state object. All other information stored in member variables of actions will represent modes of operations which do not change during execution and merely provide for generic functionality and code reuse. Since all constituents of object action-hierarchies represent pure functionality and store their state in associated heuristic state objects, object action-hierarchies as a whole again represent true functionality. By not having internal states stored in any action object and hence object action-hierarchies, they can share heuristic state objects as a whole or individually even on the object level (in contrast to sharing implementation code only by instantiating several action objects according to action classes).

The UML box named `Action` in Figure 5.8 presents the member variables of the root action class, `Action`. The information stored in the member variables comprises a "pointer" to the associated heuristic state object in the form of an index with respect to the list of heuristic state objects stored in the global agent state object. This index is stored in the member variable named `heuristicStateIndex`. Member variables named `terminationCriterionIndices` and `function-ApproximatorIndices` contain "pointers" to the objects realizing termination criteria and function approximators that are utilized. The "pointers" again come in the form of a list of indices. The indices for the objects realizing termination criteria and function approximators are with respect to the list of such objects as stored in the associated heuristic state object. This pointer implementation via indices is illustrated in Figure 5.7. There, dashed lines represent the "pointers" implemented by indices which are called *index-based pointers*. An index-based pointer to an object realizing a termination criterion is labeled "HS_i-TC_j" ($i, j \in \mathbb{N}^+$) ("HS" is a short cut for "Heuristic State", "TC" for "Termination Criterion") indicating that it points to the j-th object as stored in the i-th heuristic state object of the global agent state object. Index-based pointers to objects realizing function approximators are labeled "HS_i-FA_j" and work analog ("FA" is a short cut for "Function Approximator"). The reason for this kind of pointer implementation is to support sharing heuristic state objects and to prevent from implicitly corrupting "real" pointers to heuristic state objects. If "real" pointers are used, corruption can happen for example, if a heuristic state object pointed to by several action objects is silently removed in one action object application undetected by the other action objects involved in sharing this heuristic state object. These other action objects will thereafter work with invalid "real" pointers.

The root action class also stores an identifier (member variable named `identifier` in Figure 5.8; all member variables names given in parentheses in this paragraph refer to this figure), and flags that do regulate general modes of operation such as whether to operate in learning mode (`doLearning`), whether to actually trigger updates to the global best search state object (`updateGlobalBest-SearchState`), and whether to initialize the objects realizing termination criteria that are used at the begin of an action object application (`doInitializeTerminationCriteria`, cf. Subsection 5.2.7). The methods provided by class `Action` are for accessing the associated heuristic state object (`getHeuristicState`), to access the global current search state object (`getSearchState`), the global best search state object (`getBestSearchState`), and the problem instance object (`access-ProblemInstance`). These methods are virtual because they might have to be overwritten in subclasses (cf. Subsection 5.2.6). Further methods can be used to reset the associated heuristic state object (`resetHeuristicState`) and to trigger update of the globally best search state object, if a new best search state has been found during action object application (`updateGlobalBest-SearchState`). The last method will not only change the best search state object stored in the global agent state object but will also store heuristic information such as counters indicating when

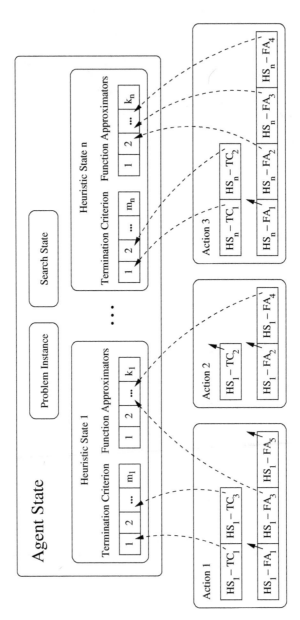

Figure 5.7:   Pointer implementation

this last update occurred. It therefore is virtual and typically has to be overwritten by subclasses corresponding to more specialized heuristic state types with additional kinds of counters.

Action objects represent functionality and can be *applied* or *executed*. This is done by *calling* an action object via calling a method named `apply`, also denoted by `apply`-*method*. Note that only the action objects contained in another action object can be called by the latter. Vice versa, since action objects are always organized in an tree-like object action-hierarchy, calling an action object by another action object means that the former its contained by the latter. So, notions call and contain can be used interchangeable. Any action object has a method `apply` which must be defined in its (concrete) class. The `apply`-method is declared in the root action class `Action`. It is a pure virtual method which must be defined by concrete action classes which is illustrated in Figure 5.8 by putting the method name `apply` in parentheses in the UML box named `Action` representing root action class and without parentheses in any of the UML boxes representing its concrete subclasses. Any UML boxes there without a method named `apply` represent subclasses which consequently are abstract. Since actions operate on the whole LSA state, method `apply` is called for action objects with a pointer to the global agent state object as parameter. The global agent state object then can be changed by the `apply`-method. *Execution of an object action-hierarchy* starts by calling the `apply`-method of the root-action object of the object action-hierarchy which in turn will call the `apply`-methods of the action objects it holds on to directly. The top-down distribution of calls to `apply`-methods finally ends at the leave objects of the object action-hierarchy which do not call any other `apply`-methods anymore. As an example, consider Figure 4.2 which shows an hierarchy-tree that is supposed to illustrate an action-hierarchy implementing a round-robin application scheme of two ILS metaheuristics (cf. Subsection 4.1.3). The nodes of the depicted hierarchy-tree represent action-objects, arrows indicate containment (or rather pointers with the direction of the pointers reversed). When calling the `apply`-method of the root-action object represented by UML box named "RoundRobin" it in turn will call the `apply`-methods of the two action objects it contains in turn which implement two ILS as represented by the two UML boxes named "SimpleILS" each. Each such action object then calls the `apply`-method for the action objects it contains and which implement a perturbation, a local search procedure and an acceptance criterion. The action objects implementing local search procedure are represented by the nodes named "LsProcedure" again call an action object implementing a local search step. In this example, the two action objects implementing perturbations and the two local search steps are elementary action objects. The two action objects implementing the two acceptance criteria are leave action objects.

As this example has shown, among the leave action objects of any object action-hierarchy will be at least some elementary action objects. All action objects will change their associated heuristic state objects when updating their heuristic state information. Elementary action objects furthermore are responsible for changing the global current search state object. Since actions essentially are extended local search operators and local search operators are implemented as methods of search state objects (cf. Section 2.8 and Subsection 5.2.2), elementary action objects will simply call an appropriate method for the global current search state object that implements the local search operator the action realized by the elementary action object extents. This way, they manipulate the global current search state object representing the current search state. As such, elementary action objects interface from the action part of the GAILS framework to the problem-specific part by simply acting as wrapper for methods for search state objects implementing local search operators as can be seen in Figure 5.1. In addition to calling an appropriate search state object method, elementary action objects merely have to update any heuristic information changed by the search state manipulation method called such as the change in cost induced or number of steps done in case a local search procedure implementation was applied.

The local search operators implemented by methods of search state objects typically comprise local search steps or procedures. As was mentioned before (cf. Subsection 5.2.2), they can also be

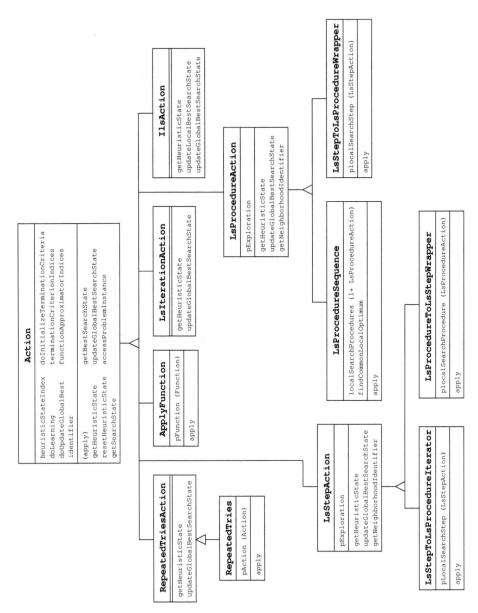

Figure 5.8: Action class hierarchy

more fine-grained and provide means for enumerating a neighborhood. These methods then can, for example, be used by an elementary action object to implement a best or first improvement local search step with the original cost function substituted by another, perhaps represented by an object realizing a function approximator. Methods of search state objects can implement complete metaheuristics such as ILS also. This permits to integrate successful implementations of local search methods as a whole which then can be used as they are. Such, it gives great flexibility for making GAILS algorithm instantiations efficient and additionally this possibility further supports reuse of code.

During the execution of an object action-hierarchy, the search state objects such as global current and best search state objects and hence the represented search states are changed. Also, heuristic information has to be updated in associated heuristic state objects. Both kinds of updates must be propagated to other action objects of an object action-hierarchy. Propagation of search state object changes happens implicitly via the common and unique global current and best search state objects in the global agent state object, but heuristic information must be conveyed in both directions through an object action-hierarchy also. At any level of an object action-hierarchy, an individual action object that calls other action objects must provide up to date information about the search progress that happened outside the called action object – for any required information. Also, any information that changed during the execution of a called action object – either heuristic or search state specific information – and which is needed by other action objects must be transferred back. For example, an action object implementing an acceptance criterion for an ILS needs as input a history of search state objects to decide which search state object to set as new global current search state object and hence where to continue the search from. Other examples for necessary information transfer are changed objects realizing function approximators and step counts used for summing up steps made by several individual action objects.

The problem is how to efficiently and reliably convey heuristic information between action objects. The solution to this problem is to share heuristic information by means of sharing commonly accessible heuristic state objects that store the heuristic information to be shared analog to the global current search state object. Recall that sharing heuristic information is possible only since it is not stored in action objects but in extra heuristic state objects and since all heuristic state objects are stored in the global agent state object. The decision to share heuristic state objects will quickly be justified by working out the disadvantages of the alternative approach of storing heuristic information in each action object individually and by working out the advantages of sharing heuristic state objects:

- If the heuristic information and hence the state of an action object is stored by each action object itself in member variables, access methods for these member variables are required which will induce additional implementation overhead. Other drawbacks of this approach are additional memory consumption caused by redundant storage of heuristic information in each action object, extra computation time needed to perform necessary synchronization of the distributed and redundantly stored heuristic information, and an additional implementation effort for realizing a reliable synchronization. The latter two disadvantages are due to the fact that heuristic information such as common counters or perhaps termination criteria typically are to be exchanged between several action objects that are distributed over an object action-hierarchy and which do not have mutual and direct access. Any heuristic information will be only accessible and hence can be synchronized to other action objects, if any action objects involved in synchronization are mutually accessible. If heuristic information is to be synchronized between two action objects at lower levels of the object action-hierarchy in different branches, it has to be passed circumstantially via the root-action object since they are not directly mutually accessible for each other. This is illustrated in

Figure 5.9. It shows two hierarchy-trees illustrating the same object action-hierarchy. The nodes represent action objects and the containment relationship is indicated by the arcs of the hierarchy-trees. The global agent state objects, each containing a problem instance object and a global current search state object, are depicted below the hierarchy-tree labeled "AgentState", "ProblemInstance", and "SearchState", respectively. The communication of elementary action objects (represented by the leave nodes of the hierarchy-trees) with the search state object is depicted by dotted lines. In part a) of Figure 5.9, heuristic information is passed on via two paths indicated by solid and dashed lines between the nodes representing action objects involved and labeled from "1" to "9". Both paths involve several action objects of higher levels of the object action-hierarchy, always including the root-action object.

- In contrast, action objects can easily share information by sharing heuristic state objects. This way, no redundant storage of heuristic information in individual or even all action objects is needed. Even if action objects do not share heuristic state objects, synchronization can be regulated directly between any two action objects without having to go via the root-action object of an object action-hierarchy or any other action object on the way. Part b) of Figure 5.9 illustrates this procedure. There, the heuristic information needed to be passed on from action objects represented by nodes labeled "1" and "4" to action objects represented by nodes labeled "7" and "9", respectively, only needs to take the detour via heuristic state objects represented by boxes named "HeurState$_1$" and "HeurState$_2$" and which are directly accessed by action object pairs (1,7) and (4,9), respectively.

Action object executions rely on and are altered by the heuristic information that is contained in their associated heuristic state object. They need certain heuristic information as input via their associated heuristic state object and they will change certain heuristic information such as counters or a search state object history which in turn are used by other action objects, e.g. to compute termination. Since heuristic state objects are organized in a class hierarchy with each heuristic state class also providing some form of heuristic state type (cf. the notes at the begin of this section) one can say that action objects need a certain type of associated heuristic state object to operate on. They need to operate on a type that provides at least the needed amount of heuristic information they use. Otherwise, an action object cannot operate properly. Accordingly, to ensure operation on associated heuristic state objects of fitting types it is required to constrain action object application. One way to facilitate such constraints is to group and type action objects according to heuristic state classes by means of an appropriately designed class hierarchy for action objects. Figure 5.10 – which is an extension of Figure 5.5 — illustrates this action object grouping and typing according to heuristic state classes. Most heuristic state classes are assigned an action class in a one-to-one correspondence according to the following naming scheme (and indicated by horizontal arrows $\longleftrightarrow$ in this figure): Action class XYZAction is assigned to heuristic state class XYZHeuristic-State with XYZ for example being LsStep or Ils. For any action class XYZAction or any object instantiation xyzAction of an action class XYZAction, heuristic state class XYZHeuristicState is called the *corresponding heuristic state class* or *type*. Inheritance among represented action classes is indicated by dashed lines. The needed constraints now can be realized by requiring that the type of the associated heuristic state object for any action object is the corresponding heuristic state type of the action object. The implications of this requirement are demonstrated next.

Any action object that changes the same set of heuristic information has the same corresponding heuristic state type and because of the one-to-one mapping of heuristic state classes and action classes also the same action type. Action objects can be seen as black boxes and therefore can be used interchangeably without harm concerning the set of heuristic information they have to change in their associated heuristic state object, if they have the same type. For example, an action object implementing an ILS needs one action object implementing a perturbation that processes

a)

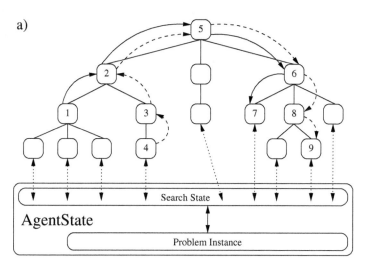

b)

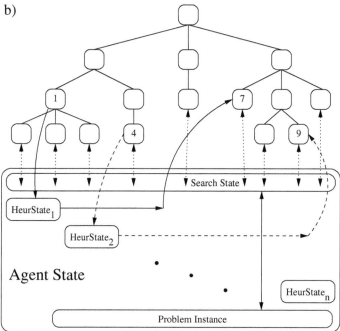

Figure 5.9: Action-hierarchy information passing

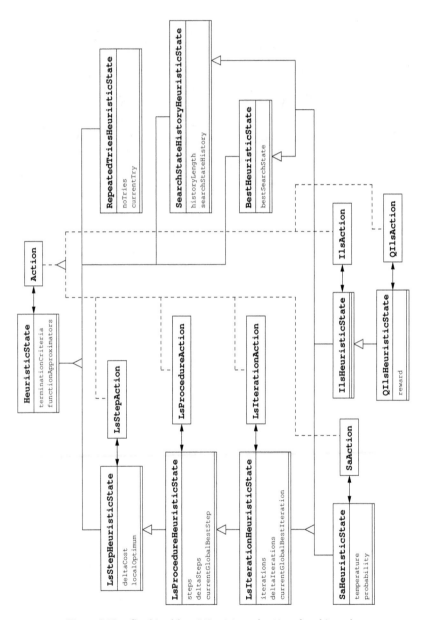

Figure 5.10:　Combined heuristic state and action class hierarchy

exactly the information that is contained in a heuristic state object of type `LsStepHeuristicState`. Other action objects required are one that implements a local search procedure and which must operate on an associated heuristic state object of type `LsProcedureHeuristicState` and one action object that operates on an associated heuristic state object of type `IlsHeuristicState` and that implements an acceptance criterion. An action object implementing an ILS can be instantiated by giving to it action objects such that these are set up to operate on associated heuristic states of the proper heuristic state types. Due to the one-to-one correspondence of heuristic state classes to action classes, this can easily be enforced. When instantiating an action object, it can be required in the class declaration and method definition that all action objects it needs for its operation and which it has to call, i.e. it contains, have to be of a certain action type and hence work on associated heuristic state objects of a certain, namely the corresponding, heuristic state type. In case of an action object implementing an ILS, it simply is needed to require action objects in its class declaration and definition of the action types just mentioned. The action object implementing an ILS calls the provided action objects in turn in the `apply`-method and takes care of updating the heuristic information in the various associated heuristic state objects including its own, if they are not shared in the first place. This is only possible, however, since the action object implementing an ILS knows exactly which corresponding heuristic state types the associated heuristic state objects have that are used by the action objects it calls. Hence, it knows which information to update and how to access it and that is allowed to do so. Typing of action objects and using these action types in class declarations and definitions this way can be seen as a contract. Typing of action objects by means of heuristic state types not only seems to be necessary but also facilitates building object action-hierarchies reliably.

Generally, when typing action objects according to heuristic state types, an action object that contains and calls other action objects then can state, by requiring certain action types for the called action objects in its class declaration and definition, which amount and kind of heuristic information the contained action objects must work on without requiring a special inner working. Due to the typing, the calling action object knows which heuristic information to provide, retrieve and to pass around via the associated heuristic state objects of the action objects it calls. Action objects with different inner workings but same input-output-interface as determined by their corresponding heuristic state type can then arbitrarily and easily be replaced. For examples, different object action-hierarchies implementing an ILS varying only in the kind of perturbation used can be instantiated quickly, given different action objects implementing different perturbation versions. In almost the same manner, not only different action object versions for the same problem type can be exchanged arbitrarily, but analogously, action objects for other problem types can be plugged in. This way, object action-hierarchies such as one implementing an ILS can be reused effectively across several problem types, since only elementary action objects such as ones implementing perturbations and local search procedures are problem type dependent and have to be exchanged in principle. All other action objects such as the ones implementing the acceptance criterion can be reused as is. Figure 4.2 nicely depicts the reuse of object action-hierarchies. The two nodes of the hierarchy-tree presented in this figure named "SimpleILS" represent object action-hierarchies implementing an ILS. They only differ in the action objects implementing their acceptance criterion and the one implementing the local search step their local search procedure implementing action object uses. This is illustrated by differently labeled nodes representing the respective action objects in Figure 4.2. By exchanging the action objects represented by nodes named "TSP-DoubleBridge", "TSP-2opt-LsStep", and "TSP-3opt-LsStep" to ones that also implement a perturbation and local search step but operate on a different problem type, the whole object action-hierarchies implementing the two ILS illustrated in Figure 4.2 can be transferred basically as is to another problem type; all other action objects are problem type independent and can be reused.

Sharing heuristic state objects is a very advantageous possibility resulting from typing action objects according to heuristic state classes. Consider an action object `xyzAction` of (concrete) action class `XYZAction`. The corresponding heuristic state class (and hence type) then is `XYZHeuristicState`. Any heuristic state object `xyzSubClassHeuristicState` instantiated from a subclass `SubClassHeuristicState` of class `XYZHeuristicState` can be used as an associated heuristic state object by action object `xyzAction` also, since any such heuristic state object `xyzSubClassHeuristicState` will include all heuristic information specified by superclass `XYZHeuristicState` and hence all heuristic information action object `xyzAction` needs for its operation. This is because subclasses expand their superclasses thus containing more heuristic information; the objects they instantiate also have the type of the superclass (cf. the notes at the begin of this section). Any additional heuristic information added by a subclass can be neglected without harm. Now, consider two or more action objects supposed to use the same heuristic state object. Each action object needs to operate on an associated heuristic state object with type of its corresponding heuristic state class. As long as the shared heuristic state object is an instantiation of a subclass of the corresponding heuristic state classes for *all* action objects involved, sharing associated heuristic state objects will work.

For example, an action object of type `IlsAction` being the root-action object of an object action-hierarchy implementation of an ILS needs for its operation an action object of type `LsStepAction` implementing a perturbation, an action object of type `LsProcedureAction` implementing a local search procedure, and an action object implementing an acceptance criterion. The latter action object needs a search state object history for its operation and hence needs to operate on the heuristic state object which is an instantiation of a subclass of class `SearchStateHistoryHeuristicState`, e.g. class `IlsHeuristicState`. All action objects listed effectively can share an associated heuristic state object of class `IlsHeuristicState` since it is a subclass of the corresponding heuristic state classes for *all* action objects involved, i.e. classes `LsStepHeuristicState`, `LsProcedureHeuristicState`, and `SearchStateHistoryHeuristicState`. In this example, the root action object will increment the iteration counter of the shared heuristic state object of class `IlsHeuristicState`, the action object implementing the local search procedure will increment the step counter of the shared heuristic state object, while the action object implementing the acceptance criterion takes care of the search state object history of the shared heuristic state object. All this will happen without need to synchronize updates between the involved action objects directly. Any potential corruption of the shared heuristic state object by interference is avoided automatically since each action object involved naturally changes heuristic information mutually exclusive. Nevertheless, the heuristic information can and will be used commonly, e.g. for computing termination.

As was briefly mentioned before, using typing of action objects corresponds to programming by contract. When instantiating an action object, its type determines which corresponding heuristic state type it requires for its associated heuristic state object. The contract includes also that before calling an action object, the heuristic information stored in the associated heuristic state object of the called action object must be up to date which has to be ensured by the calling action object. Equally, after each action object application, a called action object has to ensure up to date information in its associated heuristic state object again. A special inner working for an action object is not required, even if using an action object implementing a simple local search step instead of an action object implementing a perturbation seldom makes sense, for example. This, however, is a semantic problem and beyond the reach of an implementation framework as the one discussed here. The typing mechanism introduced rather is some kind of a-priori program verification comparable to the typing of programming languages. It eases the construction of correct GAILS algorithms instantiations, although it introduces some complexity overhead, both on the implementation and the usage side.

5.2.6   Action Typing

Typing of action objects has been identified to be crucial to enable reuse of code such as code for instantiating an object action-hierarchy implementing an ILS. The question is how to design typing of action objects in terms of a proper design of the action class hierarchy? The programming language used to implement the GAILS framework, C++, supports typing and type checking of objects at compile time and hence enables typing and type checking of action objects at compile time without having to implement costly runtime type checking [Str91]. Recall that heuristic state objects are realized by a class hierarchy. The induced heuristic state types are transferred to action classes. Each resulting action class (and hence type) belongs to exactly one heuristic state class (and hence type) according to a one-to-one relationship. Each action class in such a one-to-one relationship is realized as an abstract class which foremost misses a definition of method `apply`. Each such abstract class directly inherits from the root action class, namely class `Action`. Such abstract action classes directly involved in a one-to-one relationship are called *basis action classes* or *types*. Figure 5.10 shows the one-to-one correspondence, each UML box representing an action or a heuristic state class is named as the represented class. The inheritance relationship for actions is indicated by dashed lines. Only those basis action classes which are intended to work as superclass of concrete classes implementing local search methods are represented and thus shown. All names of abstract basis action class end with `Action`, all names of concrete actions classes, i.e. basically identifiable by an implemented apply method, without. Each concrete action class will inherit a certain basis action class and this way will commit itself by inheritance to a certain (basis) action type and to a certain corresponding heuristic state type. When instantiating an action object, this commitment is carried over to the actual object instantiations.

5.8 on page 148 shows parts of the action class hierarchy containing representations of basis action classes. The action class represented by the UML box named `RepeatedTriesAction` in this figure is an example for an abstract basis class. It is inherited by class `RepeatedTries` as represented by the UML box named equally in Figure 5.8. An action object of class `RepeatedTries` takes an action object of any action type pointed to by a pointer stored in member variable `pAction` and applies it repeatedly. This member variable is shown in Figure 5.8. The pointer type of this member variable is always given in parentheses. In general, the pointer type of member variables pointing to action objects in this and other figures that show parts of the action class hierarchy in given in parentheses behind the member variable name. The intention for this action class is to provide an easy to use means to run several applications or rather *tries* of another action object. This action object for example can be the root-action object of an object action-hierarchy which implements an LSA and hence a GAILS algorithm. Repeating applications of algorithm instantiations is needed for a proper statistical analysis in the case an algorithm such as a GAILS algorithm based on LSAs is randomized. Note that the global agent state object potentially must be reseted or rather reinitialized before a try as well as the associated heuristic state objects and the objects realizing termination criteria used by the repeatedly applied action object. The reinitialization is arranged by class `RepeatedTries`. It uses specific methods from the other classes involved such as method `resetHeuristicState` from class `HeuristicState`. The resetting methods of the classes that are involved are identifiable by prefix `reset`. The current try and the total number of tries are stored in an associated heuristic state object of type `RepeatedTriesHeuristicState` in member variables `currentTry` and `noTries` there as can be seen in the UML box named `RepeatedTriesHeuristic-State` representing this class in figures 5.10 on page 152 and 5.5 on page 143.

Another example of a basis action class illustrated in Figure 5.8 is class `ApplyFunction`. Action objects of this class simply take an object of class `Function` which is intended to represent a mathematical function which can be evaluated. A pointer to an object of class `Function` is stored in member variable named `pFunction` as can be seen in the UML box named `ApplyFunction` representing class `ApplyFunction` in Figure 5.8. Yet other basis actions shown in this figure (the respective class names are given in parentheses and the classes are represented in this figure by UML boxes named this way) include basis action classes to be inherited for implementing local search steps (`LsStepAction`) and local search procedures (`LsProcedureAction`), action class to be subclassed for implementing actions that do some kind of iteration (`LsIterationAction`), and an action class to be subclassed for implementing a complete ILS metaheuristic (`IlsAction`). Action objects of concrete subclasses of the first two basis action classes can have several modes of operation according to the neighborhood exploration scheme they use (as stored in member variable `pExploration`), e.g. first or best improvement. These basis action classes additionally provide for an identifier of the kind of neighborhood they represent (accessible via method `getNeighborhood-Identifier`). Class `IlsAction` finally has a method `updateLocalBestSearchState` to update the locally best search state object stored in the associated heuristic state object of its type (cf. Subsection 5.2.4).

All these basis action classes overwrite the pure virtual methods for accessing associated heuristic state objects and to update the global best search state object stemming from the root action class (named `getHeuristicState` and `updateGlobalBestSearchState`, respectively, cf. Subsection 5.2.5). Root action class `Action` provides a method named `getHeuristicState` for accessing associated heuristic state objects. This method, however, will only return pointers to objects of the most general type for heuristic state objects, namely `HeuristicState`. Basis actions provide more specialized methods (indicated in the action class hierarchy figures, figures 5.8 on page 148, 5.11 on page 158, and 5.12 on page 162, by the fact that the UML boxes representing basis action classes in these figures also list such methods named `getHeuristicState`) that are able to return pointers to heuristic state objects of the proper, i.e. corresponding heuristic state type. This overwriting entailing a pointer "conversion" is necessary, since pointers to heuristic state objects are used polymorphically within the GAILS framework. Each pointer in principle can only point to objects of a certain type, called *pointer type*. If a pointer is used polymorphically, it can also point to objects of any subclass of its pointer type, since objects of subclasses have the type of the inherited superclass also (cf. the notes at the begin of this section). However, only those methods of the objects pointed to that are declared by the pointer type can be called via the pointer, even if the object pointed to supports additional methods as added by the intermediate subclasses between the pointer type and the object's class in a class hierarchy. If the class of the object pointed to is known, however, the pointer can be *casted* to a pointer with pointer type equal to the object's class or any intermediate subclass (and hence type). Then, additional object methods can be called via the casted pointer. Now, whenever an action object accesses its associated heuristic state object will it use the overwritten method for accessing its associated heuristic state object, `get-HeuristicState`, from the basis action class its own class inherits from. This overwritten method `getHeuristicState` will do a proper cast. The returned pointer will have a pointer type equal to the corresponding heuristic state class of the action object for which the access method was called and thus can be used by this action object to access all required heuristic information from its associated heuristic state object. If the cast fails, a fatal error will simply occur, indicating that an object action-hierarchy and/or associated heuristic state objects were set up incorrectly violating the contract (cf. Subsection 5.2.5).

The same overwriting and casting mechanism used for heuristic state objects is applied to adapt action objects to different problem types. For each new problem type and each basis action class which is supposed to have a concrete subclass that is problem type dependent such as ones instantiating elementary action objects for the new problem type, an intermediate action class must be inserted into the class hierarchy. The intermediate action class simply overwrites the virtual methods for accessing the global current and best search state object such as `getSearchState` and `getBestSearchState`. Any other methods for accessing problem type specific information such as

the problem instance object have to be overwritten also. In the overwritten methods, the intermediate action class will realize a casting of pointers to search state and problem instance objects
such that the resulting pointer type is the subclass of classes `SearchState` and `ProblemInstance`
that corresponds to the problem type needed. As a result, the returned pointers to search state and
problem instance objects now can be used to access any problem type specific information from
these objects.

Figure 5.11 presents problem type specific parts of the action class hierarchy. The classes that are
mentioned next are represented in this figure by UML boxes named according to their class names.
Each of the basis action classes `LsStepAction` and `LsProcedureAction` is first inherited by problem type specific subclasses `FspLsStepAction`, `TspLsStepAction`, and `TspLsProcedureAction`
which overwrite the problem type specific (access) methods (prefix `Tsp` indicates specialization to
TSP problem instances, `Fsp` to problem instances for the FSP). Only in the next class hierarchy
level, concrete action classes implementing different kinds of local search steps and procedures,
as a rule one for each kind of neighborhood, are derived. As an example for concrete problem
type specific subclasses (i.e. subclasses instantiating elementary action objects) consider classes
`Tsp3OptLsStep` and `TspPerturbationDoubleBridge` which are illustrated in Figure 5.11. All concrete action classes define a method `apply` and those that provide different exploration schemes
for the neighborhood structure they represent such as `Tsp3OptLsStep`, `Tsp2OptLsProcedure`, or
`FspLsStepExchange` can store pointers to methods, called *method pointers* that implement various exploration schemes. The method pointers are stored in member variable `currentApplyMethod`
and are adjusted with method `adjustCurrentApplyMethod`. Method `apply` then simply follows the
method pointer and executes the method currently pointed to which hopefully implements the intended neighborhood exploration scheme. For example, the 3-opt neighborhood for the TSP can be
explored in first or best improvement or random step manner, implemented in methods `first3Opt-`
`Improve`, `best3OptImprove`, and `random3OptImprove` each of class `Tsp3OptLsStep`. Some action
objects additionally can store some parameters representing a mode of operation such as a strength
parameter for perturbations which is stored in member variable `perturbationStrength` of class
`TspPerturbationDoubleBridge` for example.

The class hierarchy for action objects as just sketched cannot be used the same way as an action class hierarchy that is designed analogously to the heuristic state class hierarchy. There,
class `LsProcedureHeuristicState` inherits from class `LsStepHeuristicState` since it is an extension in terms of the amount of heuristic information stored. Analogously, one could make
class `LsProcedureAction` inherit from class `LsStepAction` for the same reason. Then, whenever
an action object of a concrete subclass of class `LsStepAction`, i.e. of type `LsStepAction`, supposed to implement a local search step is required, an action object of a concrete subclass of class
`LsProcedureAction`, i.e. of type `LsProcedureAction`, is supposed to implement a local search procedure which could be used as well. Now, if an action object a_1 requires an action object a_2 of
`LsStepAction` for its operation, this implies that only the heuristic information contained in the
associated heuristic state object of a_2 of corresponding heuristic state type `LsStepHeuristicState`
is used by action a_2 and therefore must be maintained by action a_1, nothing more, nothing less,
according to the contract mentioned in Subsection 5.2.5. Instead of using action object a_2, it can
be substituted by an action object a_3 of type `LsProcedureAction` without harm, since class `Ls-`
`ProcedureAction` inherits from class `LsStepAction` and therefore *is* of type `LsStepAction`. An
action objects a_3 silently substituting action object a_2, however, needs to operate on an associated
heuristic state object of corresponding heuristic state type `LsProcedureHeuristicState` which
stores more heuristic information than a heuristic state object of type `LsStepHeuristicState`.
For example, it might be dependent on properly updated step counters which must be ensured by
action object a_1, e.g. because the termination criterion of the action object is based on step counts.
But any additional heuristic information from class `LsProcedureHeuristicState` in comparison to

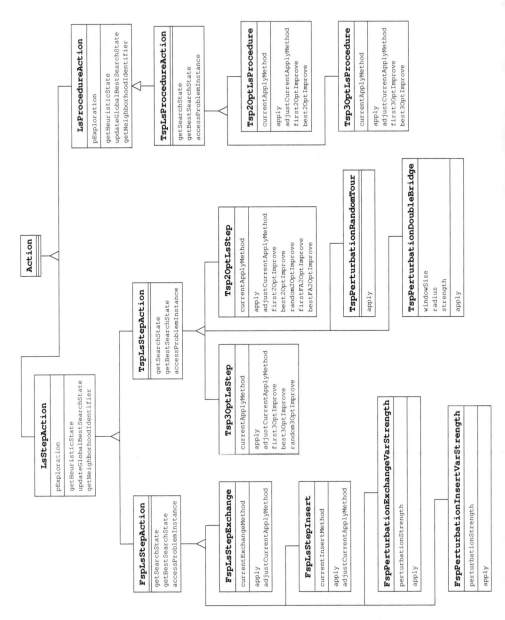

Figure 5.11:   Action class hierarchy problem type specific

class `LsStepAction` needed by an action object a_3 of type `LsProcedureAction` will not necessarily be maintained by action object a_1 according to the contract. Step counters as stored in member variable `steps` of the associated heuristic state object of action object a_3 for example will not be updated by action object a_1 simply because they are unknown to it. Action object a_1 assumes that the associated heuristic state object is of type `LsStepHeuristicState`. As a result, the substituting action object a_3 might not work properly. Nevertheless, the substitution were feasible. Preventing hazards such as the ones just described was one reason to introduce typing of action objects.

The action class hierarchy design of the GAILS framework does not allow to use an action object of type `LsProcedureAction` instead of one with type `LsStepAction`, although it might make sense. To remedy this seeming restriction, action objects that work as wrappers, also called *wrapper action objects*, can be employed. An action class instantiating a wrapper action object inherits from the basis action class in which context, i.e. on corresponding heuristic state type, the action object to be wrapped, called *wrapped action object*, is supposed to operate. A wrapper action object contains the wrapped action object by means of a pointer to the wrapped action object stored in a member variable. Both action objects potentially work on associated heuristic state objects of different corresponding heuristic state types. Instead of calling the wrapped action object directly in a certain context, the wrapper action object is called. In its `apply`-method it in turn calls the wrapped action object. Before and after calling the wrapped action object, however, the wrapper action object has to ensure that any heuristic information that is missing in any of the two associated heuristic state objects is updated or adjusted in some meaningful way. If the associated heuristic state object of the wrapper action object contains some heuristic information that cannot be stored in the associated heuristic state object of the wrapped action object and hence will not be computed and updated by the wrapped action, the wrapper action object has to come up with some meaningful derivation for the missing heuristic information after application of the wrapped action object, since this heuristic information has to be provided in the context the wrapper action object is applied according to the contract. If the wrapped action object in turn operates on an associated heuristic state object that contains some heuristic information which is unknown to the wrapper action object's associated heuristic state object and hence is not stored in it, some meaningful default values have to be provided by the wrapper action object before application of the wrapped action object in order to ensure proper execution of the wrapped action object.

For example, instead of using an action object of type `LsProcedureAction` that implements a local search procedure executing a number of local search steps, the special case of doing only one step could be used as well in principle in the form of an action object of type `LsStepAction` implementing a conventional local search step. To meet any requirements for action class `LsProcedureAction`, a wrapper class `LsStepToLsProcedureWrapper` inheriting `LsProcedureAction` has to be written whose object instantiations take an action object of type `LsStepAction` and act as action objects of type `LsProcedureAction`. This subclassing is shown in 5.8 on page 148 represented by UML boxes named respectively. In case of the mentioned example, the `apply`-method of an action object of class `LsStepToLsProcedureWrapper` simply will execute the wrapped action object implementing a local search step as stored in member variable `pLocalSearchStep` and will increment the step counters in the associated heuristic state object of the wrapper action object at most by one, since at most one step will have been done.

When wrapping an action object to a wrapper action object whose corresponding heuristic state class is a subclass of the corresponding heuristic state class of the wrapped action object, for example wrapping from class `LsStepAction` to `LsProcedureAction`, the wrapper action object and the wrapped action object can also share their associated heuristic state object. Wrapping the other way round works, too, e.g. in the form of an action class `LsProcedureToLsStepWrapper` wrapping from class `LsProcedureAction` to class `LsStepAction` as is represented by the UML box named `LsProcedureToLsStepWrapper` in Figure 5.8. A pointer to the wrapped action object of type `LsProcedureAction` is stored in member variable `pLocalSearchProcedure`. This way of wrapping, however, does not work with shared associated heuristic state objects. Since the corresponding heuristic state class of the wrapper action object is a superclass of the corresponding heuristic state of the wrapped action object, i.e. contains more heuristic information, the wrapped action object cannot work on the associated heuristic state object of the wrapper action object also but must use its own. The wrapper action object has to manage both associated heuristic states and copy information to ensure proper updates. Wrapper action objects of class `LsProcedureToLs-StepWrapper` have to hide some heuristic information after execution of the wrapped action object and have to make up some sensible values for the missing heuristic information before executing the wrapped action object.

A special case of a wrapper action object which can also be seen as a root-action object for an object action-hierarchy are action objects of class `LsStepToLsProceuderIterator` implementing a local search procedure. The UML box named `LsStepToLsProceuderIterator` in Figure 5.8 represents this class there. Action objects of class `LsStepToLsProceuderIterator` take a wrapped action object of type `LsStepAction` supposed to implement a local search step (accessed by a pointer stored in member variable `pLocalSearchStep`) and repeatedly applies it until a local optimum has been found (indicated by member variable `localOptimum` of the associated heuristic state object working as a flag of wrapped action object of type `LsStepAction`, cf. Figure 5.5). The wrapper or rather *iterator action object* of class `LsStepToLsProceuderIterator` counts the steps done and accordingly updates the heuristic information in its associated heuristic state object. Typically, it will share it with the wrapped action object.

To finish the treatment of the action class hierarchy, further important action classes and their intentions are described next. Action class `LsProcedureSequence` can be used to execute a list of action objects of type `LsProcedureAction` supposed to implement local search procedures in turn. Pointers to these action objects are stored in member variable `localSearchProcedures` which is a list of pointers. This class is represented in Figure 5.8 by UML box named `LsProcedureSequence`. The list nature of member variable `localSearchProcedures` is indicated by a 1+ in the parentheses. Name `LSProcedureAction` after 1+ refers to the pointer type of the pointers stored in the list. Each action object implementing a local search procedure is executed in turn and, depending on the mode of operation as expressed by member variable `findCommonLocalOptimum` working as flag, as soon as all implemented local search procedures have been executed once or as soon as a common local optimum has been reached, the sequencing of action object application stops. Before, however, any necessary heuristic information updates are performed, of course, by the iterating action object of class `LsProcedureSequence`.

Action class `SaAction` is the basis class for all concrete action classes for instantiating action objects implementing SA local search methods and which have to operate on an associated heuristic state object of type `SaHeuristicState` as is depicted in 5.10 on page 152. It provides a member variable `pLocalSearchProcedure` which stores a pointer to an action object of type `LsProcedureAction` supposed to implement a local search procedure and a member variable `pUpdateTemperature` which stores a pointer to an action object that is supposed to implement a temperature update action and which has type `SaAction` [BR03]. Action objects of concrete class `SimpleSa` which is a subclass of class `SaAction` then implement a simple and straightforward version of SA. All these classes and member variables are illustrated in 5.12 on page 162 with the same names as just used.

Action class `IlsAction` represented by the equally named UML box in Figure 5.12 is inherited by class `SimpleIls` implementing a simple standard ILS with the required respective actions (cf. Figure 4.1). Basis class `IlsAction` has corresponding heuristic state type `IlsHeuristicState` (as can be seen in Figure 5.10). It furthermore is inherited by various versions of action classes implementing acceptance criteria: One that always accepts a new search state (realized by class `AlwaysAccept`), one that only accepts better or equal cost new search states (realized by class `AcceptBetter`), and one that accepts better or equal cost new search states, or worse-cost new search states with a probability stored in member variable `epsilon` (realized by class `AcceptBetterOrEpsilon`). Other acceptance criteria are conceivable, too, but all action objects implementing acceptance criteria must operate on a search state object history and perhaps other ILS related heuristic information and thus on an associated heuristic state object of corresponding heuristic state type `Ils-HeuristicState` (cf. Subsection 5.2.5). Consequently, all such classes instantiating action objects implementing acceptance criteria will be subclasses of basis action class `IlsAction`. Note that an acceptance criterion of course is not an ILS. Inheriting from a basis action is about committing to a corresponding heuristic state type, not to a local search method type.

Finally, learning LSAs based on ILS-actions working according to the Q-learning approach of reinforcement learning are implemented by action objects of subclasses of classes `QIlsAction` and `QIls-BaseAction`. The former class is the basis action class for corresponding heuristic state type and concrete `QIlsHeuristicState` whose object instantiations store any heuristic information needed by learning LSAs. Class `QIlsBaseAction` is inserted into the action class hierarchy for code factorizing reasons. All action objects implementing a learning LSAs based on ILS-actions according to Q-learning need to maintain a list of action objects implementing ILS-actions. ILS-actions in the simplest case consist of a tuple of a perturbation and a local search procedure with a fixed acceptance criterion for all tuples, called *simple ILS-actions*. The maintained list for action objects implementing simple ILS-actions consists of a list of tuples of pointers to action objects implementing perturbations and local search procedures. The list of tuples of pointers is stored in a member variable `actions` which is a list. The pointer to the action object implementing the fixed acceptance criterion is stored in this simplest case of simple ILS-actions in member variable `pAccept`. Additionally, a pointer to an object realizing a policy (stored in member variable `pPolicy`) and a pointer to an object realizing a reward (stored in member variable `pReward`) are needed. All Q-learning algorithms furthermore need parameters α and γ which are stored in member variables `alpha` and `gamma`. An action object implementing a learning LSA furthermore has member variable `doLearning` which is a flag indicating whether the action object and hence the implemented learning LSA should operate in learning or in non-learning mode. The action object tuples implementing the simple ILS-actions are built from two lists of (pointers to) action objects implementing perturbations and local search procedures, by forming the Cartesian product in method `buildActions`. Objects realizing function approximators representing the respective action-value function with one function approximator per simple ILS-action applicable (cf. Subsection 3.3.2) are attached to each tuple of (pointers to) the action objects that implement the simple ILS-actions in method `attachFunctionApproximators` afterwards. Methods `saveFunctionApproximators` and `loadFunctionApproximators` are needed to save objects realizing function approximators that together learned an action-value function and to load learned ones, either for continuing learning or for using a learned action-value function implementation. The actual Q-learning based learning LSAs then are implemented by action objects of class `OneStepIls` which implement a one-step Q-learning simple ILS-action based learning LSA (cf. Subsection 3.2.6, [Wat89, Wat92], [SB98, p. 148]), and by action objects of class `QLambdaIls` which implement Watkin's Q(λ) algorithm for learning LSAs based on simple ILS-actions (cf. Subsection 3.2.6, [Wat89, Wat92], [SB98, pp. 182ff]). Both classes inherit from class `QIlsBaseAction`. The latter learning LSA additionally requires a λ parameter which is stored in member variable `lambda` of class `QLambdaIls`. All the classes and member variables mentioned in this paragraph are represented by UML boxes with same names in Figure 5.12. The list of tuples of pointers stored in member variable `actions` is followed by parentheses. The `1+` inside the parentheses signals the list nature. The following tuple of class names indicate the pointer type of the tuple components.

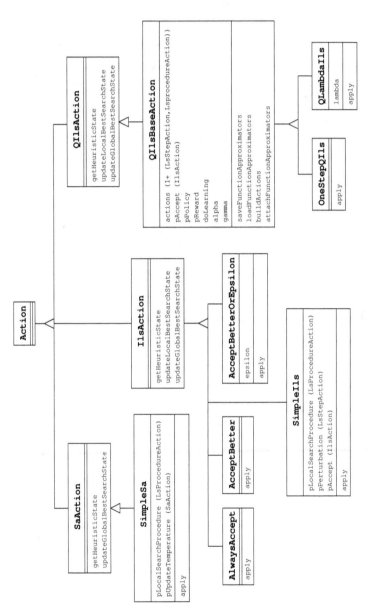

Figure 5.12: Action class hierarchy metaheuristics

5.2.7 Termination Criterion

Not all local search methods stop automatically such as a local search procedure. Nevertheless, termination must be ensured somehow. This can be done by constraining the runtime, or the number of steps or iterations to do. The same is true for action objects implementing local search methods. Action objects could implement their termination criteria directly, but this way, termination criteria implementations cannot be shared. Since many action objects employ similar or even identical termination criteria and each action object must employ at least one termination criterion massive code duplication would be the result. This is cumbersome, in particular, if the computation of whether to terminate or not, called *termination computation* or *computing termination*, is complicated or if accessing needed information such as step or iteration counts is not possible directly. Instead, termination criteria and termination computation can be encapsulated in objects. These termination criterion objects can be instantiated from the classes of a class hierarchy enabling code reuse. Additionally, termination criterion objects can be shared by several action objects and an action object can also use several termination criteria object at once easily. For example, if an action object has different phases during its execution, several termination criteria objects for ending each phase can be used. Examples of termination criteria are reaching a maximum number of steps or iterations allowed, exceeding a maximum amount of time allowed to run or reaching some other threshold value for some kind of heuristic information or even search state feature values.

In general, termination criteria represent special pieces of information describing when to stop an action application. Some functionality for the termination computation is required, but termination criteria are not actions, though. They also do not have an inner state, only modes of operation, specified for example by some threshold value such as a maximum number of steps allowed. Termination criteria are rather some kind of heuristic information that influence action application and which can change or rather be exchanged during action application. Additionally, termination criteria must not change the LSA state. Consequently, termination criterion objects are not action objects and action objects store their termination criterion objects in their associated heuristic state object. Since action objects can also share their associated heuristic state objects and action objects can employ several termination criterion objects simultaneously, each heuristic state object can hold on to several termination criteria objects (cf. Subsection 5.2.4).

Termination criterion objects are realized by the termination criterion class hierarchy with root termination criterion class `TerminationCriterion` which is illustrated in 5.13 on page 165. The individual classes are represented by UML boxes with the same names as will be used as class names throughout this subsection, only substring `TerminationCriterion` is abbreviated to `TermCrit` and substring `HeuristicState` is abbreviated to `HeurState` in the UML box names. The latter abbreviation is used for figures 5.14 on page 168, 5.16 on page 172, and 5.17 on page 173 also. Using a termination criterion object is done in three steps. First, it can be initialized with virtual method `initialize`. This typically has to be done at the beginning of the `apply`-method of the action object using it. For example, initialization is needed, if each call of an `apply`-method is allowed to run for a certain amount of time and timers might have to be reset at the begin of each such call. Initialization can, but need not, be used by an action object. If not applicable at all to a termination criterion object, the initialization method is defined to do nothing in the root termination criterion class `TerminationCriterion`. In the next step, method `computeTermination` computes whether to signal termination or not, i.e. computes termination. This virtual method has to implement how to use the information stored in the termination criterion object such as a threshold value in order to compute a boolean value which signals termination or not. Per default, it will compute a positive signal. It must be overwritten by any concrete subclass. Also, this method accesses all other information needed for computation, e.g. heuristic information to be compared with a threshold value such as step or iteration counts. The global agent state object thereby must not be changed. Nevertheless, computing termination can involve any information accessible via the global agent state object. Method `computeTermination` therefore needs access to the global agent state object which is provided (as pointer to it) as parameter. The result of the computation is stored internally in member variable `hasTerminated`. Finally, the result can be retrieved with method `terminate`. The member variables just mentioned are declared in the root termination criterion class represented by UML box named `TerminationCriterion` in Figure 5.13. A short cut for subsequently executing termination computation, and storing and accessing the result at once is method `checkTermination`. The rational for separating computing termination and accessing the truth value is as follows. It might be necessary to position the termination computation at a certain point during action object execution. The result, however, might be needed only at some later point and perhaps several times. Since termination criterion objects can be shared among actions, this way, a termination computation result can be shared among action objects, also.

Using a termination criterion object properly involves some form of contract. Any action object that employs a termination criterion object has to assert that before starting the termination computation, the global agent state object as a whole is up to date. If no termination criterion is needed, a dummy termination criterion object that always yields true or false as result of its termination computation can be used. The respective classes are depicted in Figure 5.13 represented by UML boxes named `TrueTermCrit` and `FalseTermCrit`. These classes simply overwrite virtual method `computeTermination` from the root termination criterion class, `TerminationCriterion`, always storing true or false, respectively. Other examples of termination criteria from before indicate termination as soon as a maximum amount of time has been elapsed or a search state feature value has reached a certain value. These are realized by classes `TerminationCriterionRuntime` and `TerminationCriterionMaxFeatureValue`, respectively and are also shown in Figure 5.13. Termination criterion objects of the former termination criterion class store a pointer to a timer object in member variable `pTimer` and the time threshold value in member variable `maxRuntime`. Member variable `reset` works as flag and indicates whether to reset the time in method `initialize` or not. Termination criterion objects of class `TerminationCriterionMaxFeatureValue` store the index and key of the feature to watch with respect to the feature vectors used in the features history of the global agent state object (cf. Subsection 5.2.1) in member variables `featureIndex` and `featureKey`, respectively. The feature value threshold is stored in member variable `maxFeatureValue`. Both termination criteria classes overwrite virtual `computeTermination` to fit their needs.

Other examples of termination criteria from before indicate termination as soon as a maximum number of steps or iterations have been done. These are realized by classes `LsProcedureHeuristicStateTerminationCriterionMaxSteps` and `LsIterationHeuristicStateTerminationCriterionMaxIterations`, respectively, which are also represented by identically named UML boxes in Figure 5.13. These termination criteria rely on specific heuristic information and hence respective termination criterion objects need to access a heuristic state object of a certain type, in these cases of type `LsProcedureHeuristicState` and `LsIterationHeuristicState`, respectively. Classes `LsProcedureHeuristicStateTerminationCriterionMaxSteps` and `LsIterationHeuristicStateTerminationCriterionMaxIterations` therefore have to implement and adapt heuristic state object access analog to the proceeding done within the action class hierarchy (cf. Subsection 5.2.6). For factorization reasons, class `HeuristicStateTerminationCriterion` is inserted into the termination criterion class hierarchy responsible for implementing accessing general heuristic state objects and for providing a general heuristic state object access interface. This interface basically consists of virtual method `getHeuristicState` which is defined by class `HeuristicStateTerminationCriterion`. Objects of this class additionally store a heuristic state index in member variable `heuristicStateIndex` which is with respect to the list of heuristic state objects stored in the global agent state object and which works as index-based pointer to a heuristic

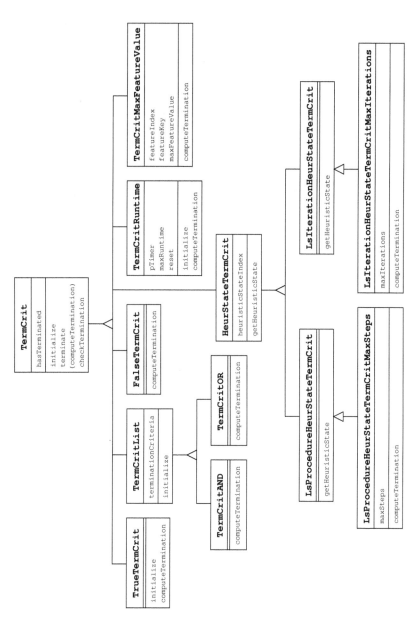

Figure 5.13: Termination criterion class hierarchy

state object. Method `getHeuristicState` returns a pointer to a heuristic state object of the root heuristic state type, `HeuristicState`. As done for the action class hierarchy which was split according to different problem types, the termination criterion class hierarchy is split according to different heuristic state types after class `HeuristicStateTerminationCriterion` yielding subclasses `LsProcedureHeuristicStateTerminationCriterion` and `LsIterationHeuristicState-TerminationCriterion` which overwrite virtual method `getHeuristicState`. The new versions of method `getHeuristicState` do a cast of the pointer returned by the original version to yield a pointer with pointer type corresponding to the respective more specialized heuristic state type before returning the pointer. In case of the two subclasses just mentioned the pointer types are `LsProcedureHeuristicState` and `LsIterationHeuristicState`, respectively. The pointer thereafter can be used to access any additional subclass-specific heuristic information. As can be seen in Figure 5.13, represented by respectively named UML boxes, classes `LsProcedureHeuristicState-TerminationCriterionMaxSteps` and `LsIterationHeuristicStateTerminationCriterionMax-Iterations` then subclasses `LsProcedureHeuristicStateTerminationCriterion` and `LsIteration-HeuristicStateTerminationCriterion`, respectively, and use the new version of method `get-HeuristicState` of the latter two termination criterion classes for accessing step or iteration counters, respectively.

It is conceivable to connect several termination criteria logically according to an AND or OR connection to yield a new one. OR-connecting several termination criteria means to stop as soon as *one* of the connected termination criteria holds true, for example to stop when an upper limit of steps, iterations, or time has been exceeded. AND-connecting them will indicate termination as soon as *all* limits have been reached, not only one. Class `TerminationCriterionList` realizes means to store and handle a list of termination criterion objects. Pointers to the termination criterion objects are stored in member variable `terminationCriteriaList` which is a list (see Figure 5.13). Class `TerminationCriterionList` overwrites virtual method `initialize` in order to initialize the whole list of termination criterion objects by initializing each termination criterion objects individually by calling their methods `initialize`. Subclasses `TerminationCriterion-OR` and `TerminationCriterionAND` then realize the respective logical combinations of termination criterion results in method `computeTermination` by triggering termination computation for all list members and connecting the results appropriately according to the logical combination they represent. Lists of termination criterion objects can easily be shared, too, in particular if the action objects involved in sharing also operate on the same shared associated heuristic state object. For example, consider a termination criterion object realizing an OR-connected list of other termination criterion objects that realize stopping after a maximum number of steps or iterations, or a maximum amount of time in the context of an object action-hierarchy implementing an ILS as is illustrated in Figure 4.1. All individual action objects of the object action-hierarchy implementing the ILS can and will work on the same associated heuristic state object for this example (cf. Subsection 5.2.6). Clearly, any heuristic information not changed by the execution of an action object involved will not trigger termination anywhere. The action object implementing a local search procedure will not change any iteration counters and hence will not trigger termination because of reaching a maximum number of iterations allowed. But it may cause termination because a maximum number of steps has been exceeded. In this case, since the termination criterion object is the same for all action objects together forming the object action-hierarchy implementing the ILS, the action object implementing a local search procedure will stop and, even if still some iterations could be done in principle, the root-action object of the object action-hierarchy implementing the ILS will stop, too, since it shares the termination criterion list and hence their contained termination criterion objects *and* the associated heuristic state object. This way, the signal to terminate automatically is propagated up the object action-hierarchy.

5.2.8   Function Approximator

Classes of the function approximator class hierarchy with root function approximator class `Function-Approximator` realize the learning components of the GAILS framework in the form of function approximators that are used to represent strategy functions. As can be seen in Figure 5.14 (substring `FunctionApproximator` of UML box names is abbreviated to `FunctionApprox` there), subclasses of the root function approximator class are devised according to different kinds of function approximators such as SVMs (classes `LibSvmFunctionApproximator` and `SVR_L_inc_Function-Approximator`) and Regression Trees (class `RegressionTreeFunctionApproximator`). Policies can be realized as preference structures using a list of function approximators with one function approximator per action in case of using an action-value function (cf. Subsection 3.3.2). As such, no special treatment of different kinds of strategy and value functions by means of subclasses is necessary in principle. Since policy representation is rather the responsibility of a specific learning LSA, it is recommendable to leave the organization of strategy functions to action classes implementing learning LSAs (cf. Subsection 5.2.6).

The subclasses of class `FunctionApproximator` hide any details such as whether a realized function approximator really works incrementally or in pseudo incremental mode, e.g with block-wise or *batch* learning (cf. Subsection 3.3.3). Together with providing a fixed interface by the root function approximator class, this makes function approximator objects completely interchangeable. The interface for function approximator objects thereby basically consists of pure virtual methods `learn` and `compute`. Both take a feature vector of real values as input represented by an object of class `Features`. Method `learn` additionally requires a target value (real value) which makes a complete training example. Each training example can be used to update the internal model of a function approximator object in the form of appropriate data structures which represent the current approximation function (cf. Section 2.6). Method `compute` computes a target value according to the internal model of a function approximator object. Any concrete subclass of class `Function-Approximator` realizing an actual function approximator has to overwrite pure virtual methods `learn` and `compute` as can be seen in Figure 5.14 indicated by putting these method names in parentheses in the UML box representing the root function approximator class and repeating them in the UML boxes representing concrete function approximator classes realizing actual function approximators.

Method `learn` has variants for processing several training examples at once. To support block-wise or batch learning, several training examples can be stored temporarily before actually updating the internal model. The maximally allowed number of stored training examples for function approximator objects is contained in member variable `learningFrequency`. Pure virtual method `triggerLearning` can be used to explicitly start the computation of a new internal model, i.e. trigger learning. It also has to be overwritten by any subclass of root function approximator class `FunctionApproximator`. Other member variables store input and output filenames for loading and storing internal models (named `inputFilename` and `outputFilename`, respectively), the expected dimension of the input feature vector (named `inputDimension`), and a flag indicating whether an internal model has already been built (named `hasModel`). All these member variables and methods are depicted in Figure 5.14 also.

Function approximator objects of class `DumpFunctionApproximator` from Figure 5.14 can be used to write training examples to a file instead of building an internal model. This happens transparently for the action object using the function approximator in method `learn`. Method `compute` simply returns a random number. The values stored in member variables named `precision` and `width` of class `DumpFunctionApproximator` determine the respective modifications for writing floating point numbers.

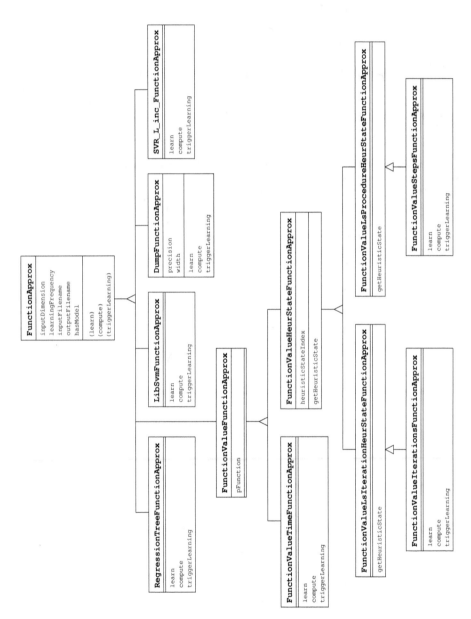

Figure 5.14: Function approximator class hierarchy

Instead of using a learned internal model to compute an approximated target value for a given feature vector, a target value can also be computed by a fixed function with the input value for computing a target value, called *input value*, stemming from any other resource available. Class `FunctionValueFunctionApproximator` and its subclasses are intended for this purpose (see Figure 5.14). Function approximator objects of this type store in member variable `pFunction` a pointer to an object of class `Function`, called *function object*, which realizes the function. All concrete subclasses of class `FunctionValueFunctionApproximator` provide a `learn` and `compute` method such that method `learn` simply does nothing and `compute` delegates the computation of a target value to the stored function object. Subclasses of class `FunctionValueFunctionApproximator` are designed according to the source for the input value. Function approximator objects of class `FunctionValueTimeFunctionApproximator` provides the currently elapsed time as input value, function approximator objects of subclasses of class `FunctionValueHeuristicStateFunctionApproximator` provide heuristic information input value. The heuristic state object index with respect to the list of heuristic state objects stored in the global agent state object is stored in member variable `heuristicStateIndex` there. The heuristic state objects indexed by member variable `heuristicStateIndex` can be retrieved with method `getHeuristicState`. Analog to the proceeding in subsections 5.2.6 and 5.2.7 for the action and termination criterion class hierarchies, the function approximator class hierarchy after class `FunctionValueHeuristicStateFunctionApproximator` is split according to different heuristic state types. In Figure 5.14, two branches are depicted. One branch consists of UML boxes representing classes `FunctionValueLsIterationFunctionApproximator` and `FunctionValueIterationsFunctionApproximator`. The latter (concrete) class provides an iteration counter as input value. The other branch consists of UML boxes representing classes `FunctionValueLsProcedureFunctionApproximator` and `FunctionValueStepsFunctionApproximator`. The latter (concrete) class provides a step counter as input value.

5.2.9 Features

Class `Features` is intended to represent feature vectors. A feature vector consists of a number of individual features. An individual feature in turn consists of a key-value pair. The key is represented as string, the value as real value. A feature vector can also be seen as a mapping from keys to real values. A so-called *agent state feature vector* is supposed to characterize a complete LSA state. As such, any agent state feature vector in principle consists of three kinds of individual features: problem instance features, search state features, and heuristic state features (cf. subsection 5.2.3, 5.2.2, and 5.2.4, respectively). Problem instance features provide properties valid for the whole problem instance independent of any current search state. Accordingly, they do not change during the search but can induce some bias dependent on the nature of the current problem instance to solve which can be used to classify problem instances. Search state features provide up to date properties of the current search state or rather solution encoding (cf. Subsection 3.2.2). Heuristic state features basically coincide with heuristic information characterizing the search progress. Recall that feature vectors and agent state feature vectors in particular and hence features objects realizing them (called *agent state features object* in the latter case) are the interface between the action part and the learning part of the GAILS framework, since the interface methods of function approximator objects, methods `learn` and `compute`, are based on features objects. These methods furthermore are not interested in feature keys, they simply regard feature objects as a vector of real values.

Features objects realizing agent state features are stored in a features history in the global agent state object (cf. Subsection 5.2.1). They are reused by means of an update mechanism (cf. Subsection 5.3.1). Feature vectors realized as individual features objects enable to maintain compatibility between function approximator objects anytime and easily and make function approximator objects completely and transparently interchangeable. They thus really provide an easy to use and simple interface. The only restriction is to use a function approximator object consistently with the same set (and hence number) of features, i.e. the same input dimension during all learning and application.

Individual features can additionally be normalized empirically. For each individual feature, a lower and an upper bound can be stored also. These bounds can be set with methods setFeatureLower-Bound and setFeatureUpperBound, respectively, and accessed with methods getFeatureLower-Bound and getFeatureUpperBound, respectively. If a flag doActualizeBounds is set, any new assignment of values to an individual feature will be checked against the lower and upper bounds already stored for this individual feature. If the new value is lower than the stored lower bound, the lower bound is updated to the new value. The same happens analog for the upper bound. Flag doNormalize indicates whether the stored feature values are to be normalized according to the procedure proposed in Subsection 3.3.1 with the stored lower and upper bounds when accessing them.

5.2.10 Policy

Recall from Subsection 3.2.1 that policies can either be realized as look-ahead search using state-value functions – if a model of action behavior is available – or by means of preference structures – if no model can be used. Function approximators can represent a state-values function directly. In the case of an action-value function, one function approximator per action applicable will be used (cf. Subsection 3.3.2). In both cases, each applicable action will yield a preference or an equivalent value (cf. Subsection 4.4.3). These then have to be compared to pick one action to actually execute. The classes of the policy class hierarchy with root policy class Policy exactly encapsulate and represent this comparison and selection process, foremost in order to avoid code duplication.

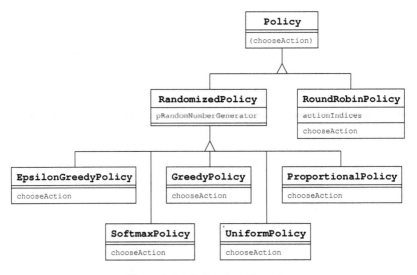

Figure 5.15: Policy class hierarchy

The policy class hierarchy is illustrated in Figure 5.15. The UML boxes there represent policy classes and UML box names coincide with the names of the policy classes they represent. Each policy class has to overwrite pure virtual method `chooseAction` from root policy class `Policy`. This method takes as input a list of preference values. The value at the i-th position belongs to the i-th action object implementing the i-th action applicable. Typically, this value stems from the i-th function approximator that approximates the action-values for the i-th action object applicable (cf. Subsection 5.2.6). Hence, the positions with respect to the input list work as action object indices. Method `chooseAction` will compare the input preference values and will return a pair consisting of the index and the respective preference value of the action object chosen. For each kind of policy, one concrete subclass is built. Since some policies involve randomization, intermediate subclass `RandomizedPolicy` is inserted which provides access to an object realizing a random number generator via the pointer stored in member variable `pRandomNumberGenerator` as can be seen in Figure 5.15. Several policies are realized such as greedy (realized by class `GreedyPolicy`), ϵ-greedy (realized by class `EpsilonGreedyPolicy`), softmax (realized by class `SoftmaxPolicy`) (cf. Subsection 3.2.1), uniform (realized by class `UniformPolicy`), proportional (realized by class `ProportionalPolicy`), and round-robin (realized by class `RoundRobinPolicy`). The uniform policy picks each action independent from its preference value with equal probability, basically yielding a random action selection. The proportional policy picks each action with probability proportional to its preference value. The round-robin policy finally picks each action in turn according to the order of the action in the list, i.e. according to their indices.

5.2.11 Reward and Move Cost

Recall from subsections 4.3.3 and 4.3.4 that several reward and move cost computations are conceivable. In order to avoid code duplication, this computation is also encapsulated in objects. Two classes for representing rewards and move cost named `Reward` and `MoveCost`, respectively, exist. They are contained in the reward class hierarchy which is depicted by Figure 5.16. Again, UML boxes represent classes. UML box and class names coincide.

Class `Reward` stores the pointer to a move cost object in member variable `pMoveCost` and computes the reward in method `computeReward`. This is done by subtracting from the pure reward, which is the reward without move costs (cf. Subsection 4.3.4) and which is computed by pure virtual method `computePureReward`, the move costs computed by method `computeMoveCost`. The resulting difference is called actual reward *actual reward* (cf. Subsection 4.3.4). This actual reward can subsequently be bounded in method `computeReward` yielding the *actual bounded reward*. Pure virtual method `computePureReward` has to be overwritten by any concrete subclass of class `Reward`. Method `computeMoveCost` simply calls method `computeMoveCost` of the stored move cost object. This method from class `MoveCost` in turn calls pure virtual method `computePureMoveCost` to compute the *pure move cost*. The pure move costs are unbounded original move costs which can further be processed in method `computeMoveCost` to yield the *move costs*. These can finally be bounded in this method, too, yielding the so-called *bounded move costs*. Pure virtual method `computePureMoveCost` must be overwritten in any concrete subclass of class `MoveCost`, virtual method `computeMoveCost` can be overwritten to introduce a further processing (and subsequently bounding) of the pure move costs. One result of this class hierarchy design as illustrated in Figure 5.16 is that move costs and pure rewards are completely independent from each other. It is possible to bound both actual rewards and move costs. Therefore, class `RewardBase` has been inserted as superclass for `Reward` and `MoveCost` has been inserted. It can store a lower and an upper bound value in member variables `lowerBound` and `upperBound`, respectively, and a flag indicating whether to apply lower and/or upper bounds in member variables `applyLowerBound` and `applyUpperBound`, respectively. Bounding is performed in method `bound` which has to be used in methods `computeReward` and `computeMoveCost`.

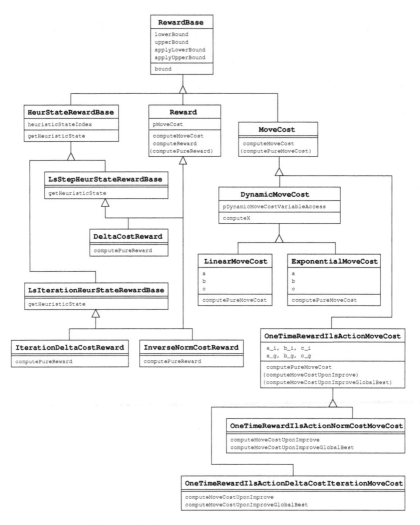

Figure 5.16:   Reward and move cost class hierarchy

Reward and move cost computation can be based on any information from an LSA state in principle. Therefore, any methods declared by class `Reward` and `MoveCost` and discussed in this subsection so far have a pointer to the global agent state object as parameter. If the information to base the computation on stems from heuristic states, some kind of adaption to heuristic state types has to be dealt with as for the action and termination criterion class hierarchies (cf. subsections 5.2.6 and 5.2.7). Figure 5.16 shows the usual subclass-split according to different heuristic state types into classes `LsStepHeuristicStateRewardBase` and `LsIterationHeuristicStateReward-Base` after an intermediate class `HeuristicStateRewardBase` used for accessing heuristic states. The latter as usual stores a heuristic state index which works as index-based pointer to a heuristic state object and is actually stored in member variable `heuristicStateIndex` and provides access to the heuristic state via method `getHeuristicState`. It then is overwritten in subclasses to provide a pointer with a pointer type of the proper heuristic state type. Finally, concrete reward classes can be realized. If they need to retrieve information from a heuristic state, they multiply inherit from class `Reward` and from a subclass of class `HeuristicStateRewardBase`. They thus have inherited one method for computing the pure reward and one for accessing a heuristic state object of the proper heuristic state type. Reward objects of class `DeltaCostReward` compute a reward which consists of the difference in cost induced by the last local search step done as resulting from an action object execution. Reward objects of class `IterationDeltaCostReward` compute a reward which consists of the difference in cost occurred during the last iteration done. In the case of an object action-hierarchy implementing an ILS such an iteration simply coincides with one application of an ILS-action. Reward objects of class `InverseNormCostReward` finally compute a reward which consists of the distance of the actual normalized cost of the current search state to the upper bound (which is 1 in the normalized case, cf. Subsection 4.3.3).

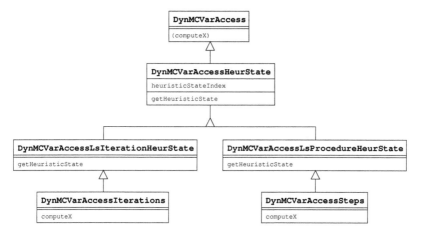

Figure 5.17: Variable access class hierarchy

Move costs can be a fixed value per move or they can differ from move to move and might be based on any information from an LSA state as well (cf. Subsection 4.3.4). Therefore, class `DynamicMoveCost` provides a method that can compute a dynamic input value for further processing by a function. The input value is computed by method `computeX` of this class which in turn uses an object of type `DynamicMoveCostVariableAccess` accessed by a pointer stored in member variable `pDynamic-MoveCostVarableAccess`. Since the dynamic input value can come from any source of information from the whole LSA state, the class hierarchy with root class `DynamicMoveCostVarableAccess` is split into subclasses according to the information source. Figure 5.17 shows a part of this class hierarchy with the well-known subclass-splits according to heuristic states type. The common class name prefix `DynamicMoveCostVariableAccess` is abbreviated to `DynMCVarAccess` for the UML box names representing the equally named classes because of space limitations. Concrete classes `LinearMoveCost` and `ExponentialMoveCost` finally realize the further processing of the input value according to a modified linear function $a \times x^b + c$ and a simple exponential function $a \times b^x + c$, respectively. The parameters are stored internally in equally named member variables (see Figure 5.16).

One-time rewards can be regarded as a form of move costs which are used to rewards special moves, for example those that are improving or the yield a new globally best LSA state (cf. Subsection 4.3.4). Class `OneTimeRewardIlsActionMoveCost` inherits from class `MoveCost` and is an abstract class for computing one-time rewards based on information that is contained in heuristic states of type `IlsHeuristicState`. The one-time reward is computed as:

$$((\text{a\_i} \cdot x^{\text{b-i}} + \text{c\_i}) \cdot I_i + (\text{a\_g} \cdot x^{\text{b-g}} + \text{c\_g}) \cdot I_g) \cdot r$$

where x ($x \in [0, 1]$) denotes the relative progression of the search relative to the known maximum runtime, I_i ($I_i \in \{0, 1\}$) is an indicator whether the last potential LSA move was an improving one, I_g ($I_g \in \{0, 1\}$) is an indicator whether the last potential move additionally yielded an overall new best LSA state, and r ($r \in \mathbb{R}$) is the immediate reward obtained during the potential LSA move. The `a_i`, `b_i`, `c_i` are the parameters for weighing the additional one-time reward for improving potential LSA moves in dependence on the current progression and `a_g`, `b_g`, `c_g` are the parameters for weighing the additional one-time reward for potential LSA moves to a new overall global best LSA states in dependence on the current progression. They are depicted as member Figure 5.17. variables in Class `OneTimeRewardIlsActionMoveCost` overwrites pure virtual method `computePureMoveCost` and computes the one-time reward in two steps, one for the improving moves and one for the moves to new globally best LSA state. The first step is computed by pure virtual method `computeMoveCostUponImprove`, the latter by pure virtual method `computeMove-CostUponImproveGlobalBest`. These methods are overwritten by subclasses `OneTimeRewardIls-ActionDeltaCostIterationMoveCost` and `OneTimeRewardIlsActionNormCostMoveCost` of class `OneTimeRewardIlsActionMoveCost`. The former class is used, if the reward signal is a simple delta cost reward signal, the latter, if it is the inverse cost reward signal (cf. Subsection 4.3.3). Figure 5.17 shows the classes involved in one-time reward computation.

5.2.12   Utilities

Several utilities are needed for a GAILS algorithm instantiation of a learning LSA to run. GAILS algorithms mostly are randomized and need random number generators (realized by class hierarchy `RandomNumberGenerator`). They need means to measure time, e.g. for termination checking, and therefore need a timer (realized by class `Timer`). They need access to any command line interface (CLI) parameters and performance measures (provided by objects of classes `ProgramParameters`, `Parameter`, and `PerformanceMeasure`), have to handle warnings and exceptions (realized by classes `Warning` and `FatalException`), might want to output information in a specific format (realized by class `StandardOutputFormat`), and might want to produce trace information for better visualization of the method calls or for debugging purposes (provided by class `Trace`). In order to enable global and anytime usage of these frequently used utilities, they are encapsulated in classes (the respective class names have been given in parentheses). The utility classes will be described briefly next:

- `RandomNumberGenerator`: The classes of the random number generator class hierarchy as illustrated in Figure 5.18 realize random number generators. Root random number generator class `RandomNumberGenerator` stores an initial seed in member variable `initialSeed`. Concrete subclasses additionally store a current state in some form. As a basic interface, root random number generator class `RandomNumberGenerator` provides methods for randomly generating numbers according to several distributions as depicted by the equally named UML box representation in Figure 5.18. The names of the methods returning random numbers according to specific distributions, called *distribution methods*, all begin with prefix `get`. The figure lists several of them with self-explanatory names. Three distribution methods, `getRandom` (realizing a Gaussian distribution), `getRandomInt` (realizing a uniform distribution over $(0, 1)$), and `getNormal` (realizing an equal likely distribution over integers from 0 to n), are dependent on the actual kind of the random number generator. All other distribution methods use these basic distribution methods to compute more specialized distributions. The three basic distribution methods are pure virtual ones and therefore have to be overwritten in any concrete subclasses of class `RandomNumberGenerator`. The same is true for the initialization method `initialize` which is purely virtual, too. Actual random number implementations comprise a Lehmer (realized by class `LehmerRandomNumberGenerator`) [Leh54, PRB69] and a Mersenne twister (realized by class `MersenneTwisterRandomNumberGenerator`) [MN98] random number generator.

- `Timer`: This class provides access to a clock which can measure net and total elapsed CPU time. The mode of time measurement, total or net time, can be set with method `setTimerType`. The clock has to be started first with method `startTimers`. Afterwards, the elapsed time according to the mode of time measurement can be queried with method `elapsedTime`.

- `ProgramParameters`, `Parameter`, and `PerformanceMeasure`: According to the CLI definition format from [VHE03] which is abutted to the interface defined in [SdBS01], class `Parameter` represents one single CLI parameter with the specification of its type and range, its command line signature, a default value, and a comment for a help print out on the command line. Class `PerformanceMeasure` represents performance measures similar to parameters with a type and a comment. Class `ProgramParameters` collects all individual parameters and performance measures (see Figure 5.2), which have to register there, parses the command line, evaluates and stores the read parameter values according to registered individual parameters, and provides them on demand.

- `Warning` and `FatalException`: Objects of class `Warning` are used to output warnings to the standard error output. It prepends a string to each output message indicating that the following message is a warning. Objects of class `FatalException` are used to indicate a fatal exception. They can also output an error message with prefix and thereafter quit the program with a provided exit code.

- `StandardOutputFormat`: A standard output format for local search methods was presented in [VHE03] which, again, is abutted to the format defined in [SdBS01]. Objects of class `StandardOutputFormat` help in outputting information in this format. Such objects can manage a data stream. This data stream can be opened externally and given to an object of class `StandardOutputFormat` or a filename can be provided which then is used to open the designated file and establish a data stream by the object itself. Class `StandardOutputFormat` provides methods that are tailored to printing several types of information and even complete objects such as objects of classes `ProgramParameters`, `SearchState`, or `PerformanceMeasure`.

- `Trace`: Class `Trace` is used to define traces which can be turned on and off individually and flexibly at compile time. They do not consume any resource if they are turned off. Class `Trace` comes with a set of macro definitions that can be used to register trace types, turn on and off some of them globally, and specify an individual trace message for a certain trace type in a method. Each individual trace message in a traced method will be output, if the respective trace type is turned on. A method signature will be prepended to identify the source of the trace message in terms of the method that issued it. Each trace output is indented according to the current stack depths of method calls, but only with respect to methods actually doing traces.

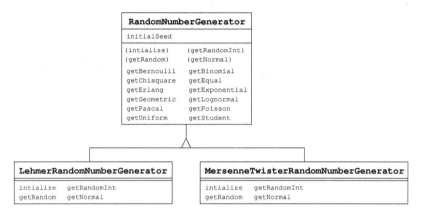

Figure 5.18: Random number generator class hierarchy

5.3   Implementation

The GAILS framework design was addressed in the previous section. This section clarifies some important implementation details or concern efficiency issues. The section does not provide comments on single methods or classes. These can be extracted from the source code.

Subsection 5.3.1 will discuss details of the implementation of the update mechanism for the features object, while the second subsection, Subsection 5.3.2, will briefly summarize, how new problem types or function approximators can be integrated into the GAILS framework.

5.3.1   Features Implementation Issues

Agent state features have to be computed very often, at least once per move for a learning LSA. Therefore, the computation of features of any kind – problem instance, heuristic state, or search state features – must be as efficient as possible in terms of memory and computation time consumption. Where possible, any individual feature from any of the two features sources that can change with each move, search and heuristic states, should be implemented incrementally. In the case of heuristic state features, these stem from heuristic information stored in heuristic state objects which have to be updated after each action object application anyway. In the case of search state features, this might require to hard-code the incremental computation of search state features in the methods of search state classes implementing local search operators which in turn unfortunately prevents from changing search state features easily. But it is expected that once a good set of search state features has been found, it will be used thereafter unchanged and that a good set can be found for all settings and instances of a problem type. Heuristic state features can be turned on and off individually at runtime and therefore easily via CLI parameters without substantial computation overhead. Each new heuristic state subclass knows which individual heuristic state features it adds and provides methods to turn them on and off. The respective methods for a heuristic information labeled XYZ are named turnOnXYZ and turnOffXYZ. Of course, for one run of a GAILS algorithm instantiation, the set of features and their position in the agent state feature vector must be the same. The same often is also true for a number of mutually connected and dependent runs, e.g. in a learning setting.

Agent state features objects including the so-called *current agent state features objects* characterizing the current state of the global agent state object and hence LSA state are stored in a features history in the global agent state object (represented by box labeled "featuresHistory" in Figure 5.2). As was mentioned in Subsection 5.2.1, the features history is organized as a ring buffer and the individual agent state features objects are reused. This ring buffer is depicted on the right hand side of Figure 5.19 labeled "Features History". Each agent state features object representing an agent state feature vector consists of a set of problem instance features, search state features, and heuristic state features stemming from the n ($n \in \mathbb{N}^+$) heuristic states in use as can be seen at the top of this figure. Updating the current agent state features object is easy. The ring buffer begin and end positions are advanced one step and the agent state features object at the new beginning position is reused for the immediately following update. This is done by retrieving the features object from the global current search state object and copying individual feature values to the proper positions in the agent state features object being updated. The proper positions have to be memorized when setting up the history for the first time when starting a GAILS algorithm instantiation, of course, and must not change implicitly during execution. Changing them does not make sense anyway, since otherwise any function approximator objects are rendered useless (cf. Subsection 5.2.9). Next, the agent state features object is presented to each of the n heuristic state objects stored in the global agent state object in turn as depicted in Figure 5.19 and indicated by arrows there. Each heuristic state object knows exactly which features are turned on and hence need updating and to which position in the agent state features object it is supposed to copy the respective value. Technically, this is done by traversing the heuristic state class hierarchy implicitly by calling superclass methods in a specific and fixed order.

5.3.2 Integrating New Problems or Function Approximators

Integrating new problem types and function approximators into the GAILS framework is easy. Compared to the effort to implement data structures (for representing a problem type instance, solution encoding, and internal models of function approximators) and to realize needed methods (which implement local search steps and procedures and other local search operators, and methods for learning and value computation for function approximators, and so on) the integration itself is negligible. The integration basically consists of writing appropriate wrapper classes for the implementations just mentioned as will be briefly explained in this subsection.

Suppose a new problem type to be integrated into the GAILS framework is abbreviated to "XYZ". To integrate the new problem type "XYZ", first a subclass XyzProblemInstance of class Problem-Instance has to be written (cf. Subsection 5.2.3). Objects of this subclass encapsulate problem instance representations. Appropriate data structures for storing the problem instance specification have to be devised and retrieval methods to be used by search state objects have to be written. The two following methods thereby have to be overwritten. Data structures have to be set up in method computeDataStructures and will be deleted by method deleteDataStructures. The former method will be triggered in method readInstance of class ProblemInstance. Problem

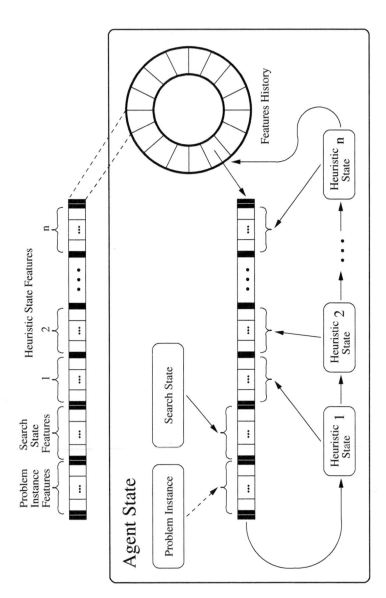

Figure 5.19:   Features update mechanism

instance feature computation has to be implemented in methods `computeLocalInstanceFeatures` and `computeGlobalInstanceFeatures`. These two methods have to be overwritten also. The first computes features local to the information connected to each solution component in the problem instance specification. Based on these local features, the second method computes instance wide, i.e. globally valid, problem instance features. Finally, methods for retrieving the problem type identifier (named `getType`) and the instance size (named `getSize`) must be provided by overwriting these virtual methods.

In addition to a problem instance representation, problem type specific solution encoding data structures and manipulation methods in the form of local search operators have to be implemented in subclass `XyzSearchState` of class `SearchState`. The data structures for encoding an initial solution and hence the initial solution encoding will be computed randomly or deterministically in methods `computeDataStructuresRandomly` and `computeDataStructures`, respectively, and can be reset, if already computed, with methods `resetDataStructuresRandomly` and `resetData-Structures` respectively. The data structures are deleted by method `deleteDataStructures`. These virtual methods from root search state class `SearchState` have to be overwritten by subclass `XyzSearchState`. In addition, methods for cloning (`clone`), copying (`operator=`) and assigning (`assignFrom`) a search state object, for comparing two search state objects (`operator==`), and for retrieving the problem type identifier (`getType`) have to be overwritten. Proper cost normalization has to be ensured via methods named `normalizeCost` and `normalizeDeltaCost` (cf. Subsection 3.3.1). Solution encodings and problem type information will be read in and output with methods `readSolution`, `printSolution`, `printTypeInformation` and `readTypeInformation`, respectively, so these methods have to be overwritten also. Next, search state features have to be contemplated and designed. Their initial computation must be taken care of in the methods that compute or reset data structures as just listed. Finally, local search operators have to be implemented. As was mentioned before (cf. subsections 5.2.2 and 5.2.5), these can comprise local search steps and procedures, and even complete metaheuristics. Each local search operator is implemented in its own method. Within this method, the solution encoding data structures will be updated according to the operator definition and the search state features, preferably incrementally, have to be computed (cf. Subsection 5.3.1).

For the learning LSA approach to work, several actions to choose from must be provided. Therefore, in the case of using several ILS-actions, several local search step or procedure methods (e.g. named `step_1`, ..., `step_N`, `procedure_1`, ..., `procedure_M`) and several perturbations methods (e.g. named `pert_1`, ..., `pert_K`) must be implemented (cf. Subsection 5.2.6). These methods can be equipped with further parameterization such as a strength parameter in case of a perturbation. To interface these local search operators implementing methods of search state class `XyzSearchState` to the action part of the GAILS framework (cf. Subsection 5.2.5), several action classes implementing elementary actions acting as wrapper must be provided (e.g. named `XyzStep1LsStep`, ..., `Xyz-Step_NLsStep`, `XyzProcedure_1LsStep`, ..., `XyzProcedure_NLsStep`, `XyzPert_1LsStep`, ..., `Xyz-Pert_KLsStep`, cf. Subsection 5.2.5). These will inherit via action class `XyzLsStepAction` from class `LsStepAction` or via class `XyzLsProcedureAction` from class `LsProcedureAction` as was discussed in Subsection 5.2.6. Note that in principle it is sufficient to implement local search operator implementing search state object methods and corresponding elementary action objects for local search steps only, since these can be upgraded easily to a local search procedure by means of action class `LsStepToLsProcedureIterator` (cf. Subsection 5.2.5). After this has been done, all object action-hierarchies can be reused as they are, only with the new elementary action objects plugged in replacing those that are specific to another problem type (cf. subsections 5.2.5 and 5.2.6).

Integrating a new function approximator of type "ZYX" is even easier. First, a new subclass `ZyxFunctionApproximator` of function approximator root class `FunctionApproximator` has to be written. This class encapsulates the data structures for its function approximator type dependent internal model. It has to provide methods for cloning (`clone`), copying (`operator=`), assigning (`assignFrom`), and perhaps for comparing (`operator==`) two function approximator objects. Since internal models have to be reused, with methods `printFunction` and `readFunction` for storing and reading in internal models have to be implemented. Sometimes it is necessary to reset an internal model with methods `resetState` and `resetResetFlag` which perhaps has to be overwritten also. Most important is, however, to implement the pure virtual methods for learning and value computation, `learn` and `compute`, respectively. In addition, purely virtual method `triggerLearning` must be implemented as was mentioned before in Subsection 5.2.8. After all these methods have been implemented, the new function approximator is ready to use and respective function approximator objects can substitute other ones.

5.4   Related Work and Discussion

The GAILS (implementation) framework is not the first attempt to use object-oriented design for generically modeling local search methods. This section reviews in Subsection 5.4.1 the most important related class libraries, toolkits, and frameworks for local search methods. Before, it will be clarified in Subsection 5.4.2 what class libraries, toolkits, and frameworks are and what their purpose is in the context of local search methods. Additionally, the GAILS framework briefly is reviewed in terms of what thereby is presented in Subsection 5.4.3.

5.4.1   Concepts of Reuse

One aim in research of local search methods is a systematic comparison of different search strategies. Amongst others, this endeavor entails the need to isolate influences on performance due to individual components of local search methods. Support for this quest is beneficial. One possibility to do so is to reduce the development time for new, perhaps prototypical, or modified local search methods by enabling fast design and implementation. Recall from the begin of Chapter 5 that such an intended rapid prototyping can be accomplished by making components of local search methods independently combinable and this way allowing for code and component reuse.

In the context of local search methods many common concepts are present. These comprise solutions, problem types, problem instances, neighborhood structures, cost functions, neighborhood exploration schemes, neighbor selection schemes, local search steps, local search procedures, individual metaheuristics, and others. These common concepts provide a generic conceptual framework for viewing local search methods which can be mapped into practical tools for the implementation of local search methods. Object-oriented programming languages such as C++ are particular suited for this venture. Object-oriented programming provides means for abstraction while enabling extensibility to the more specific. This is achieved by common and perhaps abstract superclasses reflecting common and invariant concepts and components of what is to be modeled. In object-oriented programming so-called *class libraries*, *toolkits* and (application) frameworks are used to achieve code reuse. By factoring out similarities in practice of individual occurrences of concepts and components and concentration into generic class hierarchies including generic interactions and means for extensions, complete class libraries and frameworks can be build that exactly support rapid prototyping and implementation. A side-effect of such frameworks is a unifying conceptual framework for local search: A class library for local search methods can be regarded as a taxonomy for these in the form of a hierarchy of concepts.

The rest of this subsection is devoted to introduce to general concepts of class libraries, toolkits, and various variants of object-oriented frameworks and a brief comparison thereof.

Class Libraries and Toolkits

Class libraries and toolkits according to [GHJV95, p. 26] are defined as follows:

> *A toolkit is a set of related and reusable classes designed to provide useful, general-purpose functionality. An example of a toolkit is a set of collection classes for lists, associative tables, stacks, and the like.*

Class libraries and toolkits are collections of interdependent and interacting class hierarchies. Each class hierarchy corresponds to a concept or component of a system that is supposed to be modeled and the concrete subclasses represent actual concept or component occurrences. The main objective of class libraries and toolkits is to reuse code in the form of common definitions and functionality factored out in superclasses. Accordingly, class libraries and toolkits are designed by factoring out similarities. Reuse is done by using existing concrete classes and by extending class hierarchies by means of inheritance (cf. Section 5.2).

Frameworks

Object-oriented application or implementation frameworks (or frameworks for short) are defined according to [GHJV95, p. 26] as:

> *A framework is a reusable design of all or part of a system that is represented by a set of abstract classes and the way their instances interact.*

Frameworks try to obtain a higher degree of reuse than class libraries and toolkits. They also provide reuse of design in addition to reuse of code. They therefore define a skeleton architecture and an overall structure for the collaboration of components of a system. The predetermined generic design becomes an invariant part of any actually implemented system. A framework virtually represents a partial implementation of a system in the form of an application. Frameworks can also be seen as an abstract application (hence name application framework) which can/must be tailored to an actual application. Frameworks can coarsely be distinguished into two categories: so-called *white-box* and *black-box* frameworks. The distinction is according to how a framework achieves extendibility. This question basically amounts to which kind of design patterns for reuse they mainly employ. The two categories of frameworks will be described next.

White-Box Frameworks

White-box reuse according to [GHJV95, p. 19] is done via inheritance. Notion "white" refers to visibility, since internals of inherited superclasses often are visible to subclasses. White-box frameworks mainly employ the template object-oriented design pattern [GHJV95, p. 325ff]. Figure 5.20 (adopted from [GHJV95, p. 327]) shows the general outline of this pattern. One abstract base class named `AbstractClass` provides a stable interface by defining non-changing methods such as method `templateMethod` from Figure Figure 5.20. The non-changing methods in turn call as variable parts so-called *hooks* or *hook methods*, named `primitiveOperation-1` and `primitiveOperation-2` in Figure Figure 5.20. The hook methods are pure virtual ones that must be overwritten by subclasses such as `ConcreteClass` of abstract base class `AbstractClass`. The outstanding trait of this kind

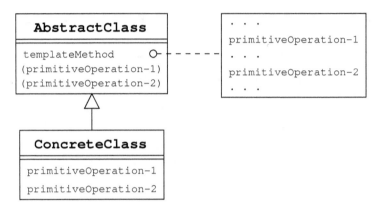

Figure 5.20:   Template design pattern

of reuse is an inversion of control according to "the Hollywood principle": "Don't call us, we'll call you" [Swe85]. Control is practiced by the method of the abstract base class that calls the virtual methods. These then can be equipped with functionality by subclasses. This kind of extension works fine if the general outline of necessary functionality is known and can be split into subtasks with known interfaces.

Black-Box Frameworks

Black-box reuse according to [GHJV95, p. 19] is done by object composition (one object holds on to another one). Black-box frameworks therefore mainly employ the strategy design pattern [GHJV95, p. 315ff]. Figure Figure 5.21 (adopted from [GHJV95, p. 316]) shows the general outline of this pattern. One class named `Context` there with its method named `contextInterface` serves as interface for the true functionality. This class and method and hence interface part remains stable. The true functionality is not implemented by subclasses of class `Context` as for the template pattern, but by another abstract class named `Strategy` in Figure Figure 5.21. Abstract base class `Strategy` is a component of the stable interface, too, and provides a pure virtual method named `algorithmInterface` in Figure Figure 5.21. It is supposed to be inherited and its subclasses must overwrite method named `algorithmInterface` there with different versions. Each subclass (named `ConcreteStrategy-A` – `ConcreteStrategy-C` in Figure Figure 5.21) this way provides a different actual functionality which is executed by the stable interface method `contextInterface` by calling method `algorithmInterface` polymorphically. New functionality and reuse is achieved by inheriting from class `Strategy` and next by composing objects anew. Because no internals of classes actually providing functionality are visible from the interface and no extensions can change internals of the interface, objects of subclasses behave as black boxes and hence this kind of reuse is labeled black-box.

Comparison

White-box and black-box reuse each have their advantages and disadvantages and so have white-box and black-box frameworks. In the following, the advantages and disadvantages of white-box and black-box frameworks, and class libraries and toolkits shall briefly be reviewed together with

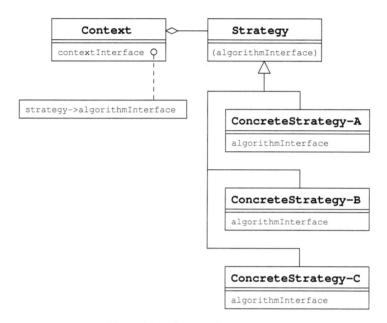

Figure 5.21: Strategy design pattern

a comparison of them.

On the one hand, class libraries and toolkits give much flexibility for users in extending them. On the other hand, there are no constraints on reuse in that no structure for usage is given or even enforced. Hence, no helping guidelines in constructing an application are given. Frameworks in contrast provide more design structure and that way help in application development (exactly because a basic design partly is already done). For both class libraries/toolkits and frameworks holds that they are far more difficult to develop than stand-alone applications. This is because they must be designed such that they can be extended and varied. Necessary extensions and variations must be anticipated in advance and provisions for the extensions and variations must be included. Frameworks naturally are most complicated in creation, since they also provide design. On the side of a framework developer they require a good understanding of the kind of applications that are to be implemented with the framework. The danger in framework design thereby is missing to anticipate needed extensions and variations which will cause problems and diminish usefulness if it happens.

White-box frameworks are more close to class libraries and toolkits than black-box ones, since they are extended by means of inheritance also, entailing more flexibility in extending them. One advantage of white-box reuse is that inheritance leaves some freedom for users of a framework for own (internal) changes and adjustments to their needs. The resulting disadvantage is a strong coupling between abstract base classes and extending subclasses. Changes in base classes often lead to necessary adjustment of *all* or substantially many subclasses, including those that have been added by other users of a framework. A problem also arises if two types of variations among subclasses are to be combined. For each combination an individual subclass of the abstract base class has to be provided which most likely will result in code duplication.

Using black-box reuse variations and extensions actually are established at runtime and therefore can be changed at runtime. It also features better encapsulation, changes are locally to the `Strategy` class hierarchy and do not affect or concern the interface as is the case for white-box reuse where the abstract base class, `AbstractClass` cannot be changed easily anymore. Users of black-box framework cannot write code easily which depends on internals of other more basic framework classes. This reduces their capability to customize a framework more than was originally anticipated by the framework developer. In addition to this, black-box frameworks need well-defined and stable interfaces which are not easy to come up with. Typically, black-box frameworks evolve over time from white-box frameworks when experience indicates how to design interfaces. Black-box frameworks are easier to use than white-box frameworks, but are harder to design properly. Finally, black-box reuse is less efficient than white-box, since delegation is used which must be organized at runtime.

5.4.2   Local Search Class Libraries, Toolkits, and Frameworks

Frameworks for local search methods basically all must try to break local search methods into independent, decoupled components that can be combined arbitrarily instead of viewing local search methods monolithically. As a result, the common central aim for all frameworks for local search methods is to separate problem type specific components (such as solution encoding, most basic solution manipulations) from in principle problem type independent components concerning search control (such as search strategies, typically those components that show up as functions or modules in pseudo-code for local search methods). Abstractly, this separation can be done by decoupling state from functionality and control. More specifically, independent abstractions for problem type related states and for search control and strategy must be provided. The second abstractions include abstraction and decoupling of states of a search and the pure search functionality also. Crucial for successful decoupling are stable and non-changing, i.e. well-defined interfaces.

The frameworks for local search methods described in the following basically all try to realize these insights. They vary to which extent they provide such abstractions and accordingly they mostly differ in which common concepts they do model via abstract base classes or not. They also differ in how they make object instantiations of concrete subclasses of these abstract base classes cooperate together in forming complete local search methods. The extent to which state and functionality can be decoupled, is determined by which concepts are abstracted by abstract base classes. As a rule of thumb, the more concepts are provided as abstract base classes, the more decoupling, but the more complicated to build actual algorithms and the less efficient these will be. Differences between frameworks often result from different perspectives on how to handle this inevitable trade-off. Decoupling state from functionality in local search methods at its core winds up to decoupling the following generic components: Solutions, neighborhood structures, neighborhood exploration schemes, neighbor selection schemes, local search procedure control, meta or high level search control, search heuristic state. Most frameworks to be presented in this subsection are written in C++ and are white-box framework. When a framework is different, this will be stated explicitly.

The *NeighborSearcher* framework [ACR98] is the first framework for local search methods. It provides abstract base classes for representing the following central concepts of local search methods: problem instance, solution encoding, local search procedure, construction methods for building solutions, and search strategies controlling local search procedures and solution construction. Objects of the latter type eventually realize local search methods, including metaheuristics. Additionally, basic search operators for manipulating solution encodings are encapsulated in separate classes as an interface for decoupling problem type specific from problem type independent aspects. Extension is achieved via subclasses in white-box reuse.

EasyLocal++ [SCL00, GS01, GS03] is also one of the first frameworks for local search methods. It is quite similar to NeighborSearcher in the central and generic concepts of local search methods that are represented by abstract bases classes and which are to be inherited to obtain realizations of more specific occurrences. The differences to NeighborSearcher mainly are due to which individual concepts exactly are supported by abstract base classes and how their interactions are organized. Fundamental types of classes are data classes, helpers, runners, and solvers. Data classes only store states in the form of solution encodings and other input and output data such as problem instance descriptions. They have no actual code except for accessing data. Helpers come in the form of neighborhood explorers for neighbor generation, neighbor selection schemes and search state update after neighbor selection. Runners hold other classes together and coordinating them this way implementing actual local search methods such as metaheuristics. Solvers finally coordinate different runners that implement different local search methods. EasyLocal++ features an extensive class hierarchies implementing several simple local search procedures and several more complex metaheuristics such as TS, SA, and others. They can readily be reused: Obtaining a local search method for a new problem type can be done by providing subclasses for problem type specific class hierarchies only. EasyLocal++ has support for proper inheritance by stating which methods must or can be redefined.

The "Heuristic OpTimization FRAMEwork" (*HotFrame* for short) [FV02] models the following concepts via class hierarchies: problem type, problem instance, solution space, solution, cost function, neighborhood structure, neighbor selection scheme, and local search procedure. Additionally, it has already built in several metaheuristics such as ILS, TS, SA, and also Genetic Algorithms (GA). In contrast to NeighborSearcher and EasyLocal++, interactions of objects are predetermined (through class interfaces and definitions) mainly by using the template mechanism of C++ [Str91]. HotFrame is very concerned with ensuring flexible design of local search methods from exiting components. It more emphasizes a building blocks principle for constructing local search methods but without neglecting decoupling at a fine degree of concepts. An effort is made for providing as many generic components as possible for not only (generic) metaheuristics but also for all other aspects of local search methods up to a detail level of support of special solution encoding such as binary vectors. Many abstractions are tried in elaborating an interface for including new problem types in order to be able to reuse as much of the built-in generic components as possible. In practice, only very problem-specific aspects have to be included when tackling a new problem type.

The *Meta-heuristics Development Framework* (MDF) [LWL04] has the following concepts modeled: solutions objective function, penalty function, move, constraints, neighborhood generator. It is foremost intended to quickly hybridize existing metaheuristics. It provides the concept of an *engine* to combine metaheuristic components and coordinate them with the goal to achieve hybridization. Coordination and the search is controlled by requests and responses. The search becomes request-driven. It extends a framework specialized to TS [LWJ03].

The *FOM* framework [PRG+03] features the concepts problem, solution, explorable solution, neighborhood, and metaheuristic and has already integrated some metaheuristics such as TS, SA, GRASP, and VNS.

Discropt [PS04] is yet another framework for local search methods that has the following concepts included: solution, neighborhood operator, search functionality and control, objective function. It concentrate on support for problem-type specific issues and runtime efficiency.

Localizer++ [MH01, MH05] is based on OPL++ which is a modeling language for constraint programming [MH00]. Localizer++ has a slightly different aim than the other frameworks presented so far in that it provides high-level abstractions to quickly realize local search steps. In case of the other frameworks, the details of solution manipulation must always be provided. After that, the framework takes over. Localizer++ rather is a class library since it does not so much care about constraining and hence supporting the development of high-level local search methods from basic local search steps. Local search methods basically must still be programmed by users themselves. This means less support for that task than in other frameworks. In return, it has features from modeling languages such as algebraic and logical operators. The structure is as is typical for constraint programming languages with a two-level architecture containing a declarative component specifying neighborhoods and hence problem type specifics, and a second component specifying local search methods that operate on declarative components. The declarative component specifies data structures by stating their invariants, i.e. the structure. Local search steps then result from changes of values to the invariant structure. Search components are procedural in nature and specify local search steps in terms of feasible updates to data structures as stated by declarative components Both declarative and search components have a single base class each. Extensions are obtained via inheritance and overriding virtual methods of these single base classes. Both types of components are interleaved during search. Altogether, Localizer++ rather is a language for specifying local search steps.

The *OpenTS* framework [Har05] is written in Java and is restricted to TS-like local search methods. Abstract base classes exist defining interfaces for the following central concepts of TS: tabu search strategy, solution, objective function, move, move manager, tabu list, tabu search listener (for implementing variants such as reactive TS, cf. Subsection 4.4.1 and [BT94, Bat96, BP97, BP99, BP01]).

A framework for memetic algorithms (cf. Subsection 4.4.1 and [MF99, Mos99]) such as TS is proposed in [Wu01]. It basically is the same framework as EasyLocal++ with the same notions.

A toolkit for heuristic search methods is *iOpt* [VDLL01, VD02, DV04]. It is written in Java and supports also population-based methods besides trajectory-based local search methods. That part of iOpt which is concerned with local search methods is called heuristic search framework (HSF). As other frameworks, it tries to capture common functionality of local search methods, called *component categories* in [VDLL01]. As a slight difference, abstraction is realized via Java interfaces. Specific occurrences of concepts such as specific metaheuristics or solution encodings are realized as Java classes adhering to the interfaces and extensions of the framework are obtained by means of subclasses. Classes for special problem domains such as scheduling exist as well as for other common solution encodings such as vectors of variables and sets of sequences. Other represented concepts comprise solution, solution construction and improvement, local search step, neighborhood, and neighborhood selection schemes. The HSF part of iOpt of iOpt can incorporate constraints [DV04] from other parts to improve local search methods by supporting check for constraints during search. Basic local search methods are already implemented as well as several metaheuristics (SA, GLS, TS).

The *ILOG Solver* [ILO04, Url05c] is a commercial class library for optimization. It is mainly concerned with exact methods for constraint logic programming but supports from Version 5.0 on implementation of local search methods by means of a class library, also. It can combine local search methods with exact methods. It is extended via inheritance and includes implementations of popular metaheuristics and neighborhoods, only adaption to new problem types is necessary in principle.

The *Templar* framework [JMRS98, Jon00] allows to create objects and embed them into other applications for distributed computation. Two basic concepts, problem and engines, are cast into classes yielding problem and engine classes. Problem classes capture problem specifics such as solution encodings, and basic local search operators. Engine classes implement local search methods such as TS, GA, and also exact algorithms. Engine objects can control other engine objects over a network of computers and this way allow for distribution and cooperation. The main aim of the Templar framework is to allow for distribution and cooperation over a network. It features no elaborate modeling of generic concepts of local search methods.

A framework for building GAs is GAlib [Url05b]. An overview over GA programming environments is given in [FTA94]. Further related work apart from class libraries, toolkits and framework is *Localizer* [MH99]. Localizer is a modeling language for writing local search algorithms with the aim to express these in a notation close to pseudo-code. *Comet* is [MH02] an object-oriented programming language designed for the implementation of stochastic search algorithms. It is a constraint-based language for modeling combinatorial optimization applications. *SALSA* [LC98] is a constraint programming language with local search support. The basic idea is to separate problem type specific neighborhoods from search control in order to yield concise specifications. Finally, *MAGMA* [MA04] is a *Multiagent Architecture for Metaheuristics*. It is only a conceptual framework for multi-agent metaheuristics, though. An overview over class libraries, toolkits, and frameworks in the context of local search and combinatorial optimization can also be found in [VW02, FVW02].

5.4.3 Discussion

This subsection concludes the section about related works and gives a quick review of which concepts of reuse are used in the GAILS framework. Afterwards, the GAILS framework will be compared with the related work from the previous subsection.

The GAILS framework employs white- as well as black-box reuse. The strategy pattern is used for example in abridged form (no context interface class, but with interface method `apply`) for building action-hierarchies by providing basis action classes according to heuristic state types as abstract base classes (cf. Subsection 5.2.6). The interface to function approximators with methods `learn` and `compute` is also designed according to the strategy pattern (cf. Subsection 5.2.8). The interplay of the `Reward` and `MoveCost` classes is organized as strategy pattern as well. There, the `Reward` class plays the role of class `ConextInterface` from Figure Figure 5.21 with method `computeReward` calling method `computeMoveCost` of class `MoveCost` which mimics class `Strategy` from Figure 5.21. The usage of the template pattern in the GAILS framework can be found for example in the construction and deconstruction process of classes of type `SearchState`. Root class `SearchState` corresponds to class `AbstractClass` in Figure 5.20 and provides among others method `initialize` calling pure virtual method `computeDataStrcuture` which must be implemented in subclasses.

The GAILS framework reuse of design is manifested foremost in the modeling of the concept of hierarchical action structures and its variation with typing action classes according to heuristic state classes. The interfaces between the three independent parts of the GAILS framework – problem-specific, action-hierarchy, and learning part – together with the fixed skeleton architecture and interfaces also can be seen as reuse of design. The abstract application that is represented by the GAILS framework is that of a learning LSA implementing a learning local search method. A specific application for example would be a learning LSA according to the GAILS standard learning scenario (cf. Subsection 4.2.3) for a specific problem type. Reuse of design also occurs in the form of reuse of action hierarchies and the hollywood principle can also be found in the execution of action-hierarchies. Altogether, the GAILS framework is an adequate application framework.

In contrast to the other frameworks just presented, only the GAILS framework clearly and centrally separates also local search control from its state (heuristic states in the GAILS framework). The organization of local search methods according to which kind of heuristic state they use is unique. The GAILS framework incorporates means for learning such as function approximators and search state abstractions in the form of features and models learning LSAs and hence reinforcement learning which no other framework attempts. A practical observation [JM97, JM02] is that most influential and crucial for performance of local search methods such as metaheuristics it is to make basic solution manipulation code – in the form of local search steps and procedures – as efficient as possible. Consequently, the GAILS framework employs local search steps and even local search procedures as most basic components of local search methods. These components are handled as black-boxes in contrast to some other frameworks where concepts for neighborhood generation and selection schemes or cost functions are made explicit. Form a user point of view, forming and reusing local search methods is well supported in the form of action-hierarchies and by means of typing according to heuristic states.

Chapter 6

Experiments

In Chapter 4, the GAILS method has been introduced and some attempts have been undertaken to analyze the proposed method theoretically. Since local search algorithms in general are highly randomized and the search spaces are too complex to investigate them theoretically beyond a certain point, empirical experiments are carried out, too. This chapter presents first empirical experiments conducted with GAILS algorithms. The experiments presented here are intended to get an impression how GAILS algorithms work and how they perform. They are intended to show that learning is possible and fruitful and to find out what is learned and what can be learned. In order to prepare the presentation of the experiments and their results, further background information about the experimental design used has to be given. This information comprises configurations used, descriptions of the problem types tackled as well as specific details for experiments conducted with GAILS algorithms.

This chapter is organized as follows. First, general topics concerning empirical experiments with metaheuristics and GAILS algorithms such as the design of experiments are covered in Section 6.1. Next, the specific GAILS algorithms used for the experiments are presented in detail in Section 6.2. Section 6.3 then briefly presents further details about the function approximators used and details about the problem type independent aspects of the features used. Section 6.4 finally presents the experiments and results for the Traveling Salesman Problem (TSP) while 6.5 presents them for the Flow Shop Problem (FSP), respectively.

6.1 Experimentation with GAILS Algorithms

This section discusses several issues related to empirical experimentation with metaheuristics and GAILS algorithms in particular. These comprise a description of how to generally conduct empirical experiments with local search methods and metaheuristics yielding some definitions of needed notions and the design of the empirical experiments conducted for this thesis.

In Subsection 6.1.1, general concepts and notions are presented and defined. The specific design of experiments is described in Subsection 6.1.2. Finally, in Subsection 6.1.3, all parameters of GAILS algorithms as used in the subsequent experiments are summarized concisely in a table with references to their description.

6.1.1 General Concepts

The goal of empirical experiments is to evaluate the *general* performance of what is being investigated. Two aspects of this statement need closer examination. These are the notions "performance" and "general" (performance). In order to evaluate performance, first some kind of performance measure has to be declared and second resulting individual measured performances have to be put into relation to each other in order to assess them. In the context of local search methods, the performance measures typically are solution quality and secondary runtime according to the goal in solving COPs as mentioned in Section 2.2. Recall from there that solution quality is to be emphasized. So for the first experiments described here the performance measure for evaluating GAILS algorithms is set to be solution quality only. The precise performance measure used is the cost of the best local optimum found during the run of an algorithm program instantiation (or *algorithm instantiations* for short). The cost might be given in absolute terms, normalized, or as percentage excess over the cost of a known global optimum or a known lower bound.

As was already mentioned, in order to be able to assess resulting performances for GAILS algorithms, these will have to be compared to other state of the art local search methods and metaheuristics in particular to find out which one(s) are performing best and how they compare. Theoretical contemplations from Subsection 4.5.3 suggest that the following GAILS algorithm options are most promising for conducting first experiments:

- Standard GAILS learning scenario using ILS actions: A learning LSA moves by means of local optima moves from one local optimum to the next (cf. Subsection 4.2.3).

- Infinite-horizon discounted return model using Q-learning in the form of one-step Q-learning or Watkin's $Q(\lambda)$.

- Using function approximators to learn and represent action-value functions and using a condensed state space in the form of features as input for the function approximators. The action-value functions thereby will be represented as one function approximator per ILS-action employed (cf. Subsection 3.3.2).

- Normalized features using lower and upper bounds for normalization (Subsection 3.3.1).

Most closely to GAILS algorithms according to the options just described are conventional ILS metaheuristics (cf. Section 2.5) using in their iterations the ILS-actions from the GAILS algorithms. For a given GAILS algorithm, these resulting conventional or normal ILS algorithms are called *corresponding normal ILS algorithms* (and the ILS algorithms themselves are called *normal ILS algorithms*). The ILS metaheuristic is a state of the art and generally very good performing metaheuristic [LMS02, JM02]. By using corresponding normal ILS for comparison enables to compare directly the differences in employing a strategy for ILS-action selection in contrast to only using one. Also, a learned strategy can be contrasted to the trivial strategies of only using one ILS-action all the time. Therefore, the experiments described here centrally compare GAILS algorithm with their corresponding normal ILS.

The second aspect that needs closer examination, besides which performance measure to use and how to assess measured performances, is how to evaluate the *general* performance of algorithms. Three observations are important in this context:

1. Local search methods are rather general templates which can lead to many different actual local search algorithms. These can in turn be parameterized manifold.

2. The resulting local search algorithms typically are randomized.

3. Measured performances for individual local search algorithms in a given parameterization cannot be obtained for all conceivable use cases, such as for all problem types and problem instances.

All these facts individually, and the more together, entail that local search methods cannot be evaluated completely in an empirical manner. Instead, empirical experiments can only yield samples of the general behavior. Crucially, some kind of generalizing inference hence has to be employed to infer from exemplary results to the general case. The impact on empirical experiments in practice of these three observations and on the experiments described here will be addressed in the following.

Local search algorithms typically are generic in nature as is exemplified by the variable parts and other variables in the pseudo code (such as the variable parts LocalSearchProcedure and Perturbation in Algorithm 2.1 and the variables α and γ in Figure 5.12). Therefore, local search algorithms can be parameterized in that different actual components can be assigned to the variable parts and actual values can be assigned to the variables. Each such assignment is called *actual parameterization*. The parameter values assigned are called *actual parameter values*. The observation is that different parameterizations yield differently performing local search algorithms. When comparing local search algorithms with each other, the question arises which parameterizations should be used for the comparison. In practice, before comparison, the best or some good performing parameterizations for each local search algorithm to be compared are searched for in preliminary tuning experiments, and next only the best parameterizations found are used for comparison. The parameterization of GAILS and ILS algorithms in actual algorithm instantiations is reflected by the fact that these are also parameterized. Parameterization for algorithm instantiations typically is realized via so-called *command line interface (CLI) parameters* that come in the form of CLI arguments. CLI parameters thereby control both assignments of parameter values to variables as well as the choices for variable parts of an algorithm. For GAILS and ILS algorithm instantiations, each actual setting of all (CLI) parameters, i.e. assignment of values to them (perhaps only set by default) is called *actual parameter setting*. Each parameterization of an algorithm corresponds to an actual parameter setting of its algorithm instantiation and vice versa. Accordingly, they are called *corresponding parameterization* and *corresponding actual parameter setting*.

The most general CLI parameters comprise filenames designating the files containing the problem instance description, the files with function approximator models to be used, files containing information for turning on and off features, and filenames designating the files that store the algorithm output and the learned function approximator models. Other CLI parameters specify how long to run an algorithm instantiation, e.g. in terms of how many iterations it is allowed to run. Any algorithm instantiation in an actual parameter setting can be run on a designated problem instance. The resulting run of an algorithm instantiation in a given actual parameter setting on a problem instance is called a *job*. The algorithm and algorithm instantiation in this context are called *underlying algorithm* and *underlying algorithm (program) instantiation*, respectively. Generally, a job is any coherent run of an algorithm instantiation in an actual parameter setting. Sometimes, for example, one job processes several problem instances at once without interruption. Such a run is also called a job. A GAILS algorithm instantiations run in an actual parameter setting on one or more specific problem instances is called *GAILS job*, an algorithm instantiation of a corresponding normal ILS is called *normal ILS job* or *ILS job*. For a given ILS-action, an ILS job is also called the *corresponding ILS job*. The underlying algorithm is called *corresponding ILS algorithm*.

Due to the randomization of GAILS and ILS algorithms, repeated single runs of identical jobs can vary in performance. The problem is how to reliably generalize from these varying performances, i.e. samples, to the general behavior. Put differently, how to assess the "real" and general behavior of a job and of an underlying algorithm? The field of statistics is exactly concerned with this problem. The way to deal with it is to run each job in several independent repetitions called *tries* and use statistical methods such as statistical tests and compute statistical characteristics (called *statistics* for short also) over the results of the individual tries such as mean (i.e. average), variance, standard deviation, minimum, maximum, and so on. The aim is to reliably – and amenable for generalization – compare job performances and hence eventually algorithm performances. This solution carries over the problem of how to assess the performance of an algorithm instantiation in general or at least for all problem instances of a given problem type, when not all problem instances can be processed. Again, several jobs are run on some problem instances that are deemed representative and next the results will be generalized by statistical means also and compared

A specialty of GAILS algorithms is that they do learn and then use what was learned. Learning and using the learned can be done interleaved in some granularity from immediately using anything learned to first only learn and afterwards only use what was learned. In consideration of the fact that runtime is an issue, it seems useful to employ the second kind of granularity first and split the learning and using process into two phases. Hence, in the experiments described here, GAILS algorithms are applied in two phases which are called *learning* or *training phase*   and *application phase*. First, a GAILS job is run in so-called *learning-* or *training mode*   where function approximators for the individual action-value functions are learned or rather trained. The training phase typically comprises several tries also in order to obtain a sufficient number of training examples and hence sufficient training. The tries of the training phase of a job thereby are not independent. Instead, the function approximator models are reused across the multiple tries. This way, learning is continued for all tries of a training phase job in a row effectively accumulating learning results over all tries. Jobs running in training mode during a training phase are called *training* or *learning phase jobs*.

After a training job, the underlying GAILS algorithm instantiation with basically the same actual parameter setting is to be run in an application phase now using the previously learned function approximator models for policy implementation and without further training them. Such jobs are called *application phase jobs*. Some parameters have different effects during training and application phase and especially learning parameters are not applicable in an application phase. Therefore, some parameters vary between otherwise identical training and application phase jobs. Given a training phase job, application phase jobs with the same underlying GAILS algorithm instantiation and actual parameter setting, except for the problem instance and application specific settings perhaps, but using the learned function approximator models are called *corresponding training phase jobs*. In fact, for each learning phase job there can be several corresponding application phase jobs that use the function approximator models learned by the training phase jobs. For a given application phase job, the training phase job that learned the function approximator models used by the application phase job is called the *corresponding training phase job*. The varying actual parameter settings for training and application phase jobs are called *training phase actual parameter settings* and *application phase actual parameter settings*, respectively. If GAILS algorithm instantiations are run on different problem instances in training and application phase, or if an additional so-called *test phase*   follows after a learning and application phase, the problem instances used for training are called *training problem instances*,   the ones used during the application phase are called *application problem instances*,   and the ones used in the test phase are called *test problem instances*. The jobs of a test phase are called *test phase jobs*,   their corresponding mode is called *test mode*. A test phase basically is an application phase, however with new, unseen problem instances. The aim is to test what was learned and for testing the transfer of the learned across problem instances.

To conclude this subsection addressing general issues in empirical experiments with local search methods, metaheuristics, and GAILS algorithms, the general parameters for empirical experiments are summarized. Randomization requires the usage of random number generators and hence a parameter with a seed for the random number generator [Leh54, PRB69, MN98]. The resulting parameter is named *seed*. In addition, several other general parameters have been mentioned or are central for forming jobs. The respective parameters are the number of tries during training phase named *tries training*, the number of tries during application phase named *tries application*, the number of tries during test phase named *tries test*, and the problem instance(s) used for training, application and testing named *input training*, *input application*, and *input test*, respectively. Since there can be more than one problem instance input to a job, several problem instances are indicated by a colon separated list of filenames. Furthermore, a termination criterion determining the maximum runtime named *termination criterion* is needed. The termination criterion usually is the same for training and application phase jobs, but can be different for test phase jobs. A parameter named *termination criterion test* determining the termination criterion for test phase jobs therefore is also employed. For the experiments described here, the termination criterion will always be measured in number of iterations allowed to run. The parameters just described together with all other parameters used in the experiments described here which will be introduced later are listed in Table 6.1 on page 200 (cf. Subsection 6.1.3).

6.1.2 Experimental Design

Recall that the aim of the invention of the GAILS method is to incorporate automated (machine) learning for exploiting search space regularities in order to improve performance or gain new insights in how to improve local search methods (cf. Subsection 4.1.1). The previous subsection addressed several important issues in the context of empirically investigating local search methods and GAILS algorithms in particular. This subsection will more specifically concentrate on how the first experiments with GAILS algorithms conducted and presented in this chapter are designed.

The first experiments must be designed such that they can help in answering the various kinds of research questions that naturally emerge in evaluating the GAILS method:

1. Does learning actually happen by learning stable policies by improving performance compared to random selection policies? Is and can something be learned by a learning LSA at all, for example in the form of stable and useful policies? What are the limits? What are the chances, what the costs?

2. If learning happens, what kind of regularities are discerned by GAILS algorithm program instantiations? Are these COP type general regularities? Are they amenable for hard-coded exploitation? More generally, how does learning work?

3. Does learning, does using learned knowledge, in fact improve performance in comparison to the best existing methods or compared to the most similar ones?

It is not likely that a newly invented method will excel over existing and finely tuned state of the art methods at once. For example, the first Otto four stroke engine that was built was also not comparable with the engines that drive nowadays Formula 1 racing cars and certainly was performing worse than competing steam engines. Yet, the method itself might be successful but has to be tuned and explored more. Therefore, primarily the inner workings of the GAILS algorithms are to be examined in the first experiments conducted and described here, aiming at firstly finding answers to the first two kinds of questions. Accordingly, the design of experiments is such that the questions of the first two kinds are likely to be answered and will be described in what follows.

As was already mentioned, it seems most beneficial to compare GAILS algorithms according to the standard learning scenario using ILS-actions with its most closely related metaheuristic which is ILS. Therefore, the basic design of the experiments described here is as follows: An experiment consists of a number of coherent jobs that are run together in order to investigate some aspect of interest. For each experiment, a set of l ($l \in \mathbb{N}^+, l > 1$) ILS-actions is given as well as a set of problem instances, divided into training, application, and test problem instances. In order to have not too many variations for the first experiments, each ILS-action includes the better acceptance criterion which has shown to produce good results in general [MO95, JM97, JM02]. For each ILS-action, a corresponding normal ILS algorithm instantiation is run. Also, a GAILS algorithm variant which picks in each iteration one of the l ILS-actions randomly is derived called *corresponding random ILS algorithm. corresponding random ILS algorithms* This type of GAILS algorithm in general is called *random ILS*. Its algorithm instantiation is also run. The jobs resulting from running normal and random ILS algorithm instantiations are called *normal* and *random ILS jobs*, respectively (and for a given ILS-action, the corresponding ILS job is called *corresponding normal ILS job*). The different ILS-actions can and will be obtained by varying either the local search procedure or the perturbation or both. The local search procedures and perturbations can be varied for example by changing their parameters such as a strength. When providing several differently parameterized perturbations and/or local search procedures, several ILS-actions resulting from the possible combinations according to a Cartesian product over perturbations and local search procedures supplied are obtained. Letting the resulting corresponding normal ILS algorithm instantiations run independently from each other basically can be seen as a tuning process for finding out which ILS-action and hence ILS parameterization performs best. Simultaneously, two GAILS algorithms, called *Q-ILS algorithms* or *Q-ILS* for short (cf. Section 6.2), using the l ILS-actions and Q-learning are also set up in several parameterizations and their algorithm instantiations are run with the corresponding actual parameter settings, first in training, next in application mode, and perhaps in addition in test mode. The jobs resulting from running Q-ILS algorithm instantiations are called *Q-ILS jobs* (and can be further divided into training and application phase jobs). Recall that the number of tries thereby can vary for training, application, and test mode as well as the number of actual parameter settings. So for both Q-ILS algorithms instantiations let p_{tr} ($p_{tr} \in \mathbb{N}^+$) be the resulting number of training phase actual parameter settings used, p_a ($p_a \in \mathbb{N}^+$) be the number of application phase actual parameter settings used, and p_{te} ($p_{te} \in \mathbb{N}^+$) be the number of training phase actual parameter settings used.

There are two types of experiments conducted here:

- *Single-instance experiments* will compare normal and random ILS and Q-ILS algorithms for single problem instances individually. In particular, no transfer of a learned action-value function across multiple problem instances will be attempted during this kind of experiment, but each training and corresponding application phase job will be run on the same single problem instance. Also, no test phase jobs are needed. This kind of experiment is used to detect whether something was learned and what. For each individual problem instance, the l normal ILS and the random ILS algorithm instantiations are run on the given problem instance yielding $l + 1$ jobs. Simultaneously, the two Q-ILS algorithm instantiations are run in training mode with p_{tr} different training phase parameter settings on the given problem instance yielding $2 \times p_{tr}$ training jobs. Each such training job will train and store an action-value function or rather its models and next, the corresponding $2 \times p_a$ application phase jobs will be run on the given problem instance using the previously learned action-value function.

- *Multi-instance experiments* will compare normal and random ILS and Q-ILS algorithms for several problem instances at once including some test problem instances that will have not been processed before in the training and application phase. Each normal and the random ILS algorithm instantiation is run on a number of n_{tr} ($n_{tr} \in \mathbb{N}^+$) training problem instances. The results of these runs are used to pick one or some best performing normal/random ILS algorithm instantiation to run it in a second stage on test problem instances. The two Q-ILS algorithm instantiations in their different training phase actual parameter settings are also run on the n_{tr} training problem instances yielding $n_{tr} \times 2 \times p_t$ training phase jobs. Afterwards, the corresponding p_a application phase jobs are run on the n_{tr} training problem instances which now serve as application problem instances using the previously learned function approximator models yielding $n_{tr} \times 2 \times p_a$ application phase jobs. Based on the result of the application phase jobs, one or some p_{te} best parameterizations for each Q-ILS algorithm are selected and the resulting algorithms instantiations are run with the respective actual parameter settings on the test problem instances also. The results of the runs of the selected normal/random ILS and Q-ILS algorithm instantiations in the test phase on the test problem instances finally are compared. Note that for the test problem instances no further training phase is started for the Q-ILS algorithm instantiations and that they typically are new and unseen problem instances.

Note that the number of tries used to run the normal and random ILS algorithm instantiations is always the same as the number of tries for the Q-ILS for application and test phase jobs. Multi-instance experiments are used to assess the ability to generalize and transfer a learned action-value function across several problem instances. Single-instance experiments are used to examine the learning behavior itself. Multi-instance experiments are more relevant for practice in that they do not need additional training for new, unseen problem instances and only will evaluate instead of also training action-values functions on them and hence will be faster and more comparable to standard or normal ILS algorithms. Sometimes, a single experiments employs sub-experiments. These sub-experiments can be single- or multi-instance experiments, also. In order to describe the mode of such sub-experiments, notions *single-instance* and *multi-instance (experiment) mode* will be used.

For each experiment conducted and described here, a table called *parameterization table* such as Table 6.8 on page 230 will be shown. This table declares the actual parameter values used and is shown first for each experiment. Only those parameters are shown that are applicable and that will be assigned an actual parameter value different from the default value declared in Table 6.1 on page 200. For some parameters, a comma separated list of actual parameter values is given indicating a set of available actual parameter values. The set of actual parameter settings is obtained by building the Cartesian product over the sets of actual parameter values over *all* parameters (including the ones not shown). Further explanations concerning how experimental results are presented are given in what follows, separately for single and multi-instance experiments.

Single-Instance Experiments Result Presentation

In the case of single-instance experiments, the performances of the normal ILS jobs for several problem instances are listed in a table such as Table 6.6 on page 229 called *ILS presentation table*. Each line corresponds to one normal ILS job. The first column named "Instance" gives respective problem instance name and in parentheses below the cost of a known global optimum or a lower bounds for this cost. The second and the third columns named "Perturbation" and "Ls-Procedure", respectively, give the perturbation and local search procedures in full parameterization that were used for the respective ILS-action. The actual perturbation and local search procedure parameter settings are given in brackets behind the name. Several parameters are separated by semicolons. The parameterization of perturbations and local search procedures is problem type specific and is explained in detail in the respective sections 6.4 and 6.5. In the example table, Table 6.7 , only one perturbation for the TSP, a random double bridge perturbation (indicated by "randDB[1]"), is used in any ILS-action with three actual parameter values sets. In contrast, the local search procedures used in this example vary over 2-opt, 2.5-opt, and 3-opt (indicated by "2opt[first]", "2.5opt[first]", and "3opt[first]", respectively) and each also has an actual parameter value (indicated by "first" for a first improvement neighbor selection scheme, cf. Section 2.3). Each job has been run for several tries on the given problem instance producing for each try a performance measure which is the cost of the best solution found during a try. These costs obtained by a job, one per try, are called *best costs*. The remaining columns, together labeled "Statistics", present some statistics computed over the best costs. The statistics are the average in column named "avg", the standard deviation in column named "σ", the averaged excess of the best costs over a known lower bound such as a global optimum in percent in column named "avg. exc", and the minimum and maximum over the best costs in columns named "min" and "max", respectively. A bold face colum name in any table indicates that the respective table is sorted according to the values displayed in this column.

The results of the training *and* the application phase jobs over several problem instances for the two Q-ILS algorithms and the single random ILS job are also presented in a table with one job per line. This kind of table is called *Q-ILS presentation table* and an example is Table 6.9 on page 231. The first column named "Instance" indicates the problem instance. The cost of a known global optimum or of a lower bound for this cost is given in parentheses below the instance name. The next two columns of Q-ILS presentation tables are named "Rank" and "Type" and indicate the rank of a job which is also used as job identifier and the type of a job. The type indicates whether it is a training or application phase Q-ILS job, or a random ILS job. An entry "apply" stands for application phase job, an entry "train" for training phase job, and an entry "rand" for random ILS job. The next columns under the heading "Parameters" lists the actual parameter values for each job for all parameters listed in the parameterization table that vary. The columns are named according to the respective shortcuts for the parameters as presented in Table 6.1. If a parameter is not applicable for a job, it will not be assigned a value which is indicated by an empty cell. This happens, for example, for application phase related parameters in the context of training phase jobs. Note that for application phase jobs, the training phase parameters belong to the corresponding training phase job.

Before the same statistics as in ILS presentation tables are presented under the heading "Statistics", the results of two statistical tests are given, first from a nonparametric alternative to the t-test, the Wilcoxon-Mann-Whitney (U) test (Wilcoxon test for short) next from the one-sided two samples t-test (t-test for short also) [Lar82, LW92, She00]. For the one-sided two samples t-test it is assumed that the samples of the two jobs to compare are distributed Gaussian and that the variances are the same [LW92, pp. 134f, pp. 139ff][She00, pp. 247ff]. The results of the one-sided two samples t-test are presented under the heading "H_0: not better", the results for the Wilcoxon test are presented under the heading "H_0 equal". For the one-sided two samples t-test the following null hypotheses H_0 is tested for each line, i.e. Q-ILS or random ILS job: The random ILS/Q-ILS job has not performed better than the best of the corresponding normal ILS jobs for the ILS-actions used expressed by the fact that the mean over the best costs of the random ILS/Q-ILS job is not better than the one for the best corresponding normal ILS job. The samples used to compute the test statistic are the best costs, so there are as many samples as there are tries for each job. The aim is to be able to reject the null hypothesis and assume the alternative hypothesis H_1 which is: The random ILS/Q-ILS job has performed better than the best corresponding normal ILS job expressed by the fact that the mean over the best costs of the random ILS/Q-ILS job is better than the one for the best normal ILS job. The columns for the t-test results under the heading "H_0: not better" are labeled with a significance level which is also the error probability for incorrectly rejecting the null hypothesis, although it was true. An entry "=" thereby indicates that the null hypothesis, H_0, based on the samples seen, *cannot* be rejected. An entry "+" in the case of the t-test indicates that the null hypothesis H_0, based on the samples seen, must be rejected and that the opposite, i.e. H_1 must be assumed – with the respective error probability of the column. In the case of the Wilcoxon test, an entry different from "=" also indicates that H_0 cannot be assumed and that H_1 must be assumed. However, the Wilcoxon test is a two-sides test so assuming H_1 can either mean that the random ILS/Q-ILS job has performed better than the best normal ILS job or the opposite.

If the mean over the best costs for a random ILS/Q-ILS job in a line is better than the one for the best corresponding normal ILS job (see columns named "avg"), the H_1 assumption will be that the random ILS/Q-ILS job has performed better than the best corresponding normal ILS job and this will be indicates by an entry "+". If the mean over the best costs for a random ILS/Q-ILS job in a line is worse than the one for the best corresponding normal ILS job, the H_1 assumption will be that the random ILS/Q-ILS job has performed worse than the best normal ILS job and this will be indicates by an entry "–". The column named "T" indicates the respective T-value, i.e. the test statistic, for the one-sided two samples t-test. Negative values for the T-value result, if the best normal ILS job is better than the compared random ILS/Q-ILS job. The labels for the next three columns under the heading "H_0 equal" also denote significance levels, this time for the Wilcoxon test [LW92, pp. 134f,pp. 139ff] comparing each line's corresponding random ILS/Q-ILS job with the best normal ILS job with the null hypothesis that both jobs perform equally good as expressed by the fact that the means over the best costs are equal.

The two types of tables just presented, normal ILS and Q-ILS presentation tables, can be used to compare the results of the normal ILS jobs and the random ILS and Q-ILS jobs for a given problem instance. Tables called *multi-instance summary tables* are used to present the results for test phase jobs of multi-instance experiments. These tables present the results for *all* test phase jobs including normal and random ILS and Q-ILS job results. As an example, see Table 6.17 on page 240. The first column of such a table is named "Instance" and shows the problem instance. The lines for each instance show the best Q-ILS, normal ILS, and random ILS jobs, sorted according to averaged best costs in order to support a direct comparison of performances. The next five columns present the results from a Wilcoxon and next a t-test as just described for the normal ILS and Q-ILS presentation tables. For each line the results are compared with the results of the best test phase job in the first line (which accordingly is left empty). The meaning of an entry "=" remains the same, and entry "–" indicates for both tests that H_O cannot be assumed and that the corresponding job performed significantly worse than the best job for this table. The remaining five columns present the statistics as for the normal ILS and Q-ILS presentation tables.

Besides tables, plots are useful for characterizing algorithm behavior. In research in local search method, one widely used type of plot shows the development of solution quality over time and is called *runtime development plot*. The left hand side of Figure 6.9 on page 239 shows such a plot. The x-axis indicates iterations and accordingly is labeled "iterations" which is true for *all* plots shown in this document. The y-axis is labeled "average cost over t tries" and indicates unnormalized solution costs of the best solution found until a given iteration number, averaged over the tries of a the jobs shown in the plot. The x-axis for runtime development plots is scaled logarithmic. Each job's runtime development of the best solution found up to a given iteration number, averaged over its tries, is shown as one line. Each job's line has a different type and/or color. A zoomed plot variant for the last phase of a search can also be supplied to provide a better view on the more interesting last phase of a search where the best solutions are found. Such a plot for example is the right hand side of Figure 6.9 on page 239. This variant is not scaled logarithmic. Runtime development plots show how the quality of the best solutions found develop over runtime and can be used to compare the performance of several jobs over time directly. In this document, runtime development plots that show the development of application phase Q-ILS jobs are used. In each such plot, the developments of the corresponding normal and random ILS jobs and of the corresponding training phase jobs are included also.

In order to have a look "inside" the Q-ILS algorithms, several plots concerning action usage for individual training and application phase jobs are available. For each training and application phase job, the following so-called *action usage plots* are available (plots for training phase jobs are marked by "(train)" following the job number in the figure captions; plots for application phase jobs are marked by "(apply)"):

- *Relative action usage* (cf. subfigure (b) of Figure 6.11 on page 242): This kind of plot maps for each iteration indicated at the x-axis the relative usage of the actions provided by a job – averaged over all tries – at this iteration to the y-axis. Each action thereby has its own line with different color and/or type. The y-axis is labeled "relative action usage" and its range is $[0, 1]$. Relative action usage plots especially for training phase jobs can be rather scattered so they sometime are smoothed using splines. In such as case, this is indicated by "(smoothed)" in the figure caption. Note that smoothing can slightly adulterate plots.

- *Accumulated action usage* (cf. right hand side of Figure 6.5 on page 232): These plots are a variant of relative action usage plots in that they show for each iteration and each action the accumulated number of action usages for this action up to the given iteration, averaged over the tries of the job. The y-axis is labeled "accumulated action usage" and its range is from 0 to the maximum average number of applications any action had in a try. The steeper the slope of a line for an action for some slice, the more frequently the respective action is used in average in the phase of the search that is represented by the slice. If the slope increases, the relative usage of the respective action will also increase. This works analog for decreasing slopes. A slope of zero of a line means that the designated action is not used at all for the interval over which the slope is zero.

- *Accumulated improving action usage* (cf. subfigure (b) of Figure 6.7 on page 233): This kind of plot is equivalent to accumulated action usage plots concerning only improving actions.

Mostly, smoothed relative action and accumulated action usage plots will be shown in the presentation of the experiments and their results later in sections 6.4 and 6.5. The corresponding plots for improving actions typically are very similar, since only those action that are executed can contribute with improving moves. Thus, they typically will not be shown unless noticeable differences occur or to investigate into the action usage of training phase plots sometimes. If accumulated action usage plots are shown, this mainly is because the relative action usage counterparts would not present the results as nicely.

Multi-Instance Experiments Result Presentation

In multi-instance experiments, a single job consists of running an algorithm instantiation in an actual parameter setting on several problem instances *at once*. For each try in the training phase, all training problem instances are processed in turn. The order in which the training problem instances are processed in each try thereby is randomized. For the corresponding application phase jobs operating on the training and also application problem instances (since these are the same) all tries for each application problem instance are processed in a row before the application training problem instances is tackled. In order to find the best performing normal and random ILS and Q-ILS actual parameter settings over *all* application problem instance, ranks are built. Separate rankings are computed for each of the normal and random ILS jobs, and the application phase Q-ILS jobs, but according to the same scheme. The tables showing the rankings are called *ILS ranking tables* and *Q-ILS ranking tables*, respectively. For each applicationg problem instance, the best costs for all jobs are sorted and enumerated. Equal best costs will have the same number and after some equal best costs the numbering continues with this number plus the number of respective equal best costs. The numbers are used as ranks and are summed up for each job and each training problem instances yielding one total sum per job. The jobs then are presented in the ranking tables sorted according to decreasing total sums. These tables (cf. Table 6.15 on page 238 and Table 6.16 on page 238) show with one job per line in the first column named "Rank" the overall ranking, in the second column named "Job" the job number. The next columns show all varied parameters in the case of Q-ILS jobs and the employed ILS-actions as combination of perturbation and local search procedure in columns named "Perturbation" and "LS", respectively, in the case of normal ILS jobs. Finally, in the last column named "Value" the ranking value in the form of the summed ranks is given. A random ILS job is indicated by entry "random ILS" over columns "Perturbation" and "LS".

Additionally, tables for the normal and random ILS jobs and application phase Q-ILS jobs equivalent to ILS and Q-ILS presentation tables used for presenting results for single-instance experiments (except for the statistical tests, though) might be shown for each application problem instance in order to have a closer look at the performance on individual application problem instances. If wished, action usage and runtime development plots can be produced for selected Q-ILS jobs also. All tables and plots together then can be used to select the normal and Q-ILS algorithm instantiation's actual parameter settings that are employed on the test problem instances in the test phase. For the results of the test phase jobs, ILS and Q-ILS presentation and multi-instance summary tables are used. Also, action usage and runtime development plots can be computed as well as a ranking according to the procedure used for the application phase jobs, now for both normal ILS and Q-ILS jobs over all test problem instances.

6.1.3 Parameter Presentation

GAILS algorithm instantiations (including instantiations for normal ILS as a special case, cf. Subsection 6.2.1) will have many (CLI) parameters stemming from different sources such as local search operators, parameters adjusting learning, and those setting up function approximators. The parameters will be introduced and explained during the sections and subsections to come covering the respective topics. They are briefly summarized in table Table 6.1 on page 200 in this section. In this table, for each parameter, its name in column named "Parameter Name", its abbreviation as used throughout plots and other tables presenting experimental result in column named "Short", its range in column named "Range", its default value in column named "Default", and a reference to its first introduction and explanation in column named "Subsection" are given. When presenting actual experiments, this table will be repeated then showing which actual parameter settings or rather values have been used for the presented experiment. In such a presentation, any parameter can have several values assigned, indicated by a comma separated list. The total number of jobs of the respective experiment then will be the number of actual parameter settings that result from building the Cartesian product over the sets of parameter values over *all* parameters. Note that for all parameters an additional "(s)" in the range column indicates that a colon separated list of several parameter can be given, for example for parameter "input" (cf. Subsection 6.1.1).

6.2 Q-learning

This section presents the two actual GAILS algorithms for solving COPs that are used and compared in the experiments described in this chapter. Recall from Subsection 6.1.1 that both work according to the standard GAILS learning scenario using ILS-actions (cf. Subsection 4.2.3) with the same fixed acceptance criterion per ILS-actions. They use Q-learning based reinforcement learning techniques. The name for GAILS algorithms according to the standard GAILS learning scenario using ILS-actions and Q-learning is *Q-ILS*. Two variants have been implemented for the first experiments. Both are off-policy temporal difference methods (cf. Subsection 3.2.5). The first variant is Q-learning in its simplest off-policy temporal difference form named *One-Step-Q-ILS* (cf. Subsection 3.2.6 and [Wat89, Wat92] [SB98, p. 148]). The respective learning strategy will be denoted by "Q(0)" in tables and plots. The second variant employs TD(λ) based off-policy Q-learning combining Q-

Parameter Name	Short	Range	Default	Subsection
Tries Training	t_{tr}	$\mathbb{N}^+$	20	6.1.1
Tries Application	t_a	$\mathbb{N}^+$	25	6.1.1
Tries Test	t_{te}	$\mathbb{N}^+$	40	6.1.1
Seed	s	$\mathbb{N}$	Current Time	6.1.1
Term. Criterion	tc	$\mathbb{N}^+$	4000	6.1.1
Term. Criterion Test	tc_{te}	$\mathbb{N}^+$	6000	6.1.1
Input Training	i_{tr}	`Filename(s)`		6.1.1
Input Application	i_a	`Filename(s)`		6.1.1
Input Test	i_{te}	`Filename(s)`		6.1.1
Learning Strategy	$strat$	$\{Q(0),Q(\lambda)\}$	Q(0)	6.2.2, 6.2.3
ILS-Action	a	`Name(s)`		6.2.1
Perturbation	$pert$	`Name(s)`		6.2.1
Local Search Procedure	ls	`Name(s)`		6.2.1
Feature Filename	$feat$	`Filename`		6.2.1
Initial Solution	$init$	`Name`		6.1.1
TSP Don't Look Bits	dlb	$\{$off, resetPert, resetAll$\}$	resetPert	6.4.1
Accept. Criterion Training	acc_t	$\{$better, ϵ-better, always$\}$	ϵ-better	6.2.1
Accept. Criterion Applicat.	acc_a	$\{$better, ϵ-better, always$\}$	ϵ-better	6.2.1
ϵ-Accept Training	ϵ_{a_t}	$[0, 1]$	0.3	6.2.1
ϵ-Accept Application	ϵ_{a_a}	$[0, 1]$	0.0	6.2.1
Policy Training	π_t	$\{$greedy, ϵ-greedy, uni$\}$	ϵ-greedy	6.2.1
Policy Application	π_a	$\{$greedy, ϵ-greedy, uni$\}$	ϵ-greedy	6.2.1
ϵ-Policy Training	ϵ_{π_t}	$[0, 1)$	0.3	6.2.1
ϵ-Policy Application	ϵ_{π_a}	$[0, 1)$	0.0	6.2.1
Learning Rate	α	$[0, 1]$	0.2	6.2.2
Discount Rate	γ	$[0, 1]$	0.5	6.2.2
Reward Signal	r	$\{c_\Delta, c_i\}$		6.2.2
Move Cost	mc	$\text{oT}[a_i;b_i;c_i;a_g;b_g;c_g]$ $a_i,b_i,c_i,a_g,b_g,c_g \in \mathbb{R}$	$\text{oT}[5;1;0;5;1;5]$	6.2.2
Start Learning	sl	$[0, 1]$	0.1	6.2.2
Lambda	λ	$[0, 1]$	0.5	6.2.3
Features History Length	fhl	$\mathbb{N}^+$	10	6.2.3
Learning Frequency	lf	$\mathbb{N}^+$	500	6.3.1
Function Approximator	fa	$\{\text{svr}[\nu], \text{regTree}[a;b;c]\},$ $\nu \in [0,1], a,b,c \in \mathbb{N}^+$	svr[0.01] svr[0.01]	6.3.1
Normalization Type	$norm$	$\{$theo, emp$\}$	emp	6.3.2
Bounds Tolerance	bt	$[0, 1]$	0.1	6.3.2
Number ΔCost Steps	$n_{\Delta c_s}$	$\mathbb{N}^+$	1000	6.3.3
Number ΔCost Iterations	$n_{\Delta c_i}$	$\mathbb{N}^+$	50	6.3.3
Number Rewards	n_r	$\mathbb{N}^+$	50	6.3.3

Table 6.1: Parameters for Q-ILS algorithms

learning with eligibility traces and is named $Q(\lambda)$-ILS (cf. Subsection 3.2.6 and [Wat89, Wat92] [SB98, pp. 182ff]). It is denoted by "$Q(\lambda)$" in plots and tables. The parameter determining which Q-ILS algorithm to choose in experiments is named *learning strategy*.

The One-Step-Q-ILS algorithms will be presented in Subsection 6.2.2 with its pseudo-code while the $Q(\lambda)$-ILS algorithm will be presented in Subsection 6.2.3 with its pseudo-code. Before, in Subsection 6.2.1, the common application phase variant is pictured.

6.2.1   Q-ILS-Application

Both Q-ILS algorithm variants behave the same during the application phase. This application mode variant will be presented first. Algorithm 6.1 shows the pseudo-code for the algorithm being the application mode of Q-ILS algorithms, called *Q-ILS-Application*. The Q-ILS-Application algorithm features the following components and variables ($\rightsquigarrow$ indicates a potentially randomized "mapping"):

- n ($n \in \mathbb{N}^+$) is the number of ILS-actions supplied.

- $(\mathsf{Pert}_i, \mathsf{LsProc}_i)$ ($\mathsf{Pert}_i : S \rightsquigarrow S, \mathsf{LsProc}_i : S \rightsquigarrow S$) is a tuple that represents an ILS-action. The component Pert_i of the tuple represents a perturbation, LsProc_i represents a local search procedure. The tuple itself is complemented with a fixed acceptance criterion to actually yield a complete ILS-action (cf. Subsection 4.1.2).

- s and s' ($s, s' \in S$) are LSA states (including a search state part with solution encoding and a heuristic state part (cf. Section 2.8)).

- $\vec{s}$ ($\vec{s} \in \mathbb{R}^m, m \in \mathbb{N}^+$) is the feature vector (or features for short) for corresponding LSA state s and m is the number of individual features computed.

- $\mathsf{ComputeFeatures}$ ($\mathsf{ComputeFeatures}: S \rightarrow \mathbb{R}^m$) is a function responsible for the computation of features,

- $\mathsf{GenerateInitialLSAState}$ ($\mathsf{GenerateInitialLSAState}: \emptyset \rightsquigarrow S$) generates an initial LSA state s.

- $\mathsf{AcceptanceCriterion}$ ($\mathsf{AcceptanceCriterion}: S^2 \rightsquigarrow S$) represents the acceptance criterion which complements each tuple of perturbation and local search procedure supplied to form a complete ILS-action.

- Q_i ($Q_i : \mathbb{R}^m \rightarrow \mathbb{R}$) embodies the function approximator which approximates the action-value function for action i (cf. Subsection 6.1.1 and Subsection 3.3.2).

- $\mathsf{InitializeFunctionApproximators}$ ($\mathsf{InitializeFunctionApproximators}: \emptyset \rightsquigarrow (\mathbb{R}^m \rightarrow \mathbb{R})^n$) initializes the function approximators (i.e. their models), perhaps randomly, one for each of the ILS-actions.

- i ($i \in \{1, \dots, n\}$) is an index for an individual ILS-action. The nomenclature is that i also denotes the indexed action.

- Policy ($\mathsf{Policy}: \mathbb{R}^n \rightsquigarrow \{1, \dots, n\}$) is a function that computes the strategy followed, based on action-values.

- $\mathsf{TermCrit}$ ($\mathsf{TermCrit}: S \rightarrow \{true, false\}$) denotes the termination criterion (cf. Section 5.2 and Subsection 5.2.7) used.

Procedure *Q-ILS-Application* $((\text{Pert}_1, \text{LsProc}_1), \ldots, (\text{Pert}_n, \text{LsProc}_n))$

    s = GenerateInitialLSAState

    $(Q_1, \ldots, Q_n)$ = InitializeFunctionApproximators

    Repeat

        $\vec{s}$ = ComputeFeatures(s)

        i = Policy($Q_1(\vec{s}), \ldots, Q_n(\vec{s})$)

        s' = Pert$_i(s)$

        s' = LsProc$_i(s')$

        s = AcceptanceCriterion(s, s')

    Until TermCrit(s) = true

End

Algorithm 6.1:   Q-ILS algorithm in non-learning mode

After this presentation of components and variables of the Q-ILS-Application algorithm, its mode of operation will be explained in details next. First note that the best search state found so far is recorded implicitly and will be the result of an execution of the Q-ILS-Application algorithm (cf. the ILS algorithm presentation in Section 2.5 and Algorithm 2.1).

The Q-ILS-application algorithm starts as a normal ILS by generating an initial LSA state with component GenerateInitialLSAState, possibly randomized. Generating an initial LSA state foremost means to generate an initial search state (cf. Section 2.5) and to initialize incorporated heuristic states with appropriate initial values. Since function approximators belong to heuristic states also (cf. Subsection 5.2.4), initializing heuristic states includes initializing contained function approximators as well. Since this an important part of the initialization process of a Q-ILS-Application algorithm, it has been made explicit in component InitializeFunctionApproximators in the pseudo-code of Algorithm 6.1. The type of the function approximators used and their parameterization is given implicitly and is the same for all function approximators used (possible type and their parameters are discussed in Subsection 3.3.4 and Subsection 6.3.1). For each action i employed one function approximator Q_i is initialized by loading a previously learned and stored model from a file and installing it (cf. subsections 6.1.1 and 3.3.2).

After the initialization of a Q-ILS-Application algorithm, the **Repeat** – **Until** or *main loop* is passed repeatedly until the termination criterion represented by TermCrit is met. The termination criterion is dependent on the current LSA state and more precisely typically on the heuristic state part by allowing to run an algorithm for a maximum number of iterations, steps, or time or a logical combination thereof. In the case of the Q-ILS algorithms presented throughout this document, the termination criterion allows to run for a maximum number of iterations of the main loop (cf. Subsection 6.1.1). In each iteration of the main loop first the features $\vec{s}$ for the (whole) current LSA state s are computed by function ComputeFeatures. This can possibly also be done incrementally in the local search procedures and perturbations of the various ILS-actions employed. Based on the features $\vec{s}$, the action-values $Q_1(\vec{s}), \ldots, Q_n(\vec{s})$ for the individual ILS-actions are computed. Note again that all function approximators Q_i together realize the (complete) action-value function (cf. subsections 6.1.1 and 3.3.2). The evaluation of the action-value function for ILS-action i based on the features $\vec{s}$ for the current LSA state s is indicated by $Q_i(\vec{s})$ yielding the respective action-value. The action-value itself is also denoted by $Q_i(\vec{s})$. Based on a vector $(Q_1(\vec{s}), \ldots, Q_n(\vec{s}))$ of action-values for the current LSA state, one action-value for each ILS-action, the policy component Policy selects the next ILS-action i to carry out. Typical policies used in the experiments described here are greedy, ϵ-greedy, or uniform. They are denoted by "greedy", "ϵ-greedy", and "uni", respectively, in plots and tables. The selected ILS-action i then is applied by first applying its perturbation component Pert$_i$ to the current LSA state s and next applying the local search procedure component LsProc$_i$ to the resulting intermediate LSA state s', together yielding a new local optimum again. Finally, the acceptance criterion AcceptanceCriterion decides where to continue the search from based on the old current LSA state s and the potential new one s'. A history larger than two LSA states is not employed here (cf. Section 2.5). The acceptance criterion decides and sets the new current LSA state s for the next iteration. Possible acceptance criteria are better, ϵ-better, and always, represented by "better", "ϵ-better", and "always", respectively, in plots and tables. The better acceptance criterion only accepts a new LSA state, if its cost are better or at least equal to the cost of the current LSA state. An ϵ-better acceptance criterion additionally accepts worse cost LSA states with a probability of ϵ. The always acceptance criterion finally always accepts a new LSA state. Note again that the acceptance criterion is common for all tuples or perturbation and local search procedures (i.e. effectively for all ILS-actions). Therefore, ILS-actions from now on also denote compositions of perturbation and local search procedures.

The Q-ILS-Application algorithm basically is the same as the standard or normal ILS algorithm, only a policy picks each next ILS-action from several given ones instead of only using one. As a minor difference, no (LSA) state history is used. Algorithm Q-ILS-Application therefore becomes a normal ILS algorithm by providing only one single ILS-action and setting the policy to a uniform one. For efficiency reason, features computation and function approximator usage could be skipped, too. The parameters needed for instantiating a Q-ILS-Application and an ILS algorithm in addition to those from Subsection 6.1.1 are described briefly next. The parameter names are given in parentheses, their shortcuts, ranges, and potential default values can be consulted in Table 6.1 on page 200. The first parameters are of course the ILS-actions used (*ILS-action*). These can also be given as two lists of perturbations (*perturbation*) and local search procedures (*local search procedure*). In this case, actual ILS-actions are built as Cartesian product. Note that actions, perturbations, and local search procedures are COP type dependent and that it is assumed that all provided perturbation and local search procedure components of ILS-actions are already parameterized (such as with a strength for the perturbation). This is indicated by writing parameter values separated by semicolons in brackets behind a perturbation or local search procedure name (cf. Subsection 6.1.2). Besides, it has to be decided which individual features are turned on and will be computed. This will be indicated by a so-called *feature file* which indicates which individual features are to be computed. The name of the feature file is given as a CLI parameter (*feature filename*). A further parameter is how to generate initial search states or rather solution for individual problem types (*initial solution*). This parameter is COP type specific, of course. A common acceptance criterion which can vary for training and application mode (*acceptance criterion training* and *acceptance criterion application*, respectively), and an ϵ value for a possible ϵ-better acceptance criterion, varying for training and application mode also (*ϵ-accept training* and *ϵ-accept application*, respectively) have to be supplied also. Finally, a policy has to be determined for training and application mode (*policy training* and *policy application*, respectively) and ϵ parameter values for parameterizing an ϵ-greedy policy have to be given for training and application mode (*ϵ-policy training* and *ϵ-policy application*, respectively).

Note that a ϵ-better acceptance criterion for $\epsilon = 0.0$ behaves as a better acceptance criterion and for $\epsilon = 1.0$ behaves as an always accepting acceptance criterion. Note also that an ϵ-greedy policy with $\epsilon = 0.0$ behaves as a greedy policy.

The next two subsections are concerned with the learning variants of Q-ILS algorithms. The directly following subsection, Subsection 6.2.2, presents one-step Q-learning for Q-ILS and discusses the resulting algorithm One-Step-Q-ILS including all adjustable parameters. Subsequently, in Sub-

Procedure *One-Step-Q-ILS* $(\alpha, \gamma, (\text{Pert}_1, \text{LsProc}_1), \ldots, (\text{Pert}_n, \text{LsProc}_n))$

    $s = \text{GenerateInitialLSAState}$

    $(Q_1, \ldots, Q_n) = \text{InitializeFunctionApproximators}$

    Repeat

        $\vec{s} = \text{ComputeFeatures}(s)$

        $i = \text{Policy}(Q_1(\vec{s}), \ldots, Q_n(\vec{s}))$

        $s' = \text{Pert}_i(s')$

        $s' = \text{LsProc}_i(s)$

        $r = \text{Reward}(s, s')$

        $\vec{s}' = \text{ComputeFeatures}(s')$

        $Q_i(\vec{s}) = Q_i(\vec{s}) + \alpha \cdot (r + \gamma \cdot \max_{j \in \{1, \ldots, n\}} Q_j(\vec{s}') - Q_i(\vec{s}))$

        $s = \text{AcceptanceCriterion}(s, s')$

    Until $\text{TermCrit}(s) = \text{true}$

End

Algorithm 6.2:   One-step Q-ILS algorithm in learning mode

section 6.2.3, Watkin's Q(λ)-learning for Q-ILS will be discussed, again including all adjustable parameters.

6.2.2 One-Step Q-learning

One-Step-Q-ILS is a GAILS algorithm that is based on one-step Q-learning and the GAILS standard learning scenario using ILS-actions (cf. Subsection 4.2.3, Subsection 3.2.6, and [Wat89, Wat92] [SB98, p. 148]). Action class `OneStepQIls` implements the One-Step-Q-ILS algorithm (and simultaneously its application phase variant Q-ILS-Application) in the GAILS framework (cf. Subsection 5.2.6 and Figure 5.12). The pseudo-code is displayed in Algorithm 6.3. One-Step-Q-ILS in principle works as its application phase variant Q-ILS-Application. The learning extensions are explained next beginning with an overview over the additional components.

- α ($\alpha \in (0, 1]$) is the learning rate (cf. Subsection 3.2.5).

- γ ($\gamma \in (0, 1)$) is the discount factor for the infinite-horizon discounted return model that is used in the One-Step-Q-ILS algorithm (cf. Subsection 3.1.3).

- r ($r \in \mathbb{R}$) is the immediate reward obtained per each potential LSA move. It is computed from the old current LSA state (i.e. local optimum) s and the potential new current LSA state s' just after a potential LSA move.

- Reward (Reward : $S \times S \rightarrow \mathbb{R}$) is the reward signal that computes the immediate reward r for each potential LSA move (cf. Subsection 3.1.2). This computation can also include move costs, e.g. in the form of one-time rewards (cf. Subsection 4.3.4).

As is the case for the Q-ILS-Application algorithm, subsequently choosing an ILS-action i and applying its perturbation Pert_i and local search procedure LsProc_i component to the current LSA state will yield a new LSA state s'. Since the acceptance criterion must still accept s' as new current

LSA state, this newly generated LSA state is called *potential new current LSA state*. For the same reason, the LSA state transition from s to s' is called *potential LSA move*. After the local search procedure produced the next local optimum and hence potential new LSA state s' to move an LSA to, an immediate reward for this so-called *potential transition* is computed by function **Reward** based on the potential new current and the old current LSA state, s' and s, respectively. The reward computation can be done according to the reward signals presented in Subsection 4.3.3 such as the delta cost or the inverse cost reward signal. The experiments presented here will only use these two reward signals whose immediate rewards will furthermore be normalized to $[0, 1]$. Immediate rewards can include move costs such as one-time rewards. A one-time reward is computed as:

$$((a_{\mathrm{imp}} \cdot x^{b_{\mathrm{imp}}} + c_{\mathrm{imp}}) \cdot I_{\mathrm{imp}} + (a_{\mathrm{glob}} \cdot x^{b_{\mathrm{glob}}} + c_{\mathrm{glob}}) \cdot I_{\mathrm{glob}}) \cdot r$$

where x ($x \in [0, 1]$) denotes the relative progression of the search relative to the known maximum runtime (as expressed by a maximum number of iterations), I_{imp} ($I_{\mathrm{imp}} \in \{0, 1\}$) is an indicator whether the last potential LSA move was an improving one (cf. Subsection 4.3.1), I_{glob} ($I_{\mathrm{glob}} \in \{0, 1\}$) is an indicator whether the last potential move additionally yielded an overall new best LSA state, r is the immediate reward obtained during the potential LSA move, $a_{\mathrm{imp}}, b_{\mathrm{imp}},$ and c_{imp} ($a_{\mathrm{imp}}, b_{\mathrm{imp}}, c_{\mathrm{imp}} \in \mathbb{R}$) are the parameters for weighing the additional one-time reward for improving potential LSA moves in dependence on the current progression, and finally $a_{\mathrm{glob}}, b_{\mathrm{glob}},$ and c_{glob} ($a_{\mathrm{glob}}, b_{\mathrm{glob}}, c_{\mathrm{glob}} \in \mathbb{R}$) are the parameters for weighing the additional one-time reward for potential LSA moves to a new overall global best LSA states in dependence on the current progression. Parameters $a_{\mathrm{imp}}, b_{\mathrm{imp}}, c_{\mathrm{imp}}, a_{\mathrm{glob}}, b_{\mathrm{glob}},$ and c_{glob} will be abbreviated to $a_i, b_i, c_i, a_g, b_g,$ and c_g in plots and tables and be written, separated by semicolons, in brackets behind the one-time reward move cost indicator "oT" (cf. Table 6.1).

For computing the updates for the action-value functions for the potential LSA move $s \rightarrow s'$ in the effective iteration, the features $\vec{s}'$ for the potential new current LSA state s' have to be computed by function **ComputeFeatures**(s'). The features then are used to compute the action-values $Q_j(\vec{s}')$ ($j \in \{1, \ldots, n\}$) for the current LSA state. Together with the learning rate α, the discount factor γ and the action-values $Q_i(\vec{s})$ for the just taken action i in the old current LSA state s, these action-values are used to compute a training example for updating the function approximator which approximates the action-value function for action i. The training example or update is computed according to the rule for one-step Q-learning presented in Subsection 3.2.6 [SB98, p. 148]. The training example corresponds to the previous LSA state s and hence to features $\vec{s}$. The function approximator update is indicated in the pseudo-code by having $Q_i(\vec{s})$ on the left-hand side of the respective update equation. Finally, the acceptance criterion has to accept or reject the potential new current LSA state as before in the Q-ILS-Application algorithm.

The new parameters and ingredients for algorithm One-Step-Q-ILS in addition to algorithm Q-ILS-Application are the learning and the discount rate for one-step Q-learning, α and γ, named *learning rate* and *discount rate*, respectively, and the reward signal, named *reward signal*. Shortcuts, ranges, and potential default values for these parameters can be seen in Table 6.1 on page 200. The two reward signals which can be computed, delta cost and inverse cost reward signal, are denoted by c_Δ and c_i, respectively. In addition, learning will not be started until some LSA moves have been executed. This is because in the beginning of a search, new globally best LSA states are found often and improvements turn out to be comparably huge, regardless of the real quality of the ILS-action taken. Since the rewards depend on improvements in costs for LSA moves or on the costs of LSA states visited, the rewards for the first LSA moves and hence ILS-actions taken are not representative and only refer to low cost regions of the search space. They therefore should not be used for updating action-value functions. If doing so nevertheless, the resulting first training examples will most likely be misleading and it will take a function approximator some time to unlearn them by when processing more representative training examples coming later during the search. So, a parameter named *start learning* indicates, relative to the maximum runtime, when to start learning in each try. This parameter is also listed in table Table 6.1.

6.2.3   Q(λ)-learning

The last Q-ILS algorithm to be presented in this section is Q(λ)-ILS. It also is based on ILS-actions and the standard GAILS learning scenario and employs Q(λ)-learning according to Watkin's Q(λ)-learning (cf. Subsection 3.2.6 and [Wat89, Wat92] [SB98, pp. 182ff]). Action class QLamb-daIls implements the Q(λ)-ILS algorithm (and simultaneously its application phase variant Q-ILS-Application) in the GAILS framework (cf. Subsection 5.2.6 and Figure 5.12). The pseudo-code of Q(λ)-ILS is displayed in Algorithm 6.3.

The Q(λ)-ILS algorithm extends the One-Step-Q-ILS algorithm. The further learning extensions to both algorithms are explained in the following beginning with an overview over the additional components.

- λ ($\lambda \in [0, 1]$) is the parameter from Q(λ)-learning algorithms (cf. Subsection 3.2.6 and [Wat89, Wat92] [SB98, pp. 182ff]) that in principle determines how far reaching updates for other than the old current LSA state (or rather its features) are in each potential LSA move.

- *FeaturesHistory* (*FeaturesHistory* $\subseteq \mathbb{R}^m \times \{1, \ldots, n\}$) represents the features history (cf. Subsection 5.2.1).

- e ($e: \mathbb{R}^m \times \{1, \ldots, n\} \to \mathbb{R}_0^+$) is the function that represents the eligibilities for tuples of features of LSA states and the action taken in the LSA states. Together with the features history, *FeaturesHistory*, e implements eligibility traces (cf. Subsection 3.2.6).

After the initialization of the function approximators and the LSA state as before in algorithm One-Step-Q-ILS, the features history denoted by *FeaturesHistory* is initialized. For the Q(λ)-ILS algorithm, eligibilities are not stored for search state action tuples but for features action tuples. This is because an LSA state not only changes with its search state part but also with its heuristic state part and so do its features. Furthermore, updates for action-values functions are computed based on the features of LSA states anyway. For these reasons and in order to save memory in computing eligibilities for establishing eligibility traces, a features history is used (cf. Subsection 5.2.1). Note that only those tuples of features and actions that have a non-zero eligibility value assigned will have an effect in computing updates and hence have to be stored. Due to the rapidly decreasing eligibilities for features action tuples that have not been "visited" recently – geometric weighing eligibilities according to $\gamma \cdot \lambda$ for each potential LSA move make eligibilities quickly approach zero – only a limited number of eligibilities for features action tuples need to be recorded. These therefore can be organized using a list or set of features action tuples such as *FeaturesHistory*. Typically, the length of a features history therefore is limited in practice, but as an additional parameter, it can be assigned a maximum length. The function assigning eligibilities to features action tuples only need to be recorded for the entries of *FeaturesHistory* (and in practice are simply associated directly to the elements of *FeaturesHistory*). Anytime a non-greedy action has been chosen, the eligibility traces must be reset and accordingly the features history has to be emptied (cf. Subsection 3.2.6 and [Wat89, Wat92] [SB98, pp. 182ff]).

After initializing the features history, the features $\vec{s}$ for the first current LSA state s are computed as well as its action-values $Q_1(\vec{s}), \ldots, Q_n(\vec{s})$ in order to compute the next action to carry out by policy component Policy (which in training mode typically is ϵ-greedy). The resulting features action tuple is added to the features history. Note that as an invariant, at each begin of the main

Procedure $Q(\lambda)$-*ILS* $(\alpha, \gamma, \lambda, (\text{Pert}_1, \text{LsProc}_1), \ldots, (\text{Pert}_n, \text{LsProc}_n))$

    s = GenerateInitialLSAState

    $(Q_1, \ldots, Q_n)$ = InitializeFunctionApproximators

    FeaturesHistory = $\emptyset$

    $\vec{s}$ = ComputeFeatures(s)

    i = Policy$(Q_1(\vec{s}), \ldots, Q_n(\vec{s}))$

    FeaturesHistory = *FeaturesHistory* $\cup \{(\vec{s}, i)\}$

    Repeat

        s' = Pert$_i(s)$

        s' = LsProc$_i(s')$

        r = Reward(s, s')

        $\vec{s}'$ = ComputeFeatures(s')

        i' = Policy$(Q_1(\vec{s}'), \ldots, Q_n(\vec{s}'))$

        i^* = $\text{argmax}_{j \in \{1, \ldots, n\}} Q_j(\vec{s}')$

        If $(Q_{i'}(\vec{s}') = Q_{i^*}(\vec{s}'))$

            i^* = i'

        δ = $r + \gamma \cdot Q_{i^*}(\vec{s}') - Q_i(\vec{s})$

        $e(\vec{s}, i) = e(\vec{s}, i) + 1$

        For all $(\vec{s}^*, j) \in$ *FeaturesHistory*

            $Q_j(\vec{s}^*) = Q_j(\vec{s}^*) + \alpha \cdot \delta \cdot e(\vec{s}^*, j)$

        If $(s' = \text{AcceptanceCriterion}(s, s'))$

            For all $(\vec{s}^*, j) \in$ *FeaturesHistory*

                If $(i' = i^*)$

                    $e(\vec{s}^*, j) = \gamma \cdot \lambda \cdot e(\vec{s}^*, j)$

                Else

                    $e(\vec{s}^*, j) = 0$

                    FeaturesHistory = $\emptyset$

            FeaturesHistory = *FeaturesHistory* $\cup \{(\vec{s}', i')\}$

            i = i'

            s = s'

            $\vec{s}$ = $\vec{s}'$

        Else

            $e(\vec{s}, i) = \max((s, i) - 1, 0)$

            FeaturesHistory = *FeaturesHistory* $\setminus \{(\vec{s}, i)\}$

            $\vec{s}$ = ComputeFeatures(s)

            i = Policy$(Q_1(\vec{s}), \ldots, Q_n(\vec{s}))$

            FeaturesHistory = *FeaturesHistory* $\cup \{(\vec{s}, i)\}$

    Until TermCrit(s) = true

End

Algorithm 6.3:   Q(λ)-ILS algorithm in learning mode loop, i.e. iteration, the features $\vec{s}$ of the current LSA state s have already been computed and an action i to apply in this current LSA state s has been chosen already, too, at the end of the last iteration or during the initialization phase just described. Also, the tuple $(\vec{s}, i)$ of features and action to apply has already been added to the features history, so the features history at the begin of each iteration is never empty. Action i is applied, an immediate reward r for the resulting potential LSA move is computed, and the features $\vec{s}'$ of the resulting potential new current LSA state s' are computed as before in algorithm One-Step-Q-ILS. Next, the action values $Q_1(\vec{s}'), \ldots, Q_n(\vec{s}')$ for the features of the potential new current LSA state are computed and used to select the next action i' to apply at the begin of the next iteration by policy component **Policy**. Simultaneously, a greedy action i^* for $\vec{s}'$ is computed. If the action-values $Q_{i'}(\vec{s}')$ and $Q_{i^*}(\vec{s}')$ for i' and i^* are equal, the previously chosen next action i' is set to be *the* greedy action in this iteration. Next, the so-called *one-step return* δ is computed from the immediate reward r observed, the action value $Q_{i^*}(\vec{s}')$ of the greedy action i^* for the potential new current LSA state s' as just computed, and the action values $Q_i(\vec{s})$ for the last action i applied to the old current LSA state s (cf. Subsection 3.2.6). This one-step return is one part of the computation of updates for action-values. It can be reused in the computation of the updates for all features action tuples currently contained in the features history. The eligibility for the features action tuple $(\vec{s}, i)$, i.e. consisting of the features for the old current LSA state s and the action i chosen there, is incremented by one in order to update the last entry of the features history (cf. Subsection 3.2.6). Then, for all tuples $(\vec{s}^*, j)$ in the features history, learning is triggered with the current one-step return δ weighed by the respective eligibility $e(\vec{s}^*, j)$ and a learning rate. Together they yield the proper update according to the update rule from Subsection 3.2.6 and [SB98, pp. 182ff]).

Depending on whether the acceptance criterion accepts the potential new current LSA state s' in fact as new current LSA state to proceed from or backs up to the old current LSA state, s, i.e. whether the potential LSA move is accepted or not, different further arrangements have to be done. If the potential LSA move to s' is accepted and the next action i' to take is a greedy one, i.e. $i' = i^*$, the eligibilities for the tuples in the features history are decayed by $\gamma \cdot \lambda$ (cf. Subsection 3.2.6 and [SB98, pp. 182ff]). If the potential LSA move to s' is accepted and the next action i' to take is not a greedy one, the features history has to be emptied and the eligibilities for all the features action tuples it contains are set to zero, since the eligibility traces must be deleted, if a non-greedy action has been chosen (cf. Subsection 3.2.6 and [SB98, pp. 182ff]). In any case, if the potential LSA move to s' is accepted, the next action i to take in the next iteration of the main loop is set to be the designated one i' as already chosen and the new current LSA state s for the next iteration is adjusted also to be s'. The corresponding tuple of features and action $(\vec{s}', i')$ is added to the features history, so the next iteration of the main loop can start as usual. If the potential LSA move to s' is not accepted, the LSA has to be backed up to the old and now new current LSA state s. In this case, basically any computations already made for the potential LSA move that has been rejected except for the function approximator updates and the changes to the heuristic state part such as iteration counters have to be undone. Therefore, the previous eligibility increment to tuple $(\vec{s}, i)$ which has just been added to the features history in the last iteration is removed again (but not below zero) and the tuple is removed from the features history, since the potential LSA move this tuple refers to has not been accepted. Also, if the potential new current LSA state s' has not been accepted, the features for the old and now new current LSA state s must be computed anew, since heuristic state features might have changed (this change of heuristic state features is made implicitly im the pseudo-code of Algorithm 6.3), although the search state features of course will remain the same. Nevertheless, a new action i to take at the begin of the next iteration is chosen according to the current policy component **Policy** after computing new action values $Q_1(\vec{s}), \ldots, Q_n(\vec{s})$ for the new features of the old and new LSA state. Selecting a new action instead of applying the old one again is to prevent to move to the same local optimum again. Since the features will have changed due to changes in the heuristic state part a different action might be selected this time not only by chance but because a different action became the greedy one. The tuple $(\bar{s}, i)$ for the new and old current LSA state s and new action i is added to the features history such that the main loop can start as usual. Note that the effects to the search state part of an LSA state when not accepting several potential LSA moves in a row is the same as if the first potential LSA move accepted had been selected in the first place. Nevertheless, there are changes to the heuristic state parts and the updates to the function approximators which means that learning goes on for not accepting a potential LSA move also so this does no harm.

The new parameters for the Q(λ)-ILS algorithm are λ, named *Lambda*, and the length of the features history, named *features history length*. They are also presented with shortcut, ranges, and default values in Table 6.1 on page 200.

6.3 Function Approximators and Features

The basic learning components of GAILS algorithms are function approximators. The interface of action-hierarchies to function approximators are the features (cf. Section 5.1). Recall from Subsection 5.2.9 that 3 kinds of features exist: search state, problem instance and heuristic state features. Search state and problem instances features often can be constructed systematically based on solution components and elements (cf. Section 2.1). Several variations of heuristic state features can be invented also.

In the first subsection of this section, Subsection 6.3.1, the types of function approximators used in the experiments described here and their parameters will be presented. She subsections following Subsection 6.3.1 present how search and problem instance features can be invented systematically and how to normalize them in Subsection 6.3.2 and which heuristic state features have been contrived and used for experiments in Subsection 6.3.3. Actual search state and problem instance features are COP type dependent and are presented later when actual experiment results are presented for several COPs (cf. Section 6.4).

6.3.1 Function Approximators

Basically, two main parameters in connection with function approximators have to be adjusted: the type and parameterization of the function approximator itself and an indication how often to learn. Two types of function approximators have been integrated. The parameter indicating the type and parameterization of the function approximators used is named *function approximator* and can be seen together with its shortcut, ranges, potential default value in Table 6.1 on page 200. Before presenting the two types of function approximators used, the question how often to learn is discussed.

Learning Frequency

Learning takes quite a while. Therefore, it seems advisable to at least partly reduce the time required for learning. Although some function approximators are able to learn online and incrementally after each new update that has been computed, this probably will take too much time. Instead, pseudo-online incremental learning by accumulating training examples in batches until an upper limit on accumulated training examples has been reached and only then actually trigger incremental learning can be employed (cf. Subsection 3.3.3). The upper limit on accumulated training examples is called *learning frequency* and is an additional parameter for the learning variants of Q-ILS algorithms. Again, shortcut, range, and potential default value for parameter learning frequency can be looked up in Table 6.1. Since differently many training examples will be collected for the actions employed, each time a function approximator representing an individual action-value function has accumulated enough training examples to trigger learning, learning is triggered for the function approximators representing the other individual action-values also. This prevents from having differently updated individual action-value functions for the different actions. Before writing function approximators to disk, learning is also triggered for a last time.

Support Vector Machine

The first type of function approximators that has been integrated into the GAILS framework and which is used for the experiments described in this chapter are Support Vector Machines (SVM), also called Support Vector Regression (SVR) in case a continuous function has to be learned (as is the case here). SVMs have been explained before (cf. Subsection 3.3.4). There have been several SVRs integrated into the GAILS framework in [Ban04] by adjusting existing implementations [Joa99, CB01, Rüp01, Rüp02, Rüp04, Url05g, CL04, CB04, Joa04]. Some are able to learn incrementally and one has specifically been adjusted in [Ban04] to fit the requirements for using it as a function approximator for representing action-value functions in the GAILS framework. This SVR is called ν-SVR-2L [Ban04, p. 20] since it is built upon incremental ν-SVRs proposed in [Rüp01, Rüp02].

Function approximator ν-SVR-2L has several parameters. The most important parameters for ν-SVR-2L are briefly presented next together with their values as used in the first experiments described here. Any other parameter adjustments for ν-SVR-2L not mentioned here are as in [Ban04]. The kernel that is used is the standard radial basis function kernel. Its parameter γ [Ban04, p. 20] [SS02, SS04] is always set to $\frac{1}{\#\text{Features}}$ which is suggested as a good standard value according to [CL04]. Another parameter concerns a constant C with which the slack variables for ϵ- and ν-SVRs are weighed [Ban04, p. 25ff] [SS02, SS04]. For the first experiments described in this thesis, a value of $C = 32$ was taken based on the experiences made in [Ban04]. Finally, parameter ν ($\nu \in [0,1]$) has to be set. This parameter determines a lower bound on the number of support vectors in computed models. The number of support vectors in a model must be at least ν times the number of training examples seen so far. The more support vectors are included in a model, the longer the computation of a target value for a given input feature vector will be in application mode. Also, learning the next time will also take longer. This entails that the learned models will monotonically increase with the number of training examples seen and that parameter ν has to be adjusted to the expected number of training examples for an experiment. It therefore will be varied throughout the experiments and will be the only parameter for the ν-SVR-2L function approximator that explicitly shows up in plots and tables. The ν-SVR-2L function approximator will be denoted by "svr" in plots and tables followed by an actual parameter value for parameter ν in brackets. The value for ν should not be too high in order not to slow down learning and application and not too low in order not to have too few support vectors for a useful model. As rule of thumb, it was tried to have SVR models after an One-Step-Q-ILS algorithm based training phase job with at most 1000 support vectors. Let i ($i \in \mathbb{N}^+$) be the number of iterations per try, t ($t \in \mathbb{N}^+$) the number of tries, and n ($n \in \mathbb{N}^+$) the number of problem instances processed by an application phase job. The value for ν to use then can be computed according to:

$$\nu = \frac{1000}{i \cdot t \cdot n}$$

Regression Tree

Regression trees are built step-wise depth by depth, or, in other words, level by level. During the regression tree construction, at each new node of a current depth, a *set of corresponding training examples* is given. The set itself is also called *corresponding set of training examples*, the single training examples are called *corresponding training examples*. For the root node in the first step the set of corresponding training examples is the whole set of all training examples given. For each corresponding set of training examples for a node at a current depth, a binary split of the set in two parts is conducted. This split needs not result in two parts of the same size. The split is made according to some individual input feature of the training examples. All training examples whose value for the individual feature exceeds a threshold are assigned to the first halve, all other are assigned to the second halve. The threshold and hence the split is made such that the sum of the standard deviations over the target values of the training examples in the two parts is minimized. In the next step, for each halve a new node will be set up in the next depth of the regression tree for which the parts then are the corresponding sets of training examples. The construction can go on until only corresponding training example sets of size one are left or perhaps earlier. Those nodes whose corresponding sets of training examples are not split anymore are called *leaves*. Splitting typically terminates when all leaves only have one corresponding training example. Subsequently then, the number of leaves is reduced again by pruning the regression tree by successively uniting sets of corresponding training examples for sibling leaves yielding a single leave, now with more corresponding training examples. Pruning is carried out in order to prevent from over-fitting and works as follows. Each union of leaves at a pruning step introduces an error on the training examples. Having only leaves with only one corresponding training example would be exact for the whole set of training examples and hence has error zero. Additionally, each leave gets assigned a penalty. Pruning in fact follows a scheme that tries to reduce a measure which is composed from penalties for leaves and the errors of leaves introduced by prior pruning. This measure is named *score* and pruning will try to minimize the score. The score is supposed to indicate how useful it is to prune in a certain way. Pruning has to be stopped at some time, too. A parameter named *number of leaves left* indicates how many leaves are to be left after pruning. For each leave of the final regression tree left after pruning target values have to be computed over the corresponding sets of training examples. Basically, any function approximator can be applied. Parameter named *leaves target value computation* indicates which computation scheme or function approximator to use.

The regression trees used in the experiments described here are further extended in [Gim05]. They can use simple averaging functions over the corresponding training example's target values, the k-nearest neighbor function approximator [RN03, pp. 733 – 735], and SVM/SVR to compute target values for leaves. In order to save computation, the construction process furthermore can stop before only leaves with one corresponding training example are left. A parameter named *split termination* indicates when to stop splitting. It represents the number of corresponding training examples that must at least be left in each leave. A leave whose number of corresponding training examples is less than the number indicated by split termination will not be split further. Regression trees are not incremental learning methods per se. Therefore, the regression trees implemented and used for the experiments described in this chapter are used in batch mode by accumulating training examples up to the learning frequency limit and then triggering learning. Regression trees typically store any training examples ever seen, even when run in pseudo incremental mode as just described. When approximating value functions for Q-ILS algorithms, however, the total number of training examples to accumulate quickly becomes too huge to be stored as a whole. Therefore, some kind of discarding of training examples must be employed. However, discarding training examples must ensure that effects on learning of early training examples are not forgotten altogether. A parameter called *samples per leave* indicates how many corresponding training examples at most are allowed to be left per leave. If the pruning process produced leaves with more than this allowed number of corresponding examples, some has to be discarded. The procedure works per leave by picking the allowed number of training examples randomly and considering them as centers of clusters. All other corresponding training examples are assigned to the nearest center – and hence cluster – according to Euclidean distance. Each cluster is reduced to one training example by using the feature values of the cluster's center and as target value an average over the training examples assigned to the respective cluster. The averaging is weighed additionally such that all training examples ever assigned and reduced to any of the training examples assigned to a cluster are averaged over with equal weight, too, independent when they have been added.

The regression tree function approximator will be denoted "regTree" in plots and tables followed by three parameter values for the parameters. These are represented by variables a, b, and c in Table 6.1 on page 200. The three parameters should be chosen such that not too many training examples have to be stored and hence computed in each batch. The limit is set to have no more than approximately 1000 training examples stored after each learning effort. This can be achieved by having for parameter split termination a value of 10, for parameter number of leaves left a value of 100, and for parameter samples per leave a value of 30.

6.3.2   Problem Instance and Search State Features

After the presentation of the two types function approximators used, this subsection introduces general construction schemes for search state and problem instance features.

Classification and Nomenclature of Features

There are different possible dimensions that characterize features:

1. Some individual features are dependent on LSA states (more precisely on the search and/or heuristic state parts thereof), other individual features are only problem instance specific and do not change with a change of a current LSA state. The former kind of individual features are called *(LSA) state features* whereas the latter are called *(problem) instance features*. Instance features need to be computed only once at the beginning of a search since they do not change during the search.

2. Some individual features represent a characteristic of a current LSA state or a problem instance as a whole, i.e. there is one feature value per LSA state or one per problem instance, while some individual features assign a value to each solution element of the search state part of a LSA state or to each solution component declared by a problem instance description. The former kind of individual features are called *global features* whereas the latter ones are called *local features*.

Clearly, instance features correspond to the already mentioned problem instance features (cf. Subsection 5.2.3). State features include search and heuristic state features (cf. Subsection 5.2.2 and Subsection 5.2.9). Since heuristic states typically do not have solution elements, the distinction into local and global features does not apply there. Otherwise, the two distinctive properties of individual features are orthogonal and hence four categories of individual features can be constructed. These are called *local instance features*, *global instance features*, *local state features*, and *global state features*. Except for heuristic state features most of the global state and instance features are computed based on corresponding local features. Otherwise, local features are of not too much use, since they are problem instance size dependent: they depend on and correspond to solution elements and components.

An example will clarify this. Given a TSP, one can compute for each city the maximum distance to any other city. This then would be a local instance feature, one for each city. Just as well one can compute the maximum distance to the two neighboring cities in a given tour. Although this maximum is only computed over exactly two distances to neighboring cities, there nevertheless will be one value for each city in each tour. Based one these local instance and state features, global variants can be computed, e.g by again computing a maximum, this time over all respective local feature values. For computing global features from local ones, any statistics such as the span, average, median, minimum, maximum, quartiles, and any other quantiles can be used. Since there are discrete solution and components for any COP, this procedure is generic for any COP and hence can be used to compute individual features for COPs other than TSP also. Of course, there are other individual features conceivable for any COP, but this procedure at least is a starting point which is also used for the COPs used throughout the experiments described in this chapter. The complete presentation of individual features for the individual COPs will be presented in the respective experiments sections (sections 6.4 and 6.5).

Besides search state features, heuristic state features belong to the group of state features. The heuristic features contrived and used for the first experiments with GAILS algorithms will be addressed in the next subsection (Subsection 6.3.3). Before, some issues regarding normalization of individual features which do not apply to the heuristic state features have to be addressed.

Empirical Normalization

Recall from Subsection 3.3.1 that normalization of individual feature values is needed to transfer value functions across problem instances, in particular over problem instance with different sizes. This is desirable in order to reuse what was learned and hence save computation time (cf. Subsection 4.2.1). Normalization of individual features can be done by computing lower and upper bounds LB and UB ($LB, UB \in \mathbb{R}$) on the values for individual features and by normalizing a feature value x ($x \in \mathbb{R}$) according to function $norm$ ($norm \colon \mathbb{R} \to [0, 1]$) as follows (cf. Subsection 3.3.1):

$$norm(x) := \frac{x - LB}{UB - LB}$$

Given correct bounds, the result will always be in $[0, 1]$ and scale is preserved, since this kind of normalization scales linearly. The tighter the bounds are, the better works this kind of normalization, since the more of interval $[0, 1]$ is covered and the more likely it is that normalized values are not too close to each other. Too close training examples in terms of feature values as well as costs can be a problem in practice due to the restricted precision of floating point variables. Any other interval could be used also by linear transformation without changing the principle working, but only interval $[0, 1]$ will be considered as so-called *normalization target interval* here. There are other normalization schemes conceivable, but throughout this document, any normalization refers to the linear normalization by the normalization function $norm$ to the target interval $[0, 1]$.

In order to carry out linear normalization, lower and upper bounds LB and UB for each individual feature have to be contrived and computed theoretically. These theoretically justified bounds are called *theoretical (lower/upper) bounds*. Normalized feature values computed with the help of theoretical bounds are called *theoretically normalized (individual) feature values*. The individual feature values before normalization are called *original (individual) feature values*. The process of normalization with theoretical bounds is called *theoretical normalization*. The theoretical bounds will be computed from the problem instance description and hence will be problem instance specific.

They only have to be computed ones for a search. For example, the tour cost for TSP instances can be bounded by summing for each city the minimum and maximum distances to other cities.

In practice, theoretical bounds might be too wide which foremost results in normalized feature values only covering a small fraction of the normalization target interval and having values very close to each other. Considering the cost feature for example, the coverage of normalized costs of the normalization target interval typically is very poor. Theoretical upper bounds are likely to be far too high, since they will have to be contrived for *any* candidate solution and hence for suboptimal ones, too. But Q-ILS algorithms using ILS-actions will only visit and hence compute normalized feature values for local optima. It seems unrealistic to be able to contrive theoretical upper bounds on costs for local optima. In order to do so, local optima would have to be characterized somehow, this way probably solving the whole optimization problem directly anyway. From a computational point of view, too close normalized feature values can be problematic, in particular for the cost feature. In fact, first test experiments with Q-ILS algorithms for the TSP using the ν-SVR-2L function approximator (cf. Subsection 6.3.1 and [Ban04]) have revealed that the theoretically normalized costs are very close to each other, exactly because of far too high theoretical upper bounds on costs. The result was that – even over a whole run of a job – the function approximator was not able to compute a model.

In order to circumvent this problem, a preprocessing step can be inserted that estimates bounds empirically. The empirically estimated bounds are called *empirical (lower/upper) bounds.* The so-called *empirical bounds estimation* is tailored to the Q-ILS algorithms use case using ILS-action. It was accomplished by successively running several representative ILS-actions individually as normal ILS jobs and together in random selection fashion as random ILS job for some t tries and some i iterations on a problem instance. For each local optimum encountered, feature values for all individual features are recorded and the upper and lower envelope over all runs and recorded feature values for each individual feature is computed. For a problem instance, these enveloping values then serve as empirical upper and lower bounds for an *additional* linear normalization to increase coverage over the normalization target interval and hence preventing from too close normalized feature values. The resulting normalized feature values are called *empirically normalized (individual) feature values.* The process is called *empirical normalization.* If the type of normalization is irrelevant, notion *normalized (individual) feature values* for short will be used. Note that the main purpose of using empirical bounds is to achieve a better spreading of normalized feature values and only secondly to obtain better bounds.

A problem can arise, if the recorded empirical bounds are not overly representative and hence are too narrow or perhaps too wide (which is more unlikely, though). The effect of having too narrow empirical bounds will be that an empirical normalized feature value might not stay within the designated normalization target interval. This, however, is not a problem in principle. First, the function approximators used are not restricted to a specific normalization target interval such as $[0, 1]$ (or rather $[0, 1]^m$, since there are m ($m \in \mathbb{N}^+$) individual features together forming a feature vector which is the function approximator domain). The function approximators can and will generalize to the outside of a designated normalization target interval also. Second, linear normalization as used here scales continuously and only gross outliers might pose a real problem in that the normalized feature values in such cases are far outside the normalization target interval and hence pose a problem to the function approximators. Finally, even if bounds are too tight, if observed feature values behave reasonable homogeneously and representative, outliers will also behave this way, now only partly outside the normalization target interval. Since the relations between feature values are most important in normalization and not the actual normalization target interval, it can be expected that normalized feature values will not all of a sudden be scattered arbitrarily over $\mathbb{R}$. This is true for single but also across different problem instances. In the worst case, empirical bounds will not increase the quality of bounds (since they coincide with theoretical bounds) and might only provide slightly or no better coverage over the normalization target interval. Or the bounds are too tight and will result in normalized feature values outside the normalization target interval, but nevertheless with more distance to each other as a side effect.

An attempt to prevent from too many outliers is to introduce a so-called *bounds tolerance*. The bounds tolerance b ($b \in [0, 1]$) indicates a percentage. The interval spanned by the empirical lower and upper bounds LB and UB will be increased by multiplying them by $1 - b$ and $1 + b$ respectively: $LB \cdot (1 - b)$ and $UB \cdot (1 + b)$. This will introduce a further tolerance. If theoretical bounds are tighter than the resulting empirical bounds including tolerance, the tighter theoretical bounds will be used, of course. Empirical normalization is most important for costs anyway. Experience has shown for many COPs that the cost for the best solutions found by local search methods and metaheuristics in particular virtually always quickly comes within a 10% range of the global optimum during a search. By having a bounds tolerance of 0.1 and not going beyond theoretical bounds, empirical bounds will most likely do not produce outliers downward and hence will do no harm for normalizing costs. Also, the first local optima encountered during each search typically are the worst cost local optima ever encountered during a search. The worst thing that can happen for upper bounds is that upper bounds are too tight for these worst local optima which are irrelevant anyway, since learning does not start at once (cf. Subsection 6.2.2). Almost surely the upper bounds will be far better for costs of interesting local optima than theoretical upper bounds. During a bounds estimation, the best local optimum can be at most as good as a global optimum, so it is impossible to have too wide lower bounds there, too.

As concerns empirical bounds used in single-instance experiments it is irrelevant whether the empirical bounds are recorded unnormalized or theoretically normalized. This is because the same normalized feature value will result by first normalizing an original feature value with theoretical bounds and next with empirical bounds that were recorded for theoretically normalized feature values, or by normalizing an original feature value directly according to unnormalized recorded empirical bounds. The empirical bounds recorded for theoretically normalized feature values basically are theoretically normalized feature values themselves. Having theoretical lower and upper bounds LB_T and UB_T ($LB_T, UB_T \in \mathbb{R}$) for an individual feature, normalization of a theoretically normalized original feature value with theoretically normalized empirical bounds is the same as normalizing the same original feature value with unnormalized empirical bounds, since they can be transformed into each other as follows:

$$\frac{\frac{x - LB_T}{UB_T - LB_T} - LB_{E_N}}{UB_{E_N} - LB_{E_N}} = \frac{\frac{x - LB_T}{UB_T - LB_T} - \frac{LB_{E_U} - LB_T}{UB_T - LB_T}}{\frac{UB_{E_U} - LB_T}{UB_T - LB_T} - \frac{LB_{E_U} - LB_T}{UB_T - LB_T}} = \frac{\frac{x - LB_{E_U}}{UB_T - LB_T}}{\frac{UB_{E_U} - LB_{E_U}}{UB_T - LB_T}} = \frac{x - LB_{E_U}}{UB_{E_U} - LB_{E_U}}$$

with

$$LB_{E_N} = \frac{LB_{E_U} - LB_T}{UB_T - LB_T}$$

$$UB_{E_N} = \frac{UB_{E_U} - LB_T}{UB_T - LB_T}$$

where LB_{E_N} and UB_{E_N} ($LB_{E_N}, UB_{E_N} \in \mathbb{R}$) are the theoretically normalized empirical lower and upper bounds and LB_{E_U} and UB_{E_U} ($UB_{E_U}, LB_{E_U} \in \mathbb{R}$) are the unnormalized empirical lower and upper bounds. In effect, for any individual problem instance, both normalization schemes will normalize to the same normalization target interval as determined by the empirical bounds and both schemes will provide a better spreading of normalized feature values.

As regards multi-instance experiments, however, it makes a difference whether to use theoretically normalized empirical bounds or whether to normalize with unnormalized empirical bounds directly. When learning across multiple problem instances with the aim to transfer a learned action-value function, empirical bounds have to be fixed for *all* problem instances used for training and *before* training. Also, the empirical bounds used must not change during the corresponding application and test phases. When applying a learned action-value function to a new, unseen problem instance, e.g in order to compare the performance to normal ILS, having a preprocessing for each new unseen problem instance would be unfair and would distort results. Of course, preprocessing will need additional computation time, but this will have to be done for each problem instance only once and before training. Using for each problem instance its own empirical bounds, besides the unfairness, furthermore is not appropriate. In experiments with training- and application-phase jobs always working on one single and fixed problem instance, the empirical bounds will not change from training- to application-phase and hence the features for which an action-value function was learned will be normalized the same way when using the learned action-value function. If empirical bounds change from training- and application-phase to a test phase as is the case in a multi-instance experiment, a learned action-value function is not used for which it was learned; it really is not applicable anymore with the new normalization for the new unseen problem instances. Instead, so-called *common empirical bounds* for all training problem instances are built, again as a lower and upper envelope over the respective bounds for all problem instances involved in training. These common empirical bounds then are used for both the training, application and the test problem instances in training-, application, and test phase. No further unfair preprocessing will occur in the test phase. Using unnormalized empirical bounds in such a setting is not sensible, since the unnormalized empirical bounds are not transferable over multiple problem instances for the same reasons that original feature values cannot. Transfer of empirical bounds then centrally is enabled again by normalizing empirical bounds theoretically also and next applying them to already theoretically normalized feature values. As a drawback, low quality bounds for an individual feature for only one training problem instance – due to too loose or tight bounds, or not representative lower or upper bounds in general – will lower the quality of the common empirical bounds for this individual feature for all problem instances. Nevertheless, even low quality bounds will still yield some additional spreading of feature values over the normalization target interval. Hence, it can be expected that common empirical bounds still ensure comparability and hence transfer of what was learned across multiple problem instances.

Altogether, empirical bounds should provide a better spreading of normalized feature values and a better coverage of the normalization target interval, also, while simultaneously not doing any harm other than additional computational effort, but only once per training instance. Hence, this kind of normalization is used, whenever the theoretically normalized features values showed to be too close to each other and prevented function approximators from computing proper models for action-values.

In order to control empirical normalization for Q-ILS algorithms a parameter for adjusting the type of normalization named *normalization type* with eligible values "theo" for theoretical normalization and "emp" for empirical normalization is introduced. Another parameter named *bounds tolerance* indicates the bounds tolerance. In addition, information where to find empirical bounds for a problem instance in the form of a filename for a file that will contain the empirical bounds is needed. This filename will be built by appending suffix ".bounds" to the filename of the problem instance and accordingly does not show up as independent parameter. If several problem instances are processed at once by a job, they will be indicated as a colon separated list of their filenames (cf. Subsection 6.1.1). The filename of the file containing the common bounds for these problem instances then is this colon separated list appended by suffix ".bounds". Shortcuts, ranges, and default values for the normalization related parameters can be found in Table 6.1 on page 200.

6.3.3 Heuristic State Features

This subsection is devoted to the heuristic state features that are used for the experiments covered in this chapter. Heuristic state features are usually independent from any COP type and often also from GAILS-algorithms used. They are based on heuristic information or simply *are* the heuristic information.

The heuristic state features used in the context of Q-ILS algorithms are based on delta costs for steps and iterations, iterations counters and obtained immediate rewards. They are listed and explained next:

- *Average delta cost steps*: A list of the delta costs for the last $n_{\Delta c_s}$ ($n_{\Delta c_s} \in \mathbb{N}^+$) local search steps made is maintained in heuristic states. Over these delta costs stored, an average can be computed indicating how good or steep the last local search steps and also whole local search descents were. Steep descents might want to be fostered. If the costs are normalized, so will the delta costs stored in the list and hence the computed average.

- *Normalized iterations*: The normalized number of iterations is defined as:

$$\frac{\text{Current iterations count}}{\text{Maximum number of iterations allowed}}$$

  This heuristic state feature will indicate in relative terms the progression of the search. It might be useful, if action selection should be made dependent on the relative search progress in order to change the strategy in later stages of the search, e.g. to more explorative actions. It is also used in the one-time reward computation (cf. Subsection 6.2.2 and variable "x" there)

- *Delta cost iteration*: This heuristic state feature consists of the delta cost for the previous iteration, i.e. for the previous LSA move and basically is the same as the immediate rewards according to the delta cost reward signal when using ILS-actions.

- *Average delta cost iterations*: The average delta cost iterations heuristic state feature works analog to the average delta cost steps heuristic state feature, now averaging over delta costs for iterations, i.e. LSA moves, instead of over local search steps. For averaging the last $n_{\Delta c_i}$ ($n_{\Delta c_i} \in \mathbb{N}^+$) such delta costs will be recorded.

- *New global best iterations*: This heuristic state feature is the iteration count at the time when the last time an overall new best search state has been found. The count is given in normalized terms, i.e. relative to the maximally allowed number of iterations.

- *Delta to new global best iterations*: The normalized number of iterations that have passed since the last time a new overall best search state has been found is indicated by this heuristic state feature, again in relative terms.

- *Reward*: This is the last immediate reward obtained according to the used reward signal (here either delta or inverse cost reward signal, cf. Subsection 6.2.2).

- *Average rewards*: Finally, the average rewards heuristic state feature indicates the average immediate reward obtained for the previous n_r ($n_r \in \mathbb{N}^+$) LSA moves.

The resulting additional parameters for Q-ILS algorithms that are needed to make the heuristic state feature work are the list lengths $n_{\Delta c_s}$, $n_{\Delta c_i}$, and n_r stemming from the average delta cost steps, average delta cost iterations, and average rewards heuristic state feature, respectively. They are named *number $\Delta cost$ steps*, *number $\Delta cost$ iterations*, and *number rewards*, respectively. Shortcuts, ranges, and default values can be viewed in Table 6.1 on page 200.

6.4   Experiments for the TSP

This section presents the results of the first experiments undertaken with GAILS algorithms on the TSP. First, the formal problem formulation of the TSP is given and the local search steps, perturbations, and initial solution construction schemes that have been used are described in Subsection 6.4.1. Next, in Subsection 6.4.2, the developed features for the TSP are presented and the worked out theoretical normalization scheme for features for the TSP will be explained. Finally, the experiments and their results are presented including result tables and plots in Subsection 6.4.3, before Subsection 6.4.5 summarizes the conclusions that can be drawn from the results.

6.4.1   Description

This subsection describes the formal and search operator related aspects of the Traveling Salesman Problem (TSP) beginning with a formal problem formulation.

Problem Formulation

Recall from Section 1.1 that for each TSP problem instance a number of n ($n \in \mathbb{N}^+$) so-called *cities* are given. For each two different cities i and j ($i, j \in \{1, \ldots, n\}, i \neq j$) *travel costs* (or *costs* for short) c_{ij} ($c_{ij} \in \mathbb{R}$) are given for traveling from city i to city j. The costs can be travel time, distance, or any other kind of costs. In case of the symmetric TSP, which is tackled in the experiments described in this section, it holds $c_{ij} = c_{ji}$. The goal in solving a TSP is to find a *round trip*, also called *tour*, that, starting at a city, visits any other city exactly once and finally returns to the starting city. A round trip can be represented as an ordered list of cities which will be visited in the given order. Formally, a TSP problem instance can be described as an almost fully connected graph G with $G := (V, E)$ where the set of vertices or nodes $V = \{1, \ldots, n\}$ is equal to the set of all cities and E with $E := (V \times V) \setminus \{(e, e) \mid e \in V\}$ is the set of all connections or edges. To each connection e of E, $e = (i, j) \in E$, a cost function $c : V \times V \to \mathbb{R}$ assigns a (travel) cost. For a problem instance of size n, a round trip can be described as a permutation p ($p : \{1, \ldots, n\} \to \{1, \ldots, n\}$). The city assigned to position i of a tour will be the i-th city visited in the tour and the tour will consist of all connections that exist between neighboring cities in the permutation (including the connection between the two cities assigned to positions i and n, of course). The cost c_p ($c_p \in \mathbb{R}$) of a tour p then is computed as:

$$c_p = \sum_{i=1}^{n-1} c(p(i), p(i+1)) + c(p(n), p(1))$$

The TSP is *NP*-hard, so it is justified to apply heuristic methods and local search methods in particular to solve it. After the formal problem description of the TSP has been given, several local search steps and procedures, and perturbations that are used throughout the experiments of this section are presented. In addition, initial solution construction schemes and other necessary details such a candidate lists are covered in the remaining section.

Initial solution

Three schemes how to construct an initial solution have been implemented [Kor04]. These are:

- *Random*:   The initial solution simply consists of a random permutation of all cities.

- *Nearest Neighbor:* This scheme constructs a permutation by step-wise adding city by city until a complete tour has been built. For each new city just added its nearest neighbor – according to Euclidean distance – *not* in the partial tour constructed so far is added to be visited next in the tour.

- *Random Nearest Neighbor:* This procedure is a modification of the nearest neighbor initial solution construction in that the city to begin the construction with is chosen randomly.

The three initial solution construction schemes are denoted by "random", "nn", and "nnRand", respectively, in plots and tables.

Local Search Steps

The costs for initial solutions according to the just presented construction schemes usually are far from being acceptable. Therefore, initial solutions are further improved by local search procedures that are based on one of three neighborhood structures that have been implemented for the experiments described in this section [Kor04]. Two of the neighborhood structures belong to the k-*opt* ($k \in \mathbb{N}^+$) family of neighborhood structures for the TSP. For each k, the k-opt neighborhood of a tour consists of all tours that are different in exactly k connections. Figure 6.1 and 6.2 show schematic 2- and 3-opt connection exchanges, respectively.

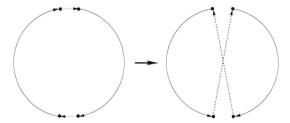

Figure 6.1: Example of a 2-opt exchange for a TSP tour

The left-hand side of these figures presents the original tour, the right-hand side the one that results from exchanging the connections that are drawn in dashed line type. The arrows indicate the original tour direction and only relevant nodes are drawn. After removing the two connections represented by the dashed lines in Figure 6.1, two new connections are inserted, again indicated by dashed lines, and the direction of the left section of the represented tour is reversed.

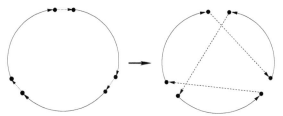

Figure 6.2: Example of a 3-opt exchange for a TSP tour

The 3-opt exchange is a little bit more complicated. Its operation can be seen in Figure 6.2.

Extending the 2-opt neighborhood structure leads to a special variant of the 3-opt neighborhood structure which is called *2.5-opt* neighborhood structure. It is illustrated in Figure 6.3. There, in addition to the exchange of two connections as for the 2-opt neighborhood structure, a city of the tour is also relocated.

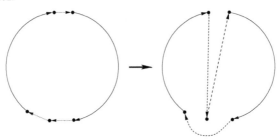

Figure 6.3: Example of a 2.5-opt exchange for a TSP tour

Local search steps can be obtained for each neighborhood structure using a first or best improvement neighbor selection scheme. Based on the resulting local search steps, local search procedures finally can be obtained (cf. Section 2.3). In the first experiments described here, only first improvement local search steps were used to form local search procedures, since they do not require to always search a whole neighborhood and hence entail some acceleration. The local search procedures just presented are abbreviated as "2opt[first]", "2.5opt[first]", and "3opt[first]" in plots and tables

Several speed-up techniques for k-opt local search procedure for the TSP have been proposed. For the experiments described here, so-called *don't look bits* (DLBs) [Ben92] have been implemented and used [Kor04]. For each city, an individual DLB is stored which generally indicates whether this city is a useful candidate for being involved in an exchange of tour connections. If set, the DLB for a city indicates that this city should not be considered for involvement in an exchange. DLBs are set for example for cities incident to connections that have just been exchanged. DLBs for some cities have to be reset after some local search steps or perturbations have been carried out, depending on whether the last exchanges make it more likely for a city to be needed in an exchange that improves tour cost. Several variants of resetting DLBs, depending on when to do the reset, have been implemented. These are: not using DLBs at all, to reset after each perturbation application only some, and to reset after each perturbation all DLBs denoted by "off", "resetPert", and "resetAll", respectively in plots and tables. The default, which is used for all experiments described here, is to reset only some DLBs after a perturbation application. The DLBs usage is controlled by a parameter named *TSP don't look bits* and is shown with shortcut, range, and default value in Table 6.1 on page 200.

Candidate Lists

Another speed-up technique is the so-called *fixed-radius* technique which restricts eligible exchanges of connections. Only connections between cities within a certain radius from each other are eligible for exchange. The radius is determined by so-called *candidate lists*: only connections with the second incident city being in the candidate list of the first one can be inserted. The radius in effect then is the candidate list length. The idea for using candidate lists for the TSP stems from the observation that most cities in a globally optimal or reasonable good solution are almost always connected to their nearest or to at least one of their nearest neighbors. The distance thereby can be Euclidean distance and the candidate list then is sorted according to Euclidean distance (and denoted by "nn" in plots and tables).

To conclude, note that when using candidate lists, it can happen in principle that connections of a globally optimal solution are not contained in the candidate lists for the cities. Since only candidate lists are searched when trying to find cities for connection exchanges, in such a case a globally optimal solution cannot be found. The risk that this happens is deemed to be minimal, though. For more information about candidate lists for the TSP in the GAILS framework see [Kor04].

Perturbations

Three perturbations have been implemented for the TSP integration into the GAILS framework [Kor04]. One perturbation is the so-called *random double-bridge* perturbation [MOF91, LMS02]. The random double-bridge perturbation is based on a 4-opt neighborhood structure and is shown in Figure 6.4. Four connections are selected randomly and removed from a tour (indicated by dotted lines in the left-hand side of Figure 6.4) resulting in four incomplete sections of a tour (labeled "A" – "D" in the right-hand side of Figure 6.4). Next, the four sections are re-assembled by inserting new connections as shown in Figure 6.4 to form a complete tour again. The new direction of the resulting tour will then be A, C, D, B. If using DLBs, all individual DLBs for all cities in a specified radius around the cities involved in the random double-bridge connections exchanges will be reset. Radius in this context is with respect to tour neighbors. In the random double-bridge perturbation version implemented for the experiments described here, the four connections are not completely selected randomly from all connections of a tour but only connections between cities within a randomly chosen section of size w ($w \in \{1, \ldots, n\}$) of a tour are eligible to be exchanged. Such a selected section is called the *window* and w is called the *window size*. Other variants of the random double-bridge perturbation that differ in how they select the connections to remove and replace have been invented and tested in [Kor04] but yielded inferior results so they are not considered here. The second perturbation employed in the experiments is called *random walk* perturbation. It randomly chooses a section or window of a tour of size w and re-sorts it randomly before inserting the re-sorted section back into the tour again. The third perturbation is called *random greedy* perturbation. This perturbation also randomly selects a section or window of a tour of size m which then is rebuilt by a candidate list based version of the nearest neighbor initial solution construction scheme. In case DLBs are used together with random walk or random greedy perturbations, the individual DLBs for the cities of the re-ordered section as well as those in a certain radius before and after the section are reset.

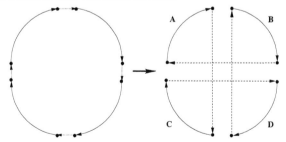

Figure 6.4: Example of a random double-bridge perturbation for a TSP tour

The three perturbations employed are denoted by "randDB" for random double-bridge, "rand-Greedy" for random greedy, and "randWalk" for random Walk in plots and tables. All are param-

eterized by a single parameter value in brackets after the name when listed. This value represents
the perturbation strength for the random double-bridge perturbation and the window size for the
random walk and random greedy perturbations. The strength parameter for the random double-
bridge perturbation basically expresses, how many times in a row the perturbation is executed
before giving back control. Strength for the random greedy and the random walk perturbations
is mostly determined by window size. In all experiments described here, the value for the window
size is set to 300 for the random double-bridge perturbation. The value for the radius for reseting
DLBs is always set to 20. In plots and tables, actions are denoted as a combination of perturbation
and local search procedure identifier, separated by a "+".

6.4.2   Features for the TSP

This subsections presents and describes all the individual features that have been implemented for
the TSP and explains how theoretical normalization is achieved. The selection and a short analysis
of the individual features actually used during the experiments for the TSP is given.

Features Description

Most of the individual features for the TSP are computed as local and global features (cf. Subsection
6.3.2). Recall that local features are computed for each city of a tour or a problem instance while
global features are computed once per tour or problem instance. The local instance features for a
city invented for the TSP are:

- *MinDist*: Minumum distance to another city.

- *MaxDist*: Maximum distance to another city.

- *MeanDist*: Mean distance to another city.

- *SpanDist*: Difference between minimum and maximum distance to another city.

- *VarDist*: Variance computed over all distances to other cities.

- *Quant25Dist*: 25%-quantile over distances to other cities.

- *MedianDist*: 50%-quantile over distances to other cities.

- *Quant75Dist*: 75%-quantile over distances to other cities.

- *Quart1SpanDist*: Difference between 25% quantile and the minimum distance.

- *RatioQuart1Dist*: Ratio between span of first quartile to whole span, i.e. $\frac{\text{Quart1SpanDist}}{\text{SpanDist}}$.

The local state features are:

- *MinDist*: Distance to nearer neighbor in a tour.

- *MaxDist*: Distance to the more distant neighbor in a tour.

- *MeanDist*: Mean distance to neighbors in a tour.

- *SpanDist*: Difference in distance to nearer and the more distant neighbor in a tour.

- *PercLeqInNNShort*: Percentage of cities in the nearest neighbors list for a city that are nearer or equally near than the nearer neighboring city in a tour.

- *PercLeqInNNMean*: Percentage of cities in the nearest neighbors list for a city that are nearer or equally near than the mean distance to the neighboring cities in a tour.

- *PercLeqInNNLong*: Percentage of cities in the nearest neighbors list for a city that are nearer or equally near than the more distant neighboring city in a tour.

- *Frust*: Frustration is the ratio of the difference between the distances to the neighbors a tour to the span over the distances to *all* cities.

Based on these local features, global features can be computed using known statistics:

- *XMinOver*: Minimum computed over values of local feature X.

- *XMaxOver*: Maximum computed over values of local feature X.

- *XMeanOver*: Mean computed over values of local feature X.

- *XQuant25Over*: Value of 25%-quantile over the values of local feature X.

- *XMedianOver*: Value of 50%-quantile over the values of local feature X.

- *XQuant75Over*: Value of 75%-quantile over the values of local feature X.

- *XSpanOver*: Difference between minimum and maximum value of local feature X.

- *XVarOver*: Variance computed over values of local feature X.

The following additional global instance feature not computed as statistics over local instance features are conceivable:

- *MeanDist*: Mean distance between any two cities.

- *Quant25Dist*: 25%-quantile over the distances between cities.

- *MedianDist*: 50%-quantile over the distances between cities.

- *Quant75Dist*: 75%-quantile over the distances between cities.

- *VarDist*: Variance of the distances between cities.

Finally, the following global state features that are not computed as statistics over local instance features have been invented:

- *Perc2Nearest*: Percentage of cities in a tour that are connected to its two nearest neighbors.

- *PercLongInNN*: Percentage of cities in a tour for which both neighbors in the tour are in the city's nearest neighbors list.

- *PercMeanInNN*: Percentage of cities in a tour for which the mean distance to its two neighbors in the tour is less than the greatest distance to a city in the city's nearest neighbors list.

- *PercOnlyShortInNN*: Percentage of cities in a tour for which only the nearer neighbor in the tour is in the city's nearest neighbors list.

- *PercNotInNN*: Percentage of cities in a tour for which no neighbor in the tour is in the city's nearest neighbors list.

- *MinDist*: Minimum distance between any two cities of a tour.

- *MaxDist*: Maximum distance between any two cities of a tour.

- *MeanDist*: Mean distance between any two cities of a tour.

- *SpanDist*: Difference between the minimum and maximum distance between any two cities of a tour.

- *Quant25Dist*: 25%-quantile of the distances between cities of a tour.

- *MedianDist*: 50%-quantile of the distances between cities of a tour.

- *Quant75Dist*: 75%-quantile of the distances between cities of a tour.

- *VarDist*: Variance of the distances between cities of a tour.

Note that the tour length is a feature also.

TSPLIB							
Problem Instances							
Name	rat575	d657	vm1084	rl1304	nrw1379	fl1577	d1655	vm1748
Cost	6773	48912	239297	252948	56638	22249	62128	336556
Name	u1817	rl1889	u2152					
Cost	57201	316536	64253					
VLSI							
Problem Instances							
Name	rbx711	rbu737	dkg813	lim963	pbd984	xit1083	dka1376	dja1436
Cost	3115	3314	3199	2789	2797	3558	4666	5257
Name	fra1488	rbv1583	rby1599	fnb1615	djc1785	dcc1911	dkd1973	djb2036
Cost	4264	5387	5533	4956	6115	6396	6421	6197
Name	dcb2086	bck2217	xpr2308	ley2323	pds2566	dbj2924	dlb3694	
Cost	6600	6764	7219	8352	7643	10128	10959	

Table 6.2: Problem instances for the TSP and costs of global optima

Features Normalization

Since there are many features available for the TSP, linear normalization using lower and upper bounds has been abbreviated. Instead of computing lower and upper bounds for each individual feature separately, normalization is done by normalizing the distances of the graph G representing a TSP problem instance by means of a lower and upper bound. These lower and upper bounds are the longest and the shortest distance in the graph. For most features for the TSP just presented this kind of normalization is compatible in that it also will produce theoretically normalized values within the interval $[0, 1]$. Note also that this kind of normalization preserves the topology of the search space with respect to the cost function (cf. Section 2.3). The lower and upper bounds for the tour length, i.e. the cost, are computed by summing up the shortest and the longest distance for each city to another city. A drawback of this approach, especially for cost normalization, is that the bounds are relatively loose. Therefore, empirical bounds according to Subsection 6.3.2 have been computed. Table 6.3 shows the parameters used for the bounds estimation.

Short	Actual Parameter Values
tc	6000
t_a	20
i_a	rat575, d657, rbx711, lim963, pbd984, pr1002, xit1083, vm1084, rl1304, dka1376, icw1483, fl1577, rbv1583, rby1599, d1655, vm1748, djc1785, rl1889, djb2036, bck2217, dcb2086, pr2392, pds2566
$pert$	randWalk[100], randGreedy[100], randDB[1], randDB[5]
ls	3opt[first], 2.5opt[first], 2opt[first]
$init$	nnRand
ϵ_{a_a}	0.0, 0.4, 1.0

Table 6.3: Actual parameter values for feature bounds estimation for the TSP

The problem instances used for the bounds estimation are all those problem instances that were picked a priori for serving as training problem instances in the experiments to come in this section. All training, application, and test problem instances used in any experiment for the TSP thereby come from two repositories: TSPLIB [Rei91, Url05f] and [Url05e]. From the second repository, only VLSI problem instances with known global optimum were taken and from the TSPLIB also only problem instances with known global optimum have been used. The number included in the instance name indicates the size of the respective problem instance. The TSP problem instances used are listed in Table 6.2 together with cost of their known global optimum. The problem instances were chosen randomly from the two repositories with the only constraint that they equally distribute over instance sizes available up to approximately 3500 cities.

All TSP experiments were run on a cluster of computers located at the Center for Scientific Computing at the University of Frankfurt [Url05a] on 2.40GHz 32 bit Intel Xeon processors with 512 KB cache and 2048 MB main memory. The extensive support and computation time provided by the Frankfurt Center for Scientific Computing is gratefully acknowledged for.

Not all invented individual features can be used because using all individual features available might overextend the function approximators realizing value functions. Consequently, a selection has to be undertaken. The empirical bounds can be used in order to help with the decision which individual features to use. Table 6.4 shows the result for this analysis for the finally selected global state features. The individual features were selected based on the analysis of the empirical bounds and also based on other, general theoretical contemplations. One of these contemplations was to not include instance features since they are only useful for biasing the search for different types of problem instances which is not to be investigated in the first experiments described here, though. Other contemplations were whether an individual feature is deemed to be predictive as concerns solution quality or possible solution quality improvements and whether individual features are mutually redundant.

Table 6.4 shows in its column named "Feature Name" the feature name, next the minimum and maximum feature values over the problem instances in columns named "Minimum" and "Maximum", in column named "∩" the coverage of the normalization target interval [0, 1] by all the feature values over all problem instances (called *coverage* for short), and finally in the last column named "∪" the percentage of the normalization target interval that was covered by values for *each* problem instance (called *common coverage*). The table is sorted according to descending coverage except for the first three individual features named "OrigTourLength", "NormTourLength", and "TourLength", which show the original, unnormalized tour length, the tour length normalized according to the computed theoretical bounds, i.e. the theoretically normalized tour length, and the tour lengths computed from normalized distances, respectively. No entry in column "Coverage" or "Common" for a feature means that the normalization target interval is not [0, 1] such as for "FrustSpanOver" and "FrustMaxOver" or that it is not applicable. The common coverage is computed as the difference between the largest lower bounds for any problem instances and the lowest respective upper bounds. Therefore, negative value or a value of zero in column "Common" indicate that the values for some problem instances are completely disjunct. The respective individual features then basically work as global instance features in that they can bias an approximation for different problem instance characteristics.

As can be seen, the coverage for some individual features and in particular for the theoretically normalized costs is very small, so empirical normalization seems necessary. Altogether the common coverage is often quite poor, so additional empirical normalization is advisable for the TSP. The individual features selected and shown in Table 6.4 are among the ones with the highest common coverage.

6.4.3 Experiments

After the preparative remarks of the previous subsection, this subsection eventually presents the results of the first experiments conducted with GAILS algorithms for the TSP. Recall that the aim of these experiments foremost is to show that and how learning takes place. The investigation is aimed at finding out about learning behavior and not to find the best performing actual parameter settings or to find a best performing action selection strategy (perhaps only as a side-effect). Due to the fact that the GAILS method and the Q-ILS algorithms are new and for the first time investigated and due to the amount and variety of information and results obtained, the analysis presented here can only be done for reasonably representative excerpts and examples. This first examination can only be a rather superficial one. Many aspects not covered here are remarkable and noticeable but cannot be presented because of lack of time and space and are left to be investigated in the future.

Several kinds of experiments were conducted for the TSP. First, the influence of changing a local search procedure with a fixed perturbation was investigated in an experiment called *local search procedure comparison experiment*. The question to answer is whether a GAILS algorithm can learn to choose a known best local search procedure: According to prior experience [Ben92, JM97, JM02, Kor04], the 3-opt first improvement local search procedure is best performing for the TSP. Based on the results of the local search procedure comparison experiment, the local search procedure was fixed while varying strengths of the random double-bridge perturbation. This experiment is called *perturbation comparison experiment*. It is supposed to help answering the question which perturbation strength performs best and how to fruitfully vary the strength during a search relating to the aforementioned balancing problem between intensification and exploration (cf. Section 2.4). The subsequent two experiments varied several different perturbations and different perturbation strengths with a fixed local search procedure. These experiments are to show how good GAILS algorithms can perform when giving them the best ILS-actions known and whether and how they

Feature Name	Minimum	Maximum	∩ (in %)	∪ (in %)
OrigTourLength	2789.0000	464441.0000	–	–
NormTourLength	0.0022	0.0168	1.46	-0.26
TourLength	4.1263	98.9004	–	–
FrustSpanOver	0.0226	1.9490	–	–
FrustMaxOver	0.0226	1.9490	–	–
MinDistSpanOver	0.0069	0.9916	98.47	55.54
SpanDistMaxOver	0.0301	1.0000	96.99	76.75
MaxDistSpanOver	0.0320	1.0000	96.80	66.99
MeanDistSpanOver	0.0282	0.9958	96.76	56.90
MaxDistMaxOver	0.0369	1.0000	96.31	66.99
SpanDist	0.0369	1.0000	96.31	66.99
PercLeqInNNLongQuant75Over	0.3000	1.0000	70.00	0.00
FrustQuant75Over	0.0000	0.6635	66.35	-0.67
Perc2Nearest	0.2417	0.8624	62.07	-21.06
MaxDistQuant75Over	0.0017	0.5508	54.90	-2.56
PercLongInNN	0.5357	1.0000	46.43	11.91
PercLeqInNNLongMeanOver	0.2919	0.6310	33.91	0.20
RatioQuart1Dist	0.0000	0.3333	33.33	-1.21
FrustMeanOver	0.0019	0.3105	30.85	4.45
Quant75Dist	0.0017	0.2556	25.39	-2.08
FrustVarOver	0.0000	0.2292	22.92	3.69
PercLeqInNNLongSpanOver	0.6000	0.8000	20.00	0.00
PercLeqInNNLongMaxOver	0.8000	1.0000	20.00	0.00
PercOnlyShortInNN	0.0000	0.1722	17.22	5.42
SpanDistQuant75Over	0.0000	0.1429	14.29	-1.51
SpanDistMeanOver	0.0031	0.1300	12.70	1.14
PercLeqInNNLongVarOver	0.0145	0.1376	12.31	3.38
MaxDistVarOver	0.0000	0.1007	10.07	1.57
VarDist	0.0000	0.0736	7.35	1.11
MeanDistVarOver	0.0000	0.0574	5.74	0.91
SpanDistVarOver	0.0000	0.0479	4.79	0.93
MinDistVarOver	0.0000	0.0386	3.86	0.62
Quart1SpanDist	0.0000	0.0169	1.69	-1.13
MaxDistMinOver	0.0000	0.0113	1.13	-0.56

Table 6.4: Feature bounds analysis for the TSP

will learn to select among them. They were conducted first as single-instance experiments and next as two multi-instance experiments and are called *single-instance* and *multi-instance experiments*, respectively. All experiments were conducted with a small but hopefully representative selection of parameter values for the parameters available. A special parameter tuning was not employed.

Some notes to the nomenclature used when describing experimental results are to be given. The notion of "problem instance" will be abbreviated to instance. Columns will directly be denoted by their labels. When saying that some jobs are "good", "best", "better", "bad", "worse", "worst", and so on, this refers to the performance of the job and could also be read as "best performing", "performed best", and so on. Individual jobs are presented with a positioning number pos ($pos \in \mathbb{N}^+$) in tables. These positions can be used to identify jobs. Instead of saying Q-ILS or ILS job at position pos, the respective job is also denoted by *Q-ILS job pos* or *ILS job pos*. The notion of policy represents the strategy, based on value functions here, which selects the actions to be applied for each LSA state. Since this notion is also used to denote parameters and actual parameter values, the original meaning as action selection strategy is denoted by *action selection strategy* when describing and analyzing experiments in what follows (and the subsequent sections also). Policy then refers to the parameters and actual parameter values that indicate how to implement a certain policy based on value functions.

Local Search Procedure Comparison

The local search comparison experiment for the TSP is supposed to compare 2-opt, 2.5-opt and 3-opt first improvement local search procedures. In order to do so, three ILS-actions were built with a common and fixed random double-bridge perturbation with a strength of one. The resulting ILS-actions for Q-ILS algorithms (cf. Subsection 6.2.1) are called *2-opt, 2.5-opt*, and *3-opt ILS-actions*, respectively. First, for each of the three ILS-actions the corresponding ILS algorithms were run (using the better acceptance criterion, since experience suggests that this acceptance criterion works good [Ben92, JM97, JM02, Kor04]). The corresponding normal ILS jobs and algorithms for the three ILS-actions are called *2-opt, 2.5-opt*, and *3-opt ILS jobs* and algorithms. The actual parameter values for the resulting normal ILS jobs can be seen in Table 6.5 (the last three colon separated instance indicate that a ranking is computed separately over these three, cf. Subsection 6.1.1). The results of the runs are shown in Table 6.6.

Short	Actual Parameter Values
i_a	d657, dka1376, fl1577, vm1748, dkd1973, d657:dka1376:fl1577
pert	randDB[1]
ls	3opt[first], 2.5opt[first], 2opt[first]
init	nn
acc_a	better

Table 6.5: Actual parameter values for normal ILS jobs of the TSP local search procedure comparison experiment

As can be seen in Table 6.6, for each instance the 3-opt ILS jobs are best performing (see column "LS" for the local search procedure employed and columns "avg" and "avg.excess" for the performance results). The differences between the best normal ILS job and the second best is huge; the 3-opt ILS job excels by far for each instance, sometimes by a factor of two and more concerning average excess. All other statistics are lower and hence best for the 3-opt ILS job, too, for each instance.

| Instance | Perturbation | LS | Statistics | | | | |
			avg	σ	avg.excess	min	max
d657	randDB[1]	3opt[first]	48995.04	40.41	0.1698 %	48913	49068
(48912)	randDB[1]	2.5opt[first]	49133.96	86.69	0.4538 %	49002	49391
	randDB[1]	2opt[first]	49289.36	98.22	0.7715 %	49119	49434
dka1376	randDB[1]	3opt[first]	4694.16	11.79	0.6035 %	4673	4732
(4666)	randDB[1]	2.5opt[first]	4725.28	13.01	1.2705 %	4694	4749
	randDB[1]	2opt[first]	4730.92	12.66	1.3913 %	4703	4759
fl1577	randDB[1]	3opt[first]	22546.60	269.54	1.3376 %	22256	23326
(22249)	randDB[1]	2.5opt[first]	22647.68	283.72	1.7919 %	22285	23412
	randDB[1]	2opt[first]	22686.72	309.16	1.9674 %	22302	23712
vm1748	randDB[1]	3opt[first]	337974.80	370.89	0.4216 %	337091	338476
(336556)	randDB[1]	2.5opt[first]	339550.00	376.32	0.8896 %	338948	340473
	randDB[1]	2opt[first]	340004.80	479.43	1.0247 %	339125	341082
dkd1973	randDB[1]	3opt[first]	6454.36	9.34	0.5195 %	6435	6473
(6421)	randDB[1]	2.5opt[first]	6516.64	12.40	1.4895 %	6496	6539
	randDB[1]	2opt[first]	6544.68	14.83	1.9262 %	6518	6579

Table 6.6:   Normal ILS job results for the local search procedure comparison experiment

Table 6.7 shows the results of a ranking according to the procedure described in Subsection 6.1.2 of the corresponding normal ILS algorithms for the three ILS-actions and a random ILS algorithm over the three ILS-actions for instances d657, dka1376, and fl1577. The ranking results further underline the superiority of the 3-opt ILS-action, in particular by the fact that the random ILS algorithm performed second best in the ranking. This is an indication that even using the 3-opt ILS-action only every third iteration is still better than using any of the two other ILS-actions all the time. These results confirm the prior experience from [Ben92, JM97, JM02, Kor04] that the 3-opt first improvement local search procedure is performing better than the 2-opt and 2.5-opt variants for the TSP.

Rank	Perturbation	LS	Value
1	randDB[1]	3opt[first]	2059
2	random ILS		2679
3	randDB[1]	2.5opt[first]	4701
4	randDB[1]	2opt[first]	5588

Table 6.7:   Ranks of normal and random ILS algorithms in the TSP local search comparison experiment

A sub-experiment in single-instance mode running Q-ILS jobs using the three 2-opt, 2.5-opt, and 3-opt ILS-actions was conducted also. The jobs were run for instances d657 and dka1376. The aim is to investigate whether the superiority of the 3-opt ILS-actions can be learned automatically by Q-ILS algorithms. Table 6.8 shows the changed and additional actual parameter values used for the single-instance mode sub-experiment. The sub-experiment comprised runs of one random ILS job, three normal ILS jobs, and 196 Q-ILS jobs (48 training phase and 144 application phase

Q-ILS jobs). Note that some parameters such as the acceptance criterion can be set individual for the training and the application phase, so one training phase job can have several corresponding application phase jobs that work with the function approximator models learned by the training phase job (cf. Subsection 6.1.1). The results for the normal ILS jobs are contained Table 6.6. The results for the best performing Q-ILS jobs and the random ILS job and are shown in Table 6.9. This table in total would show 193 jobs for each instance.

Short	Actual Parameter Values
i_{tr}	d657, dka1376
i_a	d657, dka1376
$strat$	Q(0), Q(λ)
$pert$	randDB[1]
ls	3opt[first], 2.5opt[first], 2opt[first]
$feat$	TSP-Features-Experiments-1.Manual-Selection.feat
$init$	nn
acc_a	ϵ-better
ϵ_{a_t}	0.0, 0.5, 1.0
ϵ_{a_a}	0.0, 0.5, 1.0
γ	0.3, 0.8
r	c_Δ, c_i
λ	0.4, 0.8

Table 6.8:   Changed actual parameter values for Q-ILS jobs of the TSP local search procedure comparison experiment

For instance d657, the best Q-ILS jobs are better than the best normal ILS job (average excess ranging from 0.1474 % to 0.1632 % for the best Q-ILS jobs compared to an average excess of 0.1698 % for the best normal ILS job). The differences are not statistically significant, though (according to the two statistical tests employed, cf. Subsection 6.1.2). For instance dka1376, the best Q-ILS jobs also performed better than the best normal ILS job and at least the very best Q-ILS job shows significant better performance than the best normal ILS (at the 0.1 significance level). The random ILS jobs perform as expected for both instances: they are significantly worse than the best 3-opt ILS job according to the Wilcoxon test (on a significance level of 0.1) but far better than the 2-opt and 2.5-opt ILS jobs.

In order to have a look "inside" the working of Q-ILS jobs, action usage plots for the best Q-ILS job for each instance d657 and dka1376 are shown in subfigures (a) and (b) of Figure 6.5, respectively. They reveal that the "proper", i.e. 3-opt ILS-action, was applied exclusively except for during for the rather unimportant very first iterations. Hence, the Q-ILS algorithms correctly learned that only the 3-opt ILS-action should be used.

According to the action usage plots from Figure 6.5, the respective Q-ILS jobs practically behave the same as the best normal ILS jobs but perform better. Either the usage of other ILS-actions than the 3-op-ILS-action at the very beginning of the search makes a difference, or the performance differences rather are by chance and due to the randomization inherent to algorithms. The former possibility does not seem to apply.

Instance	Rank	Parameters Type	ϵ_{aa}	ϵ_{at}	γ	λ	strat	r	H_0: equal 0.05	0.1	H_0: not better 0.05	0.1	T	Statistics avg	σ	avg. exc.	min	max
d657 (48912)	1	apply	0.0	0.5	0.8		Q(0)	c_Δ	=	=	=	=	0.8831	48984.12	46.79	0.1474 %	48913	49100
	2	apply	0.0	0.5	0.8		Q(0)	c_i	=	=	=	=	0.5743	48989.24	30.28	0.1579 %	48938	49074
	3	apply	0.0	0.5	0.3	0.4	Q(λ)	c_i	=	=	=	=	0.5144	48989.64	33.49	0.1587 %	48913	49055
	4	apply	0.0	1.0	0.3	0.4	Q(λ)	c_i	=	=	=	=	0.4509	48990.20	35.32	0.1599 %	48941	49066
	5	apply	0.0	0.5	0.3	0.8	Q(λ)	c_Δ	=	=	=	=	0.4422	48990.56	30.54	0.1606 %	48944	49053
	6	apply	0.0	1.0	0.8		Q(0)	c_i	=	=	=	=	0.3276	48991.64	32.55	0.1628 %	48928	49059
	7	apply	0.0	0.5	0.8	0.4	Q(0)	c_Δ	=	=	=	=	0.3246	48991.84	28.22	0.1632 %	48928	49047
	...	...						...	...	...	...	...	...	...	...	...	...	...
	47	rand	0.0						=	=	=	=	-1.8804	49020.32	53.71	0.2215 %	48947	49183
	...	...						...	...	...	...	...	...	...	...	...	...	...
	65	apply	0.0	0.0	0.3		Q(0)	c_Δ	=	–	=	=	-6.4360	49120.64	88.81	0.4266 %	48975	49369
	66	apply	0.5	1.0	0.3		Q(0)	c_Δ	=	–	=	=	-17.6964	49271.92	66.98	0.7359 %	49081	49352
	...	...						...	...	...	...	...	...	...	...	...	...	...
dka1376 (4666)	1	apply	0.0	0.0	0.3		Q(0)	c_i	+	+	+	+	1.6370	4689.55	7.69	0.5047 %	4672	4701
	2	apply	0.0	0.5	0.8		Q(0)	c_i	=	=	=	=	1.2359	4690.68	7.69	0.5289 %	4680	4705
	3	apply	0.0	1.0	0.8		Q(0)	c_i	=	=	=	=	0.8686	4691.60	8.84	0.5486 %	4671	4709
	4	apply	0.0	0.5	0.3	0.4	Q(λ)	c_Δ	=	=	=	=	0.8603	4691.72	7.88	0.5512 %	4674	4704
	5	apply	0.0	1.0	0.3		Q(λ)	c_i	=	=	=	=	0.3048	4693.24	9.42	0.5838 %	4676	4710
	6	apply	0.0	0.5	0.3	0.4	Q(λ)	c_Δ	=	=	=	=	0.1635	4693.68	8.74	0.5932 %	4682	4717
	7	apply	0.0	0.0	0.8		Q(0)	c_i	=	=	=	=	0.1515	4693.72	8.47	0.5941 %	4680	4714
	...	...						...	...	...	...	...	...	...	...	...	...	...
	56	rand	0.0						=	–	=	=	-2.9002	4703.68	11.42	0.8075 %	4674	4727
	...	...						...	...	...	...	...	...	...	...	...	...	...
	65	apply	0.0	1.0	0.3	0.4	Q(λ)	c_Δ	=	–	=	=	-7.5513	4724.04	15.89	1.2439 %	4687	4760
	66	apply	0.5	1.0	0.8		Q(0)	c_i	=	–	=	=	-11.1218	4729.44	10.61	1.3596 %	4702	4747
	...	...						...	...	...	...	...	...	...	...	...	...	...

Table 6.9: Q-ILS job results in the TSP local search procedure comparison experiment

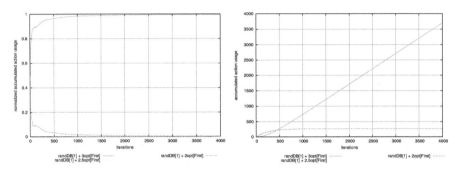

(a) Normalized accumulated action usage, instance d657

(b) Accumulated action usage, instance dka1376

Figure 6.5:   Action usage plots for the best Q-ILS jobs run on instances d657 and dka1376 in the TSP local search procedure comparison experiment

Reviewing Table 6.9 it strikes that the choice of the acceptance criterion in the application phase obviously has an enormous impact on the performance of Q-ILS jobs whereas the distribution of actual parameter values for the learning parameters does not show any trend. As can be seen in Table 6.9, the difference in performance is huge when the acceptance criterion for application phase Q-ILS jobs changes from a better acceptance criterion (indicated by a value of 0.0 in column "ϵ_{a_a}") to an ϵ-better one with $\epsilon = 0.5$ (indicated by a value of 0.5 in column "ϵ_{a_a}"). The changes for both instances occur from position 65 to 66. What cannot be seen from Table 6.9 is the fact that *all* Q-ILS jobs, irrespective whether they are training or application phase jobs, employing a better acceptance criterion are better than *all* Q-ILS jobs employing an ϵ-better acceptance criterion.

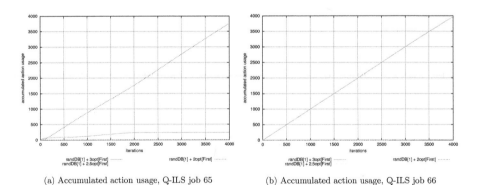

(a) Accumulated action usage, Q-ILS job 65

(b) Accumulated action usage, Q-ILS job 66

Figure 6.6:   Action usage plots for Q-ILS jobs 64 and 65 run on instance dka1376 in the TSP local search procedure comparison experiment

To assess the magnitude of the influence of the choice of the acceptance criterion, the action usage plots for the Q-ILS jobs 65 and 66 for instance dka1376 are investigated. According to subfigure (b) of Figure 6.6, Q-ILS 66 basically employs the same action selection strategy as the best Q-ILS job

1. The only parameter with effects during the application phase that has a different actual value assigned is the acceptance criterion. Hence, the difference in performance is due to the acceptance criterion which accordingly makes a substantial difference. Even Q-ILS job 65, which according to subfigure (b) of Figure 6.6 employs a totally wrong action selection strategy by always choosing the worst ILS-action possible (the 2-opt ILS-action) performs better than Q-ILS job 66. These results confirm the contemplations about acceptance criterion influence given in Subsection 4.5.2. The effects of the acceptance criterion choice entail that they potentially level out any other learning influences. The choice of the acceptance criterion is not as decisive for the training phase, though. Looking at the respective actual parameter values for the best Q-ILS jobs in Table 6.9 (column "ϵ_{a_t}"), it can be seen that these were trained with varying acceptance criteria for both instances.

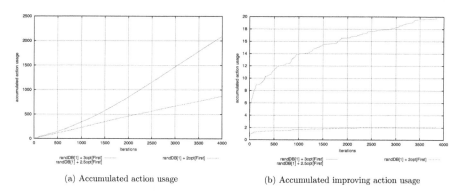

(a) Accumulated action usage   (b) Accumulated improving action usage

Figure 6.7:   Action usage plots for the corresponding training phase jobs for Q-ILS job 65 run on instance dka1376 in the TSP local search procedure comparison experiment

Concluding the analysis of the local search procedure comparison experiment for the TSP, the action usage of the corresponding training phase Q-ILS job of Q-ILS job 65 is shown in Figure 6.7. As can be seen from subfigure (a) of this figure showing the accumulated action usage during the training phase, the 3-opt ILS-action is used increasingly often towards the end of the search which is as expected. The usage of the other two LS-action is due to the ϵ-greedy policy employed with an ϵ-value of 0.3. These do not contribute substantially with improving moves the nearer to the end the search is, though. The improving moves are mostly due to the 3-opt ILS-action, especially towards the end (see accumulated improving action usage in subfigure (b) of Figure 6.7). The expectation is that the trend to use the 3-opt ILS-action is carried over to the corresponding application phase Q-ILS job 65 is not met. This puzzling phenomenon should be investigated further in future experiments and might indicate that there are some yet unknown factors which influence the transfer of a learned value function.

Perturbation Comparison

The purpose of the perturbation comparison experiment for the TSP is on the one hand to investigate the influence of the strength of a perturbation on the learning success and on the other hand to find and confirm good perturbation strengths which can be used in subsequent experiments. The investigation concentrated on varying the strength for the random double-bridge perturbation (cf. Subsection 6.4.1). Following the results of the local search procedure comparison experiment, for all experiments still to present only the 3-opt local search procedure was used to form ILS-actions.

Using the other inferior local search procedures in ILS-actions would certainly have disturbed experimental results. The ILS-actions built from the 3-opt local search procedure and the random double-bridge perturbations in different strengths g ($g \in \mathbb{N}^+$) are also named *strength g ILS-action* or *ILS-action with strength g*. The corresponding (normal) ILS jobs and algorithms for the resulting ILS-actions are called *strength g (normal) ILS job* or *algorithm*, or *(normal) ILS job* or *algorithm with strength g*. In multi-instance summary tables, strength g ILS jobs are denoted by "ILS(g)". The random ILS jobs are characterized by "ILS(R)".

Short	Actual Parameter Values
i_a	rbx711, rl1304, vm1748, d1655, bck2217
pert	randDB[1], randDB[4], randDB[7], randDB[10], randDB[15]
ls	3opt[first]
init	nn
acc_a	better

Table 6.10:   Actual parameter values for normal ILS jobs of the TSP perturbation comparison experiment

Short	Actual Parameter Values
i_{tr}	rbx711, rl1304
i_a	rbx711, rl1304
strat	Q(0), Q(λ)
pert	randDB[1], randDB[4], randDB[7]
ls	3opt[first]
feat	TSP-Features-Experiments-1.Manual-Selection.feat
init	nn
ϵ_{a_t}	0.0, 0.5, 1.0
ϵ_{a_a}	0.0, 0.5, 1.0
γ	0.3, 0.8
r	c_Δ, c_i
λ	0.4, 0.8

Table 6.11:   Actual parameter values for the single-instance mode sub-experiment of the TSP perturbation comparison experiment

The perturbation comparison experiment for the TSP is divided into three sub-experiments. The first one run and compares normal ILS jobs, the second is in single-instance mode, the third in multi-instance mode. The actual parameter values for the normal ILS sub-experiment are shown in Table 6.10.

The results for the five instances are shown in Table 6.12. The general trend discernible is that lower perturbation strengths have the tendency to perform better, although there are exceptions to this rule (for example instance rl1304). The second recognizable trend is that the best strengths for normal ILS job seem to be 4, 1, and 7, in this order. The differences in average excess between the normal ILS jobs with these three perturbation strengths to the remaining ones typically is substantially greater than the differences among these normal ILS jobs for the instances. As a consequence, the further single- and multi-instance mode sub-experiments for the perturbation comparison experiment were conducted using perturbation strengths of 1, 4, and 7.

The actual parameter values for the single instance mode sub-experiment are shown in Table 6.11. The results of the normal ILS jobs for the single instance mode sub-experiment are contained in Table 6.12, the results for the Q-ILS jobs are shown in Table 6.13 (193 entries in full table).

Instance	Perturbation	LS	Statistics				
			avg	σ	avg.excess	min	max
rbx711	randDB[4]	3opt[first]	3120.84	4.70	0.1875 %	3116	3132
(3115)	randDB[1]	3opt[first]	3121.84	5.80	0.2196 %	3115	3140
	randDB[7]	3opt[first]	3130.24	6.65	0.4892 %	3121	3142
	randDB[10]	3opt[first]	3142.36	7.18	0.8783 %	3129	3156
	randDB[15]	3opt[first]	3156.00	5.87	1.3162 %	3140	3166
rl1304	randDB[7]	3opt[first]	253412.60	349.34	0.1837 %	252948	254344
(252948)	randDB[10]	3opt[first]	253567.28	447.66	0.2448 %	252959	254736
	randDB[4]	3opt[first]	253585.28	267.97	0.2519 %	253296	254203
	randDB[1]	3opt[first]	254164.40	841.55	0.4809 %	253361	256032
	randDB[15]	3opt[first]	254318.76	448.05	0.5419 %	253513	255108
d1655	randDB[4]	3opt[first]	62449.88	170.74	0.5181 %	62255	62886
(62128)	randDB[1]	3opt[first]	62458.32	189.97	0.5317 %	62181	62947
	randDB[7]	3opt[first]	62582.12	274.81	0.7309 %	62186	63365
	randDB[10]	3opt[first]	62655.96	232.75	0.8498 %	62350	63168
	randDB[15]	3opt[first]	62790.84	145.94	1.0669 %	62526	63121
vm1748	randDB[4]	3opt[first]	337628.96	374.34	0.3188 %	337079	338289
(336556)	randDB[1]	3opt[first]	337844.44	395.17	0.3828 %	336927	338571
	randDB[7]	3opt[first]	337975.88	431.34	0.4219 %	337197	338716
	randDB[10]	3opt[first]	338981.64	581.65	0.7207 %	338089	340301
	randDB[15]	3opt[first]	340615.92	545.54	1.2063 %	339638	342063
bck2217	randDB[4]	3opt[first]	6806.96	13.93	0.6351 %	6784	6835
(6764)	randDB[1]	3opt[first]	6819.12	14.68	0.8149 %	6791	6837
	randDB[7]	3opt[first]	6822.68	14.50	0.8675 %	6790	6860
	randDB[10]	3opt[first]	6845.20	17.19	1.2005 %	6813	6878
	randDB[15]	3opt[first]	6886.48	11.94	1.8108 %	6865	6914

Table 6.12: Normal ILS job results for the TSP perturbation comparison experiment

According to the t-test, the best Q-ILS jobs are significantly better than the best normal ILS job for instance rbx711 but not for instance rl1304. In case of instance rl1304, only the very best Q-ILS job 1 is better than the best normal ILS job, all other Q-ILS job shown in Table 6.13 are only better then the second best normal ILS job. The Wilcoxon test does not show any significant differences in nether direction for the best Q-ILS jobs, though. Note that that the Wilcoxon test is a ranking-based statistical test (cf. Subsection 6.1.2). If too many samples are identical, for example when the globally optimal costs have been reached often as seems to be the case here (compare cost of global optimum with values from column "min"), the Wilcoxon test statistic will not be adequate and can be discarded for jobs with results too close to the best normal ILS job. The t-test, in contrast, is based on the averaged best costs (cf. Subsection 6.1.2) and works.

Instance	Job	Type	Parameters						H_0: equal		H_0: not better		T	Statistics				
			ϵ_{a_a}	ϵ_{a_1}	γ	λ	strat	r	0.05	0.1	0.05	0.1		avg	σ	avg. exc.	min	max
rbx711 (3115)	1	apply	0.0	0.5	0.3	0.4	Q(λ)	c_i	=	=	+	+	2.1284	3118.72	4.16	0.1194 %	3115	3133
	2	apply	0.0	1.0	0.3	0.4	Q(λ)	c_i	=	=	+	+	2.1554	3118.88	3.32	0.1246 %	3115	3126
	3	apply	0.0	0.0	0.8	0.8	Q(λ)	c_i	=	=	+	+	2.1086	3118.96	3.23	0.1271 %	3115	3127
	4	apply	0.0	1.0	0.3	0.8	Q(λ)	c_Δ	=	=	+	+	1.5952	3119.52	3.90	0.1451 %	3115	3129
	5	apply	0.0	0.5	0.3		Q(0)	c_i	=	=	+	+	1.5703	3119.56	3.86	0.1464 %	3115	3130
	6	apply	0.0	0.0	0.8		Q(0)	c_i	=	=	+	+	1.6226	3119.56	3.45	0.1464 %	3115	3129
	7	apply	0.0	1.0	0.8		Q(0)	c_i	=	=	+	+	1.5260	3119.64	3.75	0.1490 %	3115	3130
	...												...	...	...	...	...	...
	50	rand	0.0						=	=	=		−0.4837	3122.44	4.80	0.2388 %	3116	3136
	...												...	...	...	...	...	...
	65	apply	0.0	0.5	0.3		Q(0)	c_i	−	=	=		−5.6891	3131.40	6.32	0.5265 %	3119	3144
	66	apply	0.5	1.0	0.3	0.8	Q(λ)	c_Δ	−	−	=		−12.3610	3142.00	5.91	0.8668 %	3131	3152
	...												...	...	...	...	...	...
r1304 (252948)	1	apply	0.0	1.0	0.8		Q(0)	c_i	=	=	=	=	0.0042	253412.24	251.60	0.1835 %	252948	253900
	2	apply	0.0	0.0	0.8	0.4	Q(λ)	c_Δ	=	=	=	−	−0.3359	253443.60	301.53	0.1959 %	252948	254203
	3	apply	0.0	0.5	0.3		Q(0)	c_i	=	=	=	=	−0.4281	253464.12	489.99	0.2040 %	252948	255131
	4	apply	0.0	0.5	0.8		Q(0)	c_Δ	=	=	=	=	−0.5219	253465.44	366.44	0.2046 %	252948	254288
	5	apply	0.0	0.0	0.8		Q(0)	c_Δ	=	=	=	−	−0.6163	253470.08	308.86	0.2064 %	252948	254050
	6	apply	0.0	1.0	0.3		Q(0)	c_i	=	=	=	=	−0.7557	253480.56	283.11	0.2105 %	252948	253900
	7	apply	0.0	0.0	0.3	0.8	Q(λ)	c_Δ	=	=	=	=	−1.1185	253524.08	355.38	0.2277 %	252948	254203
	...												...	...	...	...	...	...
	49	rand	0.0						−	=	=		−2.4218	253872.28	882.42	0.3654 %	252948	257131
	...												...	...	...	...	...	...
	65	apply	0.0	0.0	0.3		Q(0)	c_Δ	−	=	=		−4.7875	254354.20	919.25	0.5559 %	252948	256011
	66	apply	0.5	1.0	0.3	0.8	Q(λ)	c_Δ	−	=	=		−7.9646	254362.24	483.09	0.5591 %	253550	255333
	...												...	...	...	...	...	...

Table 6.13: Q-ILS job results for the single-instance mode sub-experiment of the TSP perturbation comparison experiment

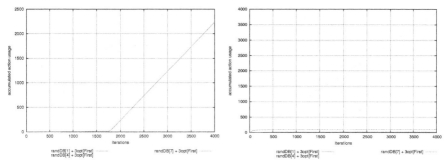

(a) Accumulated action usage, instance rbx711   (b) Accumulated action usage, instance rll1304

Figure 6.8:   Action usage plots for the best Q-ILS jobs run on instances rbx711 and rll1304 in the
TSP perturbation comparison experiment

Short	Actual Parameter Values
tc_{te}	6000
i_{tr}	rbx711:rll1304:vm1748
i_a	rbx711:rll1304:vm1748
i_{te}	d657, lim963, dka1376, u1817
$strat$	Q(0), Q(λ)
$pert$	randDB[1], randDB[4], randDB[7]
ls	3opt[first]
$feat$	TSP-Features-Experiments-1.Manual-Selection.feat
$init$	nn
ϵ_{a_t}	0.0, 0.5, 1.0
ϵ_{a_a}	0.0, 0.5, 1.0
γ	0.3, 0.8
r	c_Δ, c_i
λ	0.4, 0.8

Table 6.14:   Actual parameter values for the multi-instance mode sub-experiment of the TSP
perturbation comparison experiment

As regards the learning parameters, a systematic trend towards a good performing actual parameter
setting is not visible at once while the choice of the acceptance criterion is most important again
(see changes from Q-ILS jobs 65 to 66 for each instance and consider the fact that all Q-ILS jobs
employing the better acceptance criterion are better than those that do not, regardless whether
they are training or application phase jobs). The random ILS jobs do not perform very well, for
instance rll1304 even significantly worse than the best normal ILS job according to the Wilcoxon.
This is an indication that some regularity in the action selection strategy is better than none.

Looking at action usage plots for the best Q-ILS job for instances rbx911 and rll1304 in Figure
6.8 one can see for instance rbx711 (subfigure (a)) that the best Q-ILS job for approximately

the first 2000 iterations always applied the strength 4 ILS-actions (whose corresponding normal ILS-action performed best) and thereafter always employed the ILS-action with strength 1 (whose corresponding normal ILS performed second best). Since the best Q-ILS job performed significantly better, this learned action selection strategy seems to be better for instance rbx711 than the only use of one single ILS-action and hence learning made a difference. For instance rl1304, the best Q-ILS job basically always selected the strength 4 ILS-action (see subfigure (b) of Figure 6.8) whose corresponding normal ILS job performed third best. The differences in average excess of the normal ILS job with strength 4 to the best one with strength 7 is huge, so it comes as a surprise that basically the same job run for a second time all of a sudden yields a substantially better result (and this averaged over 30 tries)! This is an indication that the algorithm inherent randomization should not be underestimated. Perhaps, instance rl1304 simply is a very special case. This also would explain the results for the normal ILS jobs from Table 6.12 also where strengths 7 ILS-action is best in contrast to all other instances shown there.

Rank	Perturbation	LS	Value
1	random ILS		2862
2	randDB[4]	3opt[first]	3282
3	randDB[1]	3opt[first]	3903
4	randDB[7]	3opt[first]	4616

Table 6.15:   Ranks of normal and random ILS algorithms in the multi-instance mode sub-experiment of the TSP perturbation comparison experiment

Rank	Parameters						Value
	r	γ	λ	ϵ_{a_t}	ϵ_{a_a}	strat	
1	c_Δ	0.3		1.0	0.0	Q(0)	29293
2	c_i	0.8	0.4	0.5	0.0	Q(λ)	30118
3	c_i	0.3	0.8	1.0	0.0	Q(λ)	34543
4	c_i	0.3		1.0	0.0	Q(0)	34574
5	c_Δ	0.3		0.5	0.0	Q(0)	36926
6	c_Δ	0.3	0.8	1.0	0.0	Q(λ)	37398
7	c_i	0.3		1.0	0.0	Q(0)	38617
	⋮	⋮	⋮	⋮	⋮	⋮	⋮

Table 6.16:   Ranks of Q-ILS algorithms in the multi-instance mode sub-experiment of the TSP perturbation comparison experiment

To finalize the perturbation comparison experiment for the TSP, a multi-instance mode sub-experiment was run. The actual parameter values are shown in Table 6.14. The ranking over the resulting normal and random ILS algorithms is shown in Table 6.15. Surprisingly, the random ILS algorithm is ranked best over all three instances involved. One possible explanation is that different ILS-actions are best suited for different instances. Another explanation is that using different ILS-actions during a search simply is a good action selection strategy. The second explanation is challenged by the rather bad performance of the random ILS jobs during the single-instance mode sub-experiment, though. The ranks for the other normal ILS algorithms confirm the rankings for the normal ILS jobs runs found for most instances during the local search procedure comparison experiment.

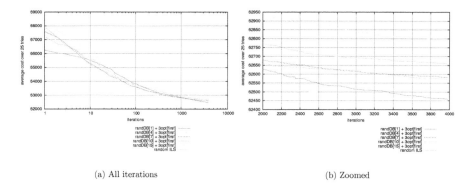

(a) All iterations   (b) Zoomed

Figure 6.9:   Runtime development plots for normal ILS jobs run on instance d1655 in the TSP perturbation comparison experiment

The ranking over the parameterized Q-ILS algorithms is shown for the top ranks in Table 6.16. The full table would contain results for 144 Q-ILS jobs. As regards the learning parameters, no regularities among actual parameter values indicating good (learning) performance are discernible at once, only the acceptance criterion influence is decisive again.

For the test phase, all normal and random ILS algorithms and the four best ranked parameterizations for the Q-ILS algorithm for each combination of actual parameter values for parameters learning strategy and reward signal (ranks 1, 2, 4, and 6) were run on the test instances. Note that they will be later referred to as *rank x Q-ILS job* or *algorithm* or *Q-ILS job* or *algorithm ranked x* ($x \in \mathbb{N}^+$) for short (the saying can also be read as "the formerly ranked x algorithm" or "job with underlying algorithm formerly ranked x").

As can be seen from Table 6.14 showing the actual parameter values, the test phase jobs are run for 6000 iteration. The reason to do so is to ensure a proper stabilization of the search process. Recall that the primary goal is to find the best solution possible (cf. Subsection 6.1.1). The goal of the experimentation here is to find the differences in the principle working of algorithms. Some jobs need more runtime than others until they do not find improvements anymore due to their general inability to do so. It would be distorting results, if runtime were too short, such that failing to find further improvements is because of inappropriately low runtime only. Runtime development plots such as the ones from Figure 6.9 for the normal ILS jobs run on instance d1615 during the perturbation comparison experiment for 4000 iterations can be used to estimated the asymptotic behavior of jobs. As can be seen from both subfigures, the asymptotic behavior has not stabilized yet, the lines are not "saturated" yet which is an indication that more runtime would probably lead to better solutions. The tendency is that allowing for too short runtime is more likely for larger and typically more difficult to solve instances.

Therefore, all test phase jobs (which tend to be larger than the d1655 one) were run for 6000 iterations for all TSP experiments. Note that a too long runtime has its dangers, too, though. Too long runtimes might blur differences between the performance of jobs, because in each try they too often come too close to the globally optimal cost and any outliers due to algorithm inherent randomization become most influential as regards averaged performance. This is indicated for example by a very small average excess and the fact that the minimum over the best costs (see column "min") for many jobs is equal to the globally optimal cost. In this case, however, the instance simply was not difficult and hence useful enough.

Instance	Rank	Type	H_0: equal		H_0: not better			Statistics				
			0.05	0.1	0.05	0.1	T	avg	σ	avg. exc.	min	max
d657	3	ILS(1)						48975.18	32.45	0.1292 %	48913	49051
(48912)	1	Q-ILS	=	=	=	=	0.7571	48981.35	40.10	0.1418 %	48915	49074
	4	Q-ILS	=	=	=	=	0.8097	48982.68	48.77	0.1445 %	48913	49097
	6	Q-ILS	=	–	=	–	1.6482	48987.95	36.74	0.1553 %	48913	49074
	1	ILS(R)	–	–	–	–	2.9403	48999.40	40.77	0.1787 %	48916	49088
	2	ILS(4)	–	–	–	–	4.0418	49019.43	61.17	0.2196 %	48938	49210
	2	Q-ILS	–	–	–	–	9.8091	49069.82	51.68	0.3227 %	48972	49195
	4	ILS(7)	–	–	–	–	17.7313	49161.28	57.91	0.5096 %	49072	49313
lim963	4	Q-ILS						2799.95	5.35	0.3926 %	2790	2812
(2789)	2	ILS(4)	=	=	=	=	0.2621	2800.25	4.88	0.4034 %	2790	2808
	1	ILS(R)	=	=	=	=	0.3841	2800.43	5.71	0.4096 %	2789	2809
	2	Q-ILS	=	=	=	=	0.3971	2800.45	5.90	0.4105 %	2789	2811
	1	Q-ILS	=	=	=	=	1.1235	2801.12	3.89	0.4347 %	2790	2808
	3	ILS(1)	=	–	=	–	1.4572	2801.62	4.92	0.4527 %	2790	2811
	6	Q-ILS	–	–	–	–	2.3769	2802.45	3.95	0.4823 %	2790	2810
	4	ILS(7)	–	–	–	–	2.5913	2803.15	5.69	0.5074 %	2791	2815
dka1376	4	Q-ILS						4687.15	7.32	0.4533 %	4669	4698
(4666)	2	Q-ILS	=	=	=	=	0.7164	4688.52	9.68	0.4827 %	4667	4707
	2	ILS(4)	=	=	=	=	1.0256	4689.07	9.35	0.4945 %	4667	4710
	6	Q-ILS	=	=	=	=	1.0581	4689.10	9.07	0.4951 %	4670	4710
	1	Q-ILS	=	=	=	=	1.0306	4689.38	11.53	0.5010 %	4669	4718
	3	ILS(1)	–	–	–	–	2.3570	4690.90	6.91	0.5336 %	4679	4706
	1	ILS(R)	=	–	–	–	1.9927	4690.95	9.59	0.5347 %	4669	4710
	4	ILS(7)	–	–	–	–	2.2536	4692.05	11.64	0.5583 %	4670	4714
u1817	1	Q-ILS						57474.72	81.69	0.4785 %	57308	57679
(57201)	2	ILS(4)	=	=	=	=	0.6181	57486.43	87.50	0.4990 %	57340	57725
	6	Q-ILS	=	=	=	=	1.0098	57496.38	108.24	0.5164 %	57310	57922
	1	ILS(R)	=	=	=	=	1.1609	57498.90	103.31	0.5208 %	57305	57814
	3	ILS(1)	=	=	=	=	1.1534	57499.97	111.79	0.5227 %	57319	57757
	4	Q-ILS	=	=	=	=	1.0861	57500.25	124.17	0.5232 %	57332	57869
	2	Q-ILS	=	=	=	=	1.1603	57500.60	114.98	0.5238 %	57344	57869
	4	ILS(7)	–	–	–	–	3.3301	57558.78	137.14	0.6255 %	57314	58072

Table 6.17: Results of test phase jobs in the multi-instance mode sub-experiment of the TSP perturbation comparison experiment

The results of the test phase jobs for the multi-instance mode sub-experiment of the perturbation comparison experiment for the TSP are presented in Table 6.17. The application phase ranking for the algorithms is only confirmed for the worst ranked strength 7 ILS algorithm (which was formerly performing best for instance rl1304, though): The strength 7 ILS jobs are always significantly the worst ones for each instance. Otherwise, on first sight, the test jobs are mixed rather arbitrarily instead of repeating any ranking trend. Only for test instances d657 and lim963, a single Q-ILS job (rank 2 and 6 Q-ILS job, respectively) is performing significantly worse than the respective best ILS or Q-ILS job while for three of the test instances a Q-ILS job is performing best, two times the one ranked 4. The random ILS jobs perform quite decent, but summarizing over all test instances they do not perform better than the strength 4 ILS job contradicting the results from the ranking for the application phase. The general tendency is that Q-ILS jobs perform better than normal ILS jobs.

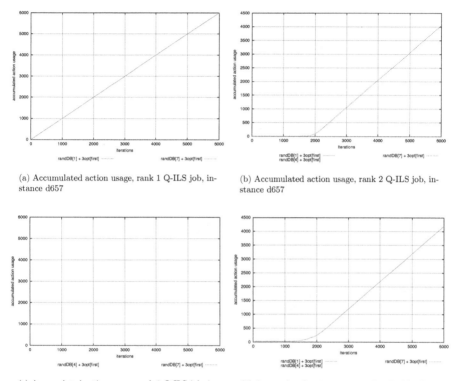

(a) Accumulated action usage, rank 1 Q-ILS job, instance d657

(b) Accumulated action usage, rank 2 Q-ILS job, instance d657

(c) Accumulated action usage, rank 1 Q-ILS job, instance u1817

(d) Accumulated action usage, rank 2 Q-ILS job, instance u1817

Figure 6.10:   Action usage plots for rank 1 and 2 Q-ILS algorithms run on test instances d657 and u1817 in the multi-instance mode sub-experiment of the TSP perturbation comparison experiment

Figures 6.10 and 6.11 show action usage plots for the rank 1, 2, and 4 Q-ILS jobs for the four test instances. For instance d657, the best normal ILS job is of strength 1. The action usage of the best Q-ILS job (rank 1 Q-ILS job) for this instance also exclusively uses the strength 1 ILS-action (see subfigure (a) of Figure 6.10) whereas the worst performing Q-ILS job (rank 2 Q-ILS job) in the beginning (first 2000 iterations) exclusively employs the ILS-action with strength 4 all of a sudden switches to exclusively using strength 7 ILS-action thereafter. The worst performing normal ILS job for instance d657 uses the strength 7 ILS-action, so the poor performance of the rank 2 Q-ILS job on this instance does not come as a surprise. Whereas the learned and transfered action selection strategy of rank 1 Q-ILS job is suitable for instance d657, the one for rank 2 Q-ILS jobs unfortunately is not. For instance u1817, the rank 1 and 2 Q-ILS jobs also performed best and worst, respectively. Looking at their action usage plots for instance u1817 in subfigure (c) and (d) of Figure 6.10, respectively, it strikes that rank 1 Q-ILS job all of a sudden only applies

the strength 4 ILS-action whereas the rank 2 Q-ILS job again changes exclusive action usage at
approximately iteration 2000 from ILS action with strength 4 to another one, now, however to the
one with strength 1.

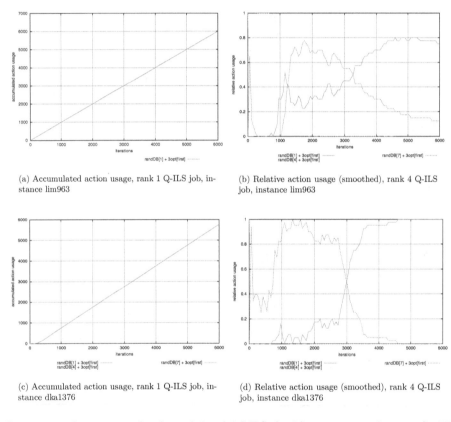

(a) Accumulated action usage, rank 1 Q-ILS job, instance lim963

(b) Relative action usage (smoothed), rank 4 Q-ILS
job, instance lim963

(c) Accumulated action usage, rank 1 Q-ILS job, instance dka1376

(d) Relative action usage (smoothed), rank 4 Q-ILS
job, instance dka1376

Figure 6.11:   Action usage plots for rank 1 and 4 Q-ILS algorithms run on test instances lim963
and dka1376 in the multi-instance mode sub-experiment of the TSP perturbation comparison experiment

For the rank 1 Q-ILS job, this switch in action usage between the instances apparently is very
useful, since the strength 4 ILS jobs is performing best for instance u1817 now and amounts to a
nice learning result: Transfer of what was learned happened. Noticeable is that the change as for
rank2 Q-ILS job obviously is instance specific, although no instance features were used. Since the
common coverage of some state features for the TSP is below zero (cf. Subsection 6.4.2 and Table
6.4) meaning that different instances will produce disjunct sets of feature values for these features,
these features can serve as instance indicators. This makes the results not overly surprising in
principle, but that it actually happened is remarkable. Obviously, some state or perhaps heuristic
features can also serve as instance dependent characteristic.

Looking at the action usage plots for rank 1 Q-ILS job for instances lim963 and dka1376 in subfigures

(a) and (c) of Figure 6.11, one can see that for those instances the strength 1 ILS-action is used exclusively again. For both instances, the strength 4 ILS job performed best among the normal ILS jobs, so using the strength 4 ILS-action, as done at the very beginning for instance dka1376 (see subfigure (c) of Figure 6.11) perhaps would have been better. Why this did not happen has to be examined further. Proper learning takes place but not under arbitrary conditions, so further influences have to be identified and investigated.

For each instance lim963 and dka1376 rank 4 Q-ILS is performing best. According to the respective action usage plots for rank 4 Q-ILS job in subfigures (b) and (d) of Figure 6.11, changing the action usage over time might be better than exclusively using only one for these instances, even if this the best performing ILS-action (according to the performance of the corresponding normal ILS job; for both instances strength 4 ILS-action). Subfigures (b) and (d) of Figure 6.11 show that the learned action selection strategy for rank 4 Q-ILS jobs is stable across instances and hence that a stable and complex action selection strategy has been learned. Consequently, even if the performance varies for different instances, transfer of what was learned is possible. Transfer is also possible independent of the instance size.

Short	Actual Parameter Values
i_{tr}	pbd984, nrw1379, rl1889, dcb2086, pds2566
i_a	pbd984, nrw1379, rl1889, dcb2086, pds2566
$strat$	Q(0), Q(λ)
$pert$	randWalk[100], randGreedy[100], randDB[1], randDB[2], randDB[4], randDB[7]
ls	3opt[first]
$feat$	TSP-Features-Experiments-1.Manual-Selection.feat
$init$	nn
ϵ_{a_t}	0.0, 0.3
ϵ_{π_a}	0.0, 0.3
γ	0.3, 0.8
r	c_Δ, c_i
λ	0.4, 0.8

Table 6.18: Actual parameter values for the TSP single-instance experiment

Second Part of Multi-Instance

In order to check the performance of greater strength ILS-actions, a second part of the perturbation comparison experiment for the TSP with single- and multi-instance mode sub-experiments was conducted. There, the strengths of the ILS-actions are 4, 7, 10, and 15 and only the better acceptance criterion or small ϵ-values for the ϵ-better acceptance criterion were used. The actual parameter values and results are shown in Appendix B.2. Summarizing, the influence of the choice of the acceptance criterion is strong again while ILS-actions with a strength greater than 7 do not seem to be of much use. The extraordinary nature of instance rl1304 is confirmed. It is the only instance where the results became substantially better for the stronger perturbations (see Table A.3). Otherwise, the superiority of strength 4 ILS-action over the strength 7 one is confirmed and the best test phase jobs are always Q-ILS ones, often significantly better than most other jobs. For inferior ILS-actions, learning perhaps can make more of a difference compared to normal ILS jobs.

Single-Instance

The single-instance and the subsequently presented multi-instance experiments vary several different perturbations and different perturbation strengths with the 3-opt local search procedure. The variation is over the best perturbations and parameterizations found. Three perturbations are used, the random double-bridge, the random walk, and the random greedy perturbation according to the results from the previous experiment and according to the results from [Kor04]. Since single-instance experiments are not as relevant for practice as multi-instance experiments – transfer of a learned value functions is ultimately aimed at (also to reduce computation overhead) – the single-instance mode experiment is analyzed only briefly.

Instance	Perturbation	LS	Statistics				
			avg	σ	avg.excess	min	max
pbd984	randWalk[100]	3opt[first]	2804.84	6.76	0.2803 %	2797	2818
(2797)	randDB[4]	3opt[first]	2805.32	7.35	0.2975 %	2797	2820
	randDB[2]	3opt[first]	2806.48	6.60	0.3389 %	2797	2815
	randDB[7]	3opt[first]	2807.32	4.68	0.3690 %	2800	2818
	randDB[1]	3opt[first]	2808.44	6.23	0.4090 %	2797	2815
	randGreedy[100]	3opt[first]	2810.76	4.82	0.4920 %	2801	2823
nrw1379	randDB[1]	3opt[first]	56811.72	54.88	0.3067 %	56711	56908
(56638)	randDB[2]	3opt[first]	56843.28	41.48	0.3624 %	56767	56920
	randGreedy[100]	3opt[first]	56899.04	50.72	0.4609 %	56773	56997
	randWalk[100]	3opt[first]	56933.80	63.34	0.5223 %	56761	57019
	randDB[4]	3opt[first]	57000.36	59.94	0.6398 %	56894	57145
	randDB[7]	3opt[first]	57244.40	72.25	1.0707 %	57079	57390
rll889	randDB[2]	3opt[first]	317829.32	562.46	0.4086 %	317166	318913
(316536)	randDB[4]	3opt[first]	318044.60	816.71	0.4766 %	316694	318860
	randDB[7]	3opt[first]	318051.72	703.39	0.4788 %	317067	319265
	randWalk[100]	3opt[first]	318487.40	1392.02	0.6165 %	316638	322345
	randDB[1]	3opt[first]	318707.76	1301.30	0.6861 %	316662	320941
	randGreedy[100]	3opt[first]	319595.00	1317.96	0.9664 %	316842	321362
dcb2086	randDB[4]	3opt[first]	6637.12	16.79	0.5624 %	6604	6674
(6600)	randWalk[100]	3opt[first]	6642.60	14.61	0.6455 %	6609	6674
	randDB[2]	3opt[first]	6647.16	17.74	0.7145 %	6610	6676
	randGreedy[100]	3opt[first]	6654.68	18.27	0.8285 %	6623	6683
	randDB[7]	3opt[first]	6656.32	16.19	0.8533 %	6631	6684
	randDB[1]	3opt[first]	6660.60	16.15	0.9182 %	6629	6695
pds2566	randWalk[100]	3opt[first]	7681.32	8.98	0.5014 %	7658	7698
(7643)	randDB[2]	3opt[first]	7688.92	11.62	0.6008 %	7666	7711
	randGreedy[100]	3opt[first]	7690.60	12.56	0.6228 %	7672	7715
	randDB[1]	3opt[first]	7690.76	11.64	0.6249 %	7667	7715
	randDB[4]	3opt[first]	7693.80	13.90	0.6647 %	7669	7716
	randDB[7]	3opt[first]	7701.44	12.47	0.7646 %	7675	7720

Table 6.19: Normal ILS job results for the TSP single-instance experiment

Instance	Rank	Type	Parameters						H_0: equal		H_0: not better		T	Statistics				
			ϵ_{π_a}	ϵ_{a_t}	γ	λ	strat	r	0.05	0.1	0.05	0.1		avg	σ	avg. exc.	min	max
pbd984	1	apply	0.3	0.3	0.8	0.4	$Q(\lambda)$	c_i	==	+	+	+	2.4317	2800.92	4.40	0.1402 %	2797	2816
(2797)	2	apply	0.0	0.0	0.8	0.8	$Q(\lambda)$	c_Δ	==	+	+	+	2.3010	2801.16	4.28	0.1487 %	2797	2811
	3	apply	0.0	0.3	0.3		$Q(0)$	c_Δ	==	==	+	+	1.9273	2801.48	5.51	0.1602 %	2797	2814
	...	...	...	...	...	...	...	...	...	...	...	...	...	...	...	...	...	...
	33	rand							==	==	==	==	0.0624	2804.72	6.83	0.2760 %	2797	2816
nrw1379	1	apply	0.0	0.3	0.8		$Q(0)$	c_i	==	==	==	==	0.9130	56799.56	37.73	0.2853 %	56736	56869
(56638)	2	apply	0.0	0.3	0.3	0.4	$Q(\lambda)$	c_Δ	==	==	==	==	0.7134	56801.36	47.55	0.2884 %	56720	56911
	3	apply	0.0	0.3	0.8		$Q(0)$	c_Δ	==	==	==	==	0.6115	56803.56	37.96	0.2923 %	56711	56902
	...	...	...	...	...	...	...	...	...	...	...	...	...	...	...	...	...	...
	23	rand							==	−	==	==	-2.0325	56842.08	50.66	0.3603 %	56741	56950
r11889	1	apply	0.3	0.0	0.8	0.8	$Q(\lambda)$	c_Δ	==	+	+	+	1.7834	317556.52	518.29	0.3224 %	316852	319998
(316536)	2	apply	0.3	0.0	0.8	0.8	$Q(\lambda)$	c_i	==	==	+	+	1.4800	317584.92	604.50	0.3314 %	316707	319209
	3	train		0.0	0.8	0.4	$Q(\lambda)$	c_i	==	==	+	+	1.3464	317598.80	645.38	0.3358 %	316662	319480
	...	...	...	...	...	...	...	...	...	...	...	...	...	...	...	...	...	...
	24	rand							==	==	==	==	0.2823	317784.04	571.61	0.3943 %	316837	319758
dch2086	1	apply	0.0	0.3	0.3	0.8	$Q(\lambda)$	c_Δ	==	+	+	+	1.7779	6629.28	14.29	0.4436 %	6604	6662
(6600)	2	apply	0.3	0.0	0.3	0.8	$Q(\lambda)$	c_Δ	==	==	==	==	0.2434	6635.96	16.92	0.5448 %	6608	6671
	3	apply	0.0	0.3	0.3		$Q(0)$	c_i	==	==	==	==	0.2357	6636.08	14.31	0.5467 %	6611	6663
	...	...	...	...	...	...	...	...	...	...	...	...	...	...	...	...	...	...
	25	rand							==	==	==	==	-1.0932	6641.64	12.06	0.6309 %	6615	6667
pds2566	1	train		0.0	0.8		$Q(0)$	c_i	==	==	==	==	0.9478	7678.50	11.86	0.4645 %	7659	7702
(7643)	2	apply	0.0	0.3	0.3	0.4	$Q(\lambda)$	c_Δ	==	==	==	==	0.4304	7680.04	11.85	0.4846 %	7661	7705
	3	apply	0.0	0.3	0.8	0.4	$Q(\lambda)$	c_Δ	==	==	==	==	0.4137	7680.08	11.99	0.4851 %	7659	7718
	4	apply	0.0	0.0	0.8		$Q(0)$	c_i	==	==	==	==	0.3833	7680.32	9.45	0.4883 %	7668	7699
	...	...	...	...	...	...	...	...	...	...	...	...	...	...	...	...	...	...
	52	rand							==	−	==	==	-1.6897	7686.08	10.85	0.5637 %	7663	7703

Table 6.20: Q-ILS job results for the TSP single-instance experiment

The actual parameter values for the single-instance experiment are shown in Table 6.18. Due to the influence of the acceptance criterion on performance discovered so far, only the better acceptance criterion is used in the application phase. The ILS-actions comprise the strength 1, 2, 4, and 7 ILS-actions which are called and denoted as before. Strength 2 ILS-action was inserted to obtain a better graduation in the low strength band. The other three ILS-actions which are the best performing ones or the best performing ones with a stronger perturbation strength. Two additional ILS-actions including of the two best performing random walk and random greedy perturbations found in [Kor04] were used. The resulting ILS-actions and corresponding normal ILS jobs are called *random walk* and *random greedy ILS-actions* and *jobs* and are denoted by "ILS(W)" and "ILS(G)", respectively, in tables. Also, the actual parameter values for the policy in the application phase is extended by an ϵ-greedy policy with an ϵ-value of 0.3 to check for influences of a small randomization in the learned action selection strategies.

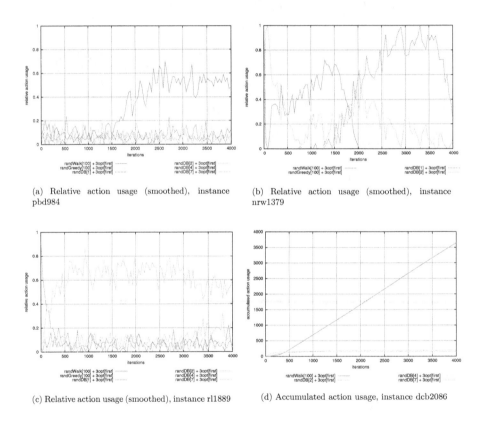

(a) Relative action usage (smoothed), instance pbd984

(b) Relative action usage (smoothed), instance nrw1379

(c) Relative action usage (smoothed), instance rl1889

(d) Accumulated action usage, instance dcb2086

Figure 6.12: Action usage plots for best Q-ILS jobs run on instances pbd984, nrw1379, rl1889, and dcb2086 in the TSP single-instance

The results for the normal ILS jobs are shown in Table 6.19, the ones for the Q-ILS jobs in Table 6.20. Six normal ILS, one random ILS, 32 training phase, and 64 application phase Q-ILS jobs were run. The full Table 6.20 hence would comprise 97 entries for each instance. The best normal

ILS jobs for each instance vary, no trend is discernible on first sight, except maybe that the normal ILS job with strength 4 is always better than the one with strength 7. For all instances the average costs of the normal ILS jobs typically are relative close to each other for each instance. The best Q-ILS jobs are better than the respective best normal ILS ones on the same instance. Only for instances pbd984 and rl1889, however, they seem to be significantly better, at least according to the t-test.

Looking at the action usage plots for the best Q-ILS job for each instance in figures 6.12 and 6.13 (subfigure (a) in the latter figure) again reveals that learning to choose the proper ILS-action is possible and happens. For all instances the best Q-ILS job basically only used one of the best performing ILS-actions (according to the performance of their corresponding normal ILS jobs). ILS-action with strength 4 and the random walk ILS-action for instance pbd984, strength 1 ILS-action for instance nrw1376, strength 2 ILS-action for instance rl1889, the random walk ILS-action for instance dcb2086, and the random walk and strength 2 ILS-action for instance pds2566 (see subfigures (a) – (d) of Figure 6.12 and subfigure (a) in Figure 6.13, respectively) For instances pbd984 and nrw1379, a change in the predominantly used action seems to be beneficial, for instance pbd984 even significantly according to the t-test. The same happens in slight variation for instance dcb2086 where towards the end the random walk and strength 2 ILS-actions (corresponding normal ILS jobs performed second and third best) are used predominantly.

Note that the best Q-ILS jobs on instances pbd984, rl1889, and pds2566 employed an ϵ-greedy policy which is the reason for the "noise" in the lower part of the relative action usage plots and the slight increase of lines in the accumulated action usage plots.

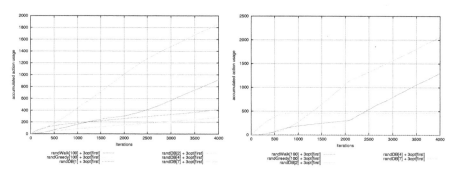

(a) Accumulated action usage, best Q-ILS jobs (training phase)

(b) Accumulated action usage, best corresponding application phase Q-ILS job

Figure 6.13: Action usage plots for the best training phase Q-ILS job run on instance pds3566 and of its best corresponding application phase Q-ILS job run in the TSP single-instance experiment

Noticeable is that the best Q-ILS job for instance pds2566 is a training phase job (using a better acceptance criterion), though. The action usage for the best of its corresponding application phase Q-ILS jobs is shown in Figure 6.13 (subfigure (b)) and shows basically the same action selection strategy and together with the fourth position of the best corresponding application phase Q-ILS job confirms the success of the action selection strategy learned. This action selection strategy switches to and alternates between the two ILS-actions whose corresponding normal ILS jobs performed best. This example also nicely illustrates how and that an action selection strategy is carried over from the learning to the application phase.

Since the results for the best Q-ILS job on instances pbd984 and nrw1379 are slightly better than the ones for the best normal ILS job on these two instances (whose ILS-actions are also used by the Q-ILS jobs; see subfigures (a) and (b) of Figure 6.12, respectively), altering the predominant usage of one ILS-actions over time might be a better action selection strategy than just using one ILS-action all the time. In particular, investigating variants of changing the predominant usage of ILS-actions towards the end might be insightful. On the other hand, a pure random action selection strategy in general does not seem to work overly well according to the results of the random ILS jobs. Also, looking at the Q-ILS presentation table of the single-instance experiment, Table 6.20, there is no indication in favor or against the thesis that an ϵ-greedy policy is better and hence a small action selection randomization is worse or better than a greedy one.

Finally note that according to the action usage plots in figures 6.12 and 6.13, the best Q-ILS jobs learned to discard the ILS-action whose corresponding normal ILS jobs performed worst altogether: They basically are not used at all or only due to ϵ-policy induced randomization.

Multi-Instance

The final experiment conducted and presented for the TSP is a multi-instance experiment. Basically the same ILS-actions and actual parameter values as for the just described single-instance experiment are used. Only, as a last test, two ϵ-better acceptance criteria with small ϵ-values of 0.1 and 0.3 for the application phase Q-ILS jobs are used in addition. The multi-instance experiment for the TSP was carried out in two parts, each only differing in the training, application, and test phase instances. The instances for the first part were picked randomly from the two repositories TSPLIB and VLSI (cf. Subsection 6.4.2) with the only constraint to cover a variety of instance sizes. The distribution of the instance size over the test instances roughly was kept the same as for the training and application instances, only slightly shifted towards using larger instances. For the second experiment part, only instances from the VLSI repository were used hoping that they are more similar to each other. In comparing the results for the two experiment parts, possibly an impression how influential problem instance consistency is might be obtained. The actual parameter values for the two parts of the multi-instance comparison experiments for the TSP are shown in tables 6.21 and 6.22, respectively.

The rankings for the normal and random ILS algorithms are shown in tables 6.23 and 6.24, respectively, the ranking for the Q-ILS algorithms are shown in tables 6.25 and 6.26, respectively. The ranking tables for the Q-ILS jobs would contain 192 lines in total. The rankings for the normal and random ILS algorithms interestingly again show the robustness of the random action selection scheme for the application phase: The random ILS algorithm is ranked 1 for the first and 2 for the second experiment part. Otherwise, the ranks of best normal ILS algorithms are rather mixed as regards their ranking for both experiments. Only the random greedy and in particular the strength 7 ILS algorithms are always the worst. According to the ranking value in column "Value", the former is not too bad, whereas the latter is far behind.

Looking at tables 6.25 and 6.26 (they would have 192 entries in total), for the Q-ILS jobs the better acceptance criterion again is far better than even an ϵ-better acceptance criterion with a very small value of 0.1: No Q-ILS job using an ϵ-greedy policy (value greater than 0.0 in column "ϵ_{a_a}") shows up among the best Q-ILS jobs in the tables for the two experiment parts. The choice of the policy during the application phase as before in the single-instance experiment does not seem to have a substantial influence. Only a slight trend towards using an ϵ-greedy policy during the application phase in contrast to following greedily the learned action selection strategy is discernible, in particular for the second experiment part. The distribution of the other learning parameters does not reveal any tendency as for the previous experiments for the TSP.

Short	Actual Parameter Values
t_{tr}	15
t_a	20
i_{tr}	rat575:lim963:vm1084:rby1599:d1655
i_a	rat575:lim963:vm1084:rby1599:d1655
i_{te}	fra1488, fnb1615, u1817, u2152, ley2323
$strat$	Q(0), Q(λ)
$pert$	randWalk[100], randGreedy[100], randDB[1], randDB[2], randDB[4], randDB[7]
ls	3opt[first]
$feat$	TSP-Features-Experiments-1.Manual-Selection.feat
$init$	nn
ϵ_{a_t}	0.0, 0.3
ϵ_{a_a}	0.0, 0.1, 0.3
ϵ_{π_a}	0.0, 0.3
γ	0.3, 0.8
r	c_Δ, c_i
λ	0.4, 0.8
fa	svr[0.004]

Table 6.21: Actual parameter values for the first part of the TSP multi-instance experiment

Short	Actual Parameter Values
i_{tr}	xit1083:rbv1583:djc1785:djb2036:bck2217
i_a	xit1083:rbv1583:djc1785:djb2036:bck2217
i_{te}	dja1436, dcc1911, bva2144, xpr2308, dbj2924, dlb3694

Table 6.22: Changed actual parameter values for the second part of the TSP multi-instance experiment

Rank	Perturbation	LS	Value
1	random ILS		5886
2	randDB[4]	3opt[first]	5888
3	randDB[2]	3opt[first]	5975
4	randWalk[100]	3opt[first]	6126
5	randDB[1]	3opt[first]	6463
6	randGreedy[100]	3opt[first]	7938
7	randDB[7]	3opt[first]	10199

Table 6.23: Ranks of normal and random ILS algorithms in the first part of the TSP multi-instance experiment

Rank	Perturbation	LS	Value
1	randWalk[100]	3opt[first]	4741
2	random ILS		5614
3	randDB[2]	3opt[first]	6010
4	randDB[4]	3opt[first]	6433
5	randDB[1]	3opt[first]	7332
6	randGreedy[100]	3opt[first]	7814
7	randDB[7]	3opt[first]	10208

Table 6.24: Ranks of normal and random ILS algorithms in the second part of the TSP multi-instance experiment

Rank				Parameters				Value
	r	γ	λ	ϵ_{a_t}	ϵ_{π_a}	ϵ_{d_a}	strat	
1	c_Δ	0.3	0.8	0.3	0.0	0.0	Q(λ)	52092
2	c_i	0.3		0.3	0.3	0.0	Q(0)	54985
3	c_i	0.8	0.8	0.3	0.3	0.0	Q(λ)	55788
4	c_Δ	0.3	0.8	0.3	0.3	0.0	Q(λ)	57025
5	c_i	0.3		0.3	0.0	0.0	Q(0)	59563
6	c_i	0.8	0.4	0.3	0.0	0.0	Q(λ)	60433
7	c_i	0.3	0.8	0.3	0.0	0.0	Q(λ)	60604
8	c_Δ	0.8	0.4	0.3	0.3	0.0	Q(λ)	62030
9	c_i	0.8		0.0	0.0	0.0	Q(0)	62619
10	c_i	0.3		0.0	0.0	0.0	Q(0)	63663
11	c_i	0.3	0.8	0.3	0.3	0.0	Q(λ)	64778
12	c_i	0.8		0.0	0.3	0.0	Q(0)	65075
13	c_i	0.3	0.8	0.0	0.3	0.0	Q(λ)	65307
14	c_i	0.8		0.0	0.3	0.0	Q(0)	65335
15	c_i	0.8	0.8	0.3	0.0	0.0	Q(λ)	65391
16	c_Δ	0.3		0.3	0.3	0.0	Q(0)	65575
	$\vdots$	$\vdots$	$\vdots$	$\vdots$	$\vdots$	$\vdots$	$\vdots$	$\vdots$

Table 6.25: Ranks of Q-ILS jobs in the first part of the TSP multi-instance experiment

Rank	r	γ	λ	ϵ_{a_t}	ϵ_{π_a}	ϵ_{a_a}	strat	Value
			Parameters					
1	c_Δ	0.8		0.3	0.3	0.0	Q(0)	50964
2	c_Δ	0.8	0.8	0.3	0.3	0.0	Q(λ)	51584
3	c_i	0.8	0.4	0.0	0.0	0.0	Q(λ)	52079
4	c_Δ	0.8	0.4	0.3	0.3	0.0	Q(λ)	52475
5	c_Δ	0.8		0.3	0.0	0.0	Q(0)	52745
6	c_i	0.8	0.4	0.3	0.3	0.0	Q(λ)	52852
7	c_i	0.8	0.4	0.0	0.3	0.0	Q(λ)	53203
8	c_Δ	0.3		0.3	0.3	0.0	Q(0)	53667
9	c_i	0.3		0.0	0.0	0.0	Q(0)	54596
$\vdots$	$\vdots$	$\vdots$	$\vdots$	$\vdots$	$\vdots$	$\vdots$	$\vdots$	$\vdots$

Table 6.26:   Ranks of Q-ILS algorithms in the second part of the TSP multi-instance experiment

For the testing phase, all normal and random ILS algorithms were run as well as the best ranked Q-ILS jobs for each combination of actual parameter values for parameters learning strategy, reward signal, and ϵ-policy (during) application (but only, if they were not ranked too bad). The actually used Q-ILS algorithms where those ranked 1, 2, 3, 4, 5, 6, and 16 for the first part of the multi-instance experiment of the TSP and those ranked 1, 2, 3, 5, 6, 7, and 9 for the second.

The results of the test phase jobs for instances fnb1615, u2152, and ley2323 for the first experiment part are presented in Table 6.27, the results for instances dcc1911, xpr2308, and dlb3594 for the second part are shown in Table 6.28. The results for the remaining instances are, due to lack of space, moved to Appendix A.2 and are shown in tables A.7 and A.8, respectively.

Regarding the first experiment part, the Q-ILS jobs typically perform best on the test instance. It seems that the bigger the instance, the more the Q-ILS jobs cluster on the first positions and the more the ILS jobs perform significantly worse than the respective best Q-ILS job. For the two largest instances, only one Q-ILS job (the ranked 6 one run on instance ley2323) is significantly worse than the respective best Q-ILS job. As concerns the ILS jobs, the picture on all test instances is reversed: only one or two ILS jobs are *not* significantly worse than the respective best Q-ILS jobs. The ranking trend for ILS algorithms from the application phase roughly is affirmed, in particular for those that were ranked worst during the application phase as is displayed by Table 6.25.

The situation for the second experiment part is as follows. For any test instance one of the Q-ILS job is performing best and is significantly better than most of the ILS jobs. The rank 2 Q-ILS algorithm performs especially well. The ranking trend for the ILS jobs from the application phase is repeated, most notably, the random ILS jobs performed generally quite good again.

In the following, several exemplary action usage plots are analyzed with the aim to investigate the transfer of learned action selection strategies across several instances and to check for any differences between experiment part one and two whose test instances supposedly are differently homogenous. Therefore, for each experiment part and each test instance actions usage plots of two selected test phase Q-ILS jobs from tables 6.27 and A.7 are compared. The Q-ILS jobs to be compared were selected as to include for each experiment part each of the two learning strategies used and each of the two policies (greedy vs. ϵ-greedy) used. The selected test phase Q-ILS jobs are those ranked 1 and 16 for the first part of the multi-instance experiment part and 2 and 9 for the second one.

Instance	Rank	Type	H_0: equal		H_0: not better		T	Statistics avg	σ	avg. exc.	min	max
			0.05	0.1	0.05	0.1						
fnb1615	16	Q-ILS						4972.82	7.31	0.3395 %	4960	4992
(4956)	2	Q-ILS	=	=	=	=	1.0411	4974.75	9.13	0.3783 %	4959	4994
	4	ILS(W)	=	–	=	=	1.1629	4974.77	7.68	0.3788 %	4961	4996
	2	ILS(4)	=	=	=	–	1.3422	4975.27	8.94	0.3889 %	4956	5000
	3	Q-ILS	=	–	=	–	1.4846	4975.40	8.18	0.3914 %	4961	4993
	4	Q-ILS	=	=	=	–	1.3763	4975.55	10.17	0.3945 %	4960	4995
	5	Q-ILS	=	–	=	–	1.5875	4975.88	9.71	0.4010 %	4956	4997
	9	Q-ILS	–	–	–	–	1.8486	4976.00	8.04	0.4036 %	4961	4990
	6	Q-ILS	–	–	–	–	2.5224	4977.48	9.08	0.4333 %	4958	4995
	1	ILS(R)	–	–	–	–	2.4712	4977.57	9.71	0.4353 %	4963	5003
	6	ILS(G)	–	–	–	–	2.6598	4977.95	9.75	0.4429 %	4960	5001
	3	ILS(2)	–	–	–	–	2.9339	4978.43	9.61	0.4525 %	4962	5002
	1	Q-ILS	–	–	–	–	2.9723	4978.98	10.85	0.4636 %	4956	5005
	5	ILS(1)	–	–	–	–	3.6768	4980.70	11.40	0.4984 %	4963	5016
	7	ILS(7)	–	–	–	–	8.9418	4990.68	10.29	0.6997 %	4963	5013
u2152	6	Q-ILS						64598.85	118.30	0.5383 %	64354	64821
(64253)	5	Q-ILS	=	=	=	=	0.3159	64607.10	115.28	0.5511 %	64413	64824
	16	Q-ILS	=	=	=	=	0.3815	64609.47	130.49	0.5548 %	64409	64967
	3	Q-ILS	=	=	=	=	0.8605	64621.22	114.24	0.5731 %	64327	64822
	9	Q-ILS	=	=	=	=	0.9625	64622.10	96.69	0.5744 %	64453	64856
	2	Q-ILS	=	=	=	=	1.2571	64632.15	118.63	0.5901 %	64453	64977
	1	Q-ILS	=	=	=	–	1.3361	64635.07	124.14	0.5946 %	64407	64837
	1	ILS(R)	=	=	=	=	1.5281	64636.85	103.64	0.5974 %	64395	64905
	4	Q-ILS	=	=	=	–	1.5812	64637.57	99.99	0.5985 %	64387	64832
	2	ILS(4)	=	–	–	–	1.7937	64643.90	106.02	0.6084 %	64452	64850
	4	ILS(W)	–	–	–	–	2.6648	64664.07	99.85	0.6398 %	64426	64909
	3	ILS(2)	–	–	–	–	3.1919	64676.57	98.61	0.6592 %	64405	64834
	5	ILS(1)	–	–	–	–	3.2446	64679.65	103.98	0.6640 %	64449	64851
	6	ILS(G)	–	–	–	–	3.9171	64692.03	92.95	0.6833 %	64455	64916
	7	ILS(7)	–	–	–	–	4.2729	64734.55	162.33	0.7495 %	64431	65056
ley2323	5	Q-ILS						8370.25	15.59	0.2185 %	8352	8425
(8352)	1	Q-ILS	=	=	=	=	0.3854	8371.50	13.34	0.2335 %	8354	8412
	16	Q-ILS	=	=	=	=	0.6929	8372.73	16.35	0.2481 %	8352	8417
	2	Q-ILS	=	=	=	=	0.7598	8373.05	17.33	0.2520 %	8355	8433
	2	ILS(4)	=	–	=	=	0.8660	8373.08	13.52	0.2523 %	8354	8415
	9	Q-ILS	=	=	=	=	0.9671	8373.62	15.63	0.2589 %	8357	8423
	3	Q-ILS	=	=	=	=	1.0597	8374.65	21.13	0.2712 %	8353	8444
	4	Q-ILS	=	–	=	–	1.3330	8374.83	15.11	0.2733 %	8356	8417
	4	ILS(W)	=	=	=	=	1.2106	8375.15	20.31	0.2772 %	8356	8438
	1	ILS(R)	=	–	=	–	1.4660	8375.25	14.91	0.2784 %	8353	8410
	6	Q-ILS	=	–	=	–	1.6127	8375.92	15.88	0.2865 %	8360	8430
	3	ILS(2)	–	–	–	–	1.7515	8376.88	18.15	0.2978 %	8354	8445
	5	ILS(1)	–	–	–	–	2.4049	8378.77	16.11	0.3206 %	8358	8419
	7	ILS(7)	–	–	–	–	3.3123	8381.25	14.08	0.3502 %	8354	8407
	6	ILS(G)	–	–	–	–	3.2567	8384.42	22.69	0.3882 %	8355	8440

Table 6.27: Results of test phase jobs in the first part of TSP multi-instance experiment

Instance	Rank	Type	H_0: equal		H_0: not better			Statistics				
			0.05	0.1	0.05	0.1	T	avg	σ	avg. exc.	min	max
dcc1911	7	Q-ILS						6421.70	7.39	0.4018 %	6406	6436
(6396)	1	ILS(W)	=	=	=	=	0.0909	6421.88	9.67	0.4045 %	6403	6448
	9	Q-ILS	=	=	=	=	0.6519	6422.95	9.61	0.4214 %	6401	6443
	2	Q-ILS	=	=	=	=	0.9184	6423.35	8.63	0.4276 %	6402	6440
	5	Q-ILS	=	=	=	=	1.0223	6423.73	10.11	0.4335 %	6400	6440
	1	Q-ILS	=	=	=	–	1.4465	6424.57	10.17	0.4468 %	6408	6445
	5	ILS(1)	=	–	–	–	1.6652	6425.00	10.12	0.4534 %	6408	6450
	3	Q-ILS	=	=	–	–	1.7779	6425.85	12.78	0.4667 %	6403	6461
	4	ILS(4)	–	–	–	–	2.1892	6425.90	9.62	0.4675 %	6405	6439
	6	Q-ILS	–	–	–	–	2.4563	6426.07	8.50	0.4702 %	6402	6446
	3	ILS(2)	–	–	–	–	2.4792	6426.32	9.19	0.4741 %	6404	6453
	6	ILS(G)	–	–	–	–	2.5670	6426.57	9.46	0.4780 %	6407	6450
	2	ILS(R)	–	–	–	–	2.8448	6427.32	10.09	0.4898 %	6405	6449
	7	ILS(7)	–	–	–	–	8.8491	6439.98	10.77	0.6875 %	6422	6463
xpr2308	2	Q-ILS						7251.70	8.05	0.4530 %	7235	7269
(7219)	9	Q-ILS	=	=	=	=	0.7047	7253.38	12.70	0.4762 %	7234	7291
	3	Q-ILS	=	=	=	=	1.0176	7253.82	10.47	0.4824 %	7234	7275
	7	Q-ILS	=	=	=	=	1.1371	7254.20	11.34	0.4876 %	7233	7279
	2	ILS(R)	=	–	=	–	1.6557	7255.43	11.73	0.5046 %	7231	7285
	4	ILS(4)	=	=	–	–	1.7028	7255.68	12.38	0.5080 %	7234	7287
	6	Q-ILS	–	–	–	–	2.7480	7256.82	8.62	0.5240 %	7238	7274
	1	ILS(W)	–	–	–	–	2.8441	7257.93	11.26	0.5392 %	7226	7279
	1	Q-ILS	–	–	–	–	3.2587	7257.95	9.07	0.5395 %	7239	7279
	3	ILS(2)	–	–	–	–	3.0825	7258.20	10.63	0.5430 %	7234	7282
	5	Q-ILS	–	–	–	–	2.9015	7258.23	11.72	0.5434 %	7234	7286
	5	ILS(1)	–	–	–	–	3.5901	7259.60	11.35	0.5624 %	7236	7282
	7	ILS(7)	–	–	–	–	3.7221	7260.82	13.25	0.5794 %	7234	7282
	6	ILS(G)	–	–	–	–	5.2705	7262.50	10.16	0.6026 %	7242	7285
dlb3694	2	Q-ILS						11022.65	12.62	0.5808 %	10981	11046
(10959)	6	Q-ILS	=	=	=	=	1.1153	11025.65	11.40	0.6082 %	10999	11046
	1	ILS(W)	–	–	–	–	1.9997	11028.08	11.62	0.6303 %	10998	11046
	5	Q-ILS	–	–	–	–	3.4518	11031.55	10.32	0.6620 %	11004	11060
	3	ILS(2)	–	–	–	–	3.3249	11031.67	11.63	0.6632 %	11009	11060
	2	ILS(R)	–	–	–	–	3.6068	11032.23	11.07	0.6682 %	11006	11063
	3	Q-ILS	–	–	–	–	3.8678	11032.77	10.71	0.6732 %	11013	11061
	7	Q-ILS	–	–	–	–	3.9691	11033.50	11.81	0.6798 %	11005	11069
	1	Q-ILS	–	–	–	–	3.5610	11033.60	14.79	0.6807 %	11004	11063
	9	Q-ILS	–	–	–	–	3.8012	11033.62	13.19	0.6809 %	11004	11064
	4	ILS(4)	–	–	–	–	4.1212	11035.33	14.80	0.6965 %	11011	11073
	5	ILS(1)	–	–	–	–	5.2158	11037.27	12.46	0.7143 %	11009	11071
	6	ILS(G)	–	–	–	–	7.9449	11043.73	11.05	0.7731 %	11023	11070
	7	ILS(7)	–	–	–	–	11.2035	11060.42	17.19	0.9255 %	11024	11093

Table 6.28: Results of test phase jobs in the second part of the TSP multi-instance experiment

Rank 1 and 9 Q-ILS jobs employed a greedy policy, rank 16 and 2 Q-ILS jobs employed an ϵ-greedy policy with an ϵ-value of 0.3.

Figure 6.14 shows the accumulated action usage plots for rank 1 Q-ILS jobs on instances fnb1615, u2152, and ley2323, of the first experiment part on the left hand side and the relative (smoothed) action usage plots for the rank 16 Q-ILS jobs on the right hand side. The action usage plots for the rank 1 Q-ILS jobs reveal that the learned action selection strategy again is instance dependent: the one for instance fnb1615 in subfigure (a) of Figure 6.14 is decisively different, unfortunately not advantageously, since the rank 1 Q-ILS job performs worst of all Q-ILS jobs for this instance. This is the case although it also mainly used the strength 4 ILS-action whose corresponding ILS job performed fourth best on this instance. The strength 4 ILS-action in particular is used towards the typically decisive end of the search exclusively. This further emphasizes the unexpectedness of the result. Q-ILS jobs ranked 1 otherwise basically only apply the strength 2 ILS-action which makes the performance good for instance ley2323 (second best job result) but bad for instance u2152 (second worst Q-ILS job result). The same phenomenon of action selection strategy variation in dependence of the instance occurs for the rank 16 Q-ILS jobs. The learned action selection strategy basically is the same for instances fnb1615 and ley2323 (the random walk ILS-action is the greedy ILS-action, the other ILS-action usages are due to the ϵ-greedy policy employed), but decisively different for instance u2152. For the latter one, the rank 16 Q-ILS jobs identifies the strength 1 ILS-action as the greedy one for most of the iterations except for the begin and the end instead of using the random walk ILS-action as greedy one all the time. Although this choice according to the results of the corresponding random walk and strength 1 ILS jobs for these instances does not seem to be sensible, rank 16 Q-ILS jobs perform stably good on all instances fnb1615, u2152, and ley2323 (and on the other ones from Table A.7 also repeating the two kinds of action selection strategies roughly, see Figure A.1). Also, the rank 16 Q-ILS jobs typically perform better or basically equal compared to best performing normal ILS jobs for each instance. The learned action selection strategy of the rank 16 Q-ILS jobs works well for several instances and hence is the proof that learned action-value function can successfully be learned and transferred to yet unseen problem instances. Remarkably, the changes in the action selection strategies over several instances for the rank 16 Q-ILS jobs do not coincide with the ones for the rank 1 Q-ILS jobs. Since the action usages and hence action selection strategies are very stable and clear for each individual instance, the differences are not deemed to be by chance.

The relative (smoothed) action usage plots for the rank 2 Q-ILS jobs for instances dcc1911, xpr2308, and dlb3694 of the second experiment part are shown on the left hand side of Figure A.3, the ones for the rank 9 Q-ILS job on the right hand side (except for instances dlb3694 where an accumulated action usage plot is shown in sub-figure (e)). These action usage plots for the rank 2 and 9 Q-ILS jobs also for each Q-ILS show similar action usages for two of the three instances and a substantially different one for a third. This time, however, the instance for which the action usage differs is the same, namely instance dlb3694. The rank 2 Q-ILS jobs are quite successful in varying their action selection strategies in dependence of the instance (also for the instances whose results are shown in Table A.8 in Appendix A.2; the respective action plots repeat the two action usage plots already shown for the rank 2 Q-ILS jobs, see Figure A.3, as was the case before for the rank 1 Q-ILS jobs in the first experiment part). For instance dlb3894, the normal ILS jobs corresponding to the greedily used random walk ILS-action by rank 2 Q-ILS job is the best performing ILS job. For instance xpr2308, strength 4 ILS job is the second best performing among the ILS jobs. Rank 2 Q-ILS switches towards the end of the search to using the strength 4 ILS-actions, while for instance dcc1911 the corresponding normal ILS jobs for the two greedily used ILS-actions by rank 2 Q-ILS job are also the best and third best performing ILS jobs. The action selection strategies for rank 9 Q-ILS job also is quite successful over the diverse instances (including the ones only shown in Appendix A.2). It never performs best, but is always among the best, maybe except for

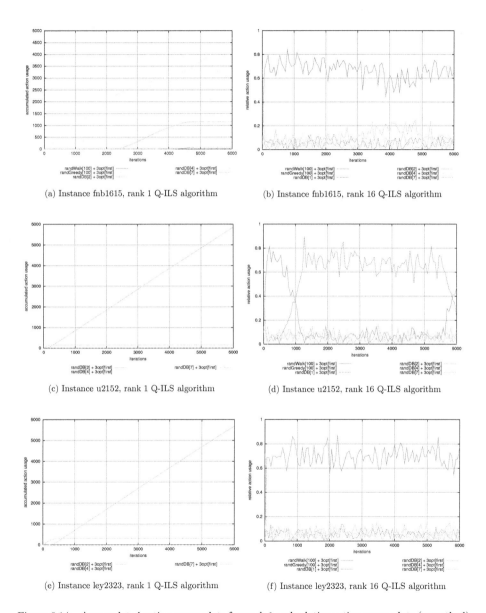

(a) Instance fnb1615, rank 1 Q-ILS algorithm

(b) Instance fnb1615, rank 16 Q-ILS algorithm

(c) Instance u2152, rank 1 Q-ILS algorithm

(d) Instance u2152, rank 16 Q-ILS algorithm

(e) Instance ley2323, rank 1 Q-ILS algorithm

(f) Instance ley2323, rank 16 Q-ILS algorithm

Figure 6.14: Accumulated action usage plots for rank 1 and relative action usage plots (smoothed) for rank 16 Q-ILS algorithm run on test instances fnb1615, u2152, and ley2323 in the first part of the TSP multi-instance experiment

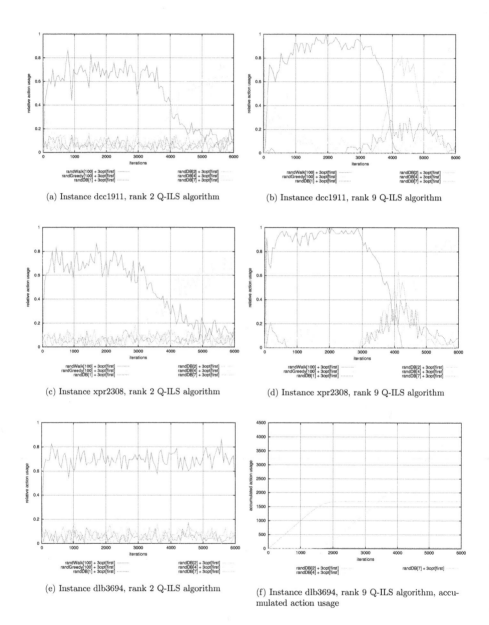

(a) Instance dcc1911, rank 2 Q-ILS algorithm

(b) Instance dcc1911, rank 9 Q-ILS algorithm

(c) Instance xpr2308, rank 2 Q-ILS algorithm

(d) Instance xpr2308, rank 9 Q-ILS algorithm

(e) Instance dlb3694, rank 2 Q-ILS algorithm

(f) Instance dlb3694, rank 9 Q-ILS algorithm, accumulated action usage

Figure 6.15:   Action usage plots for rank 2 and 9 Q-ILS algorithms run on test instances dcc1911, xpr2308, and dlb3694 in the second part of the TSP multi-instance experiment

instance dlb3694. This altogether strongly suggests that what was learned can be transferred very beneficially and should be attempted to be carried further.

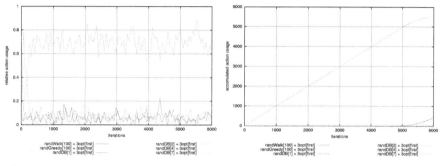

(a) Relative action usage (smoothed), rank 4 Q-ILS algorithm

(b) Accumulated action usage, rank 6 Q-ILS algorithm

Figure 6.16:   Action usage plots for rank 4 and 6 Q-ILS algorithms run on instance u2152 in the first part of the TSP multi-instance experiment

Concluding the presentation of the results for the multi-instance experiment for the TSP the following observation exemplary deserve attention. The rank 1 Q-ILS job from the first experiment part basically behaves as a strength 2 ILS job for instance ley2323, but performs far better. One quick conclusion would be to account the differences to the fact that the rank1 Q-ILS jobs employs an ϵ-greedy policy and that the differences accordingly are due to this small variation in the action usage induced. The same situation in reverse arises for instance u2152. There, rank 6 Q-ILS job is performing best. According to its action usage plot in subfigure (b) of Figure 6.16, it almost always employs the strength 2 ILS-action and uses a greedy policy. The corresponding strength 2 normal ILS jobs, though, performs significantly worse on instance u2152! Rank 4 Q-ILS job employs an ϵ-greedy policy with $\epsilon = 0.3$ and uses the obviously greedy strength 2 ILS-action only in 70 % of the time (see subfigure (a) of Figure 6.16). Here, all of a sudden, it performs almost as bad as the strength 2 normal ILS job on this instance and also significantly worse than the rank 6 Q-ILS job then. Even if one accounted the significant performance difference as regards each relation to the performance of the rank 6 Q-ILS job for instance u2152 to the fact that rank 6 Q-ILS job there uses at the very end of the search the random walk ILS-action (whose corresponding normal ILS action nevertheless also performs significantly worse and does not substantially contribute with improving moves, see Figure A.4 in Appendix A.2), still the fact remains that in one case (rank 1 Q-ILS job on instance les2323) a greedily applied ILS-action performs (probably significantly) better than a corresponding normal ILS job and in another case not (rank 4 Q-ILS job on instance u2152). An also gross difference in performance of nearly the same action selection strategy happens for rank 2 Q-ILS job on instance dlb3694. There, rank 2 Q-ILS job greedily employs the random walk ILS-action (70 % of the time due to the ϵ-greedy policy with $\epsilon = 0.3$, see subfigure (e) of Figure A.3) and is definitely significantly better than the random walk normal ILS (T-value of 3.4518 is significant with an error probability of 0.0005 in this case!).

A first explanation for this phenomenon is that these are gross outliers due to huge influence of the algorithm's inherent randomization. Another explanation could be that the proper action selection strategy differs highly between instances. For example, the results for normal ILS jobs have shown that some ILS-actions work better for some instance, others work better for others (see Table 6.19

or instance rl1304 in Table 6.12). In almost the same manner, different action selection strategies make best performing Q-ILS jobs differ greatly from instance to instance (see Figure 6.12 as a concise example). Often the best action selection strategy seems to be one that simply always applies only one single ILS-action all the time (see two best jobs from Table 6.17 and subfigure (a) of Figure 6.10). Sometimes the best action selection strategy seems to be the one that only greedily selects a single ILS-action (and employs an ϵ-greedy policy, see two best jobs from Table 6.20 and subfigure (c) of Figure 6.12). Yet sometimes the best action selection strategies seems to be the one that switches between the greedy usage of two ILS-actions (see best job for instance xpr2308 from Table 6.28 and subfigure (c) of Figure A.3). As a consequence, it can be conjectured that for each instance *individually* there might be one best action selection strategy (or even best ILS-action directly) to employ. Sometimes, some degree of instance dependent randomization might be required, sometimes not. The good averaged robust performance of the random ILS jobs, yet with outliers, for example supports the latter conjecture. These phenomena have to be investigated further to make learning success more predictable.

6.4.4   Experiments for the TSP with Regression Trees

In order to test for the influence of the choice of the function approximator on the learning success, the single- and multi-instance experiments were rerun. All actual parameter setting remained, only the function approximator was changed, now using a regression tree function approximator (cf. Subsection 6.3.1). The presentation of the results of reruns will be brief concentrating on whether significant differences that can be accounted to the changes in function approximator usage can be found. The result summary follows next, the result tables have been moved to Appendix A.3.

Single-Instance

The changes to the actual parameter values for the single-instance experiment for the TSP are shown in Table 6.29. The results of the single-instance experiment for the TSP using regression trees are contained in Appendix A.3.1.

Short	Actual Parameter Values
fa	regTree[10;100;30]

Table 6.29:   Changed actual parameter values for the regression tree version of the TSP single-instance experiment

The results for the rerun normal ILS jobs are shown in Table A.9. Comparing them with the results for the first run from Subsection 6.4.3, there are some differences in the performance of individual normal ILS jobs and also in the rankings for the individual instances. These differences in principle should not occur and are further evidence for the hypothesis that all experiments and results presented in this chapter unfortunately suffer substantially from the huge inherent randomization of the algorithms employed. They accordingly aggravate finding clear trends, unique results and making reliable predictions. Table A.10 shows the results for the Q-ILS jobs. The random ILS jobs perform slightly better this time, sometimes they occupy position 5 and 10. Trends or similarities as regards the distribution of actual parameter values for the learning parameters among the best Q-ILS jobs compared to Table 6.20 are not discernible. The best Q-ILS jobs again do not significantly perform better or worse for an instance than the best normal ILS job run on the respective instance.

Only looking at the results from tables A.10 and 6.20, the choice of the function approximator does not seem to have a substantial influence, at least none that is visible among the randomization induced noise.

Multi-Instance

The changes to the actual parameter values for the multi-instance experiment for the TSP are shown in Table 6.30. The result tables can be viewed in Appendix A.3.2.

Short	Actual Parameter Values
fa	regTree[10;100;30]

Table 6.30: Changed actual parameter values for the regression tree version of the TSP multi-instance experiments

The ranking tables for the normal ILS jobs run during the two experiment parts of the TSP multi-instance experiment using regression trees are presented in tables A.11 and A.14. They again show a very robust and good performance of the random ILS algorithms which are ranked second for both parts of the rerun. The ranking of the worst performing normal ILS algorithms from the multi-instance experiment for the TSP using support vector machines as function approximators is confirmed, but comparing tables A.11 and 6.23, and A.14 and 6.24 the ranks for the other normal ILS algorithms are basically different now. The influence of randomization or the fact that different instances have different best performing ILS-actions seems to be predominant which is conclusion that can be drawn by the repetition of the random ILS algorithm results. The ranks for the Q-ILS jobs of the application phase are displayed in tables A.12 and A.13. As before, these two tables would in total show 192 Q-ILS jobs each. Comparing them with the corresponding tables from the first multi-instance experiment for the TSP (tables 6.25 and 6.26), it can be seen that the ranking values for the second experiment part now are more close to each other whereas the ones for the best Q-ILS algorithms for the first experiment part now are better but still quite close to each other. No trend other than better using a better acceptance criterion is discernible.

The test phase for the two parts of the multi-instance experiment for the TSP with regression trees comprised running all normal ILS algorithms each time and rank 1, 2, 3, 4, 7, and 11 Q-ILS algorithms for the first experiment part and rank 1,2,3,5,6,7, and 14 Q-ILS algorithms for the second part. The results of the test phase are presented in tables A.15 and A.16. The general trend does not change compared to the results of the test phase of the first multi-experiment for the TSP described in tables 6.27 and A.7 in Subsection 6.4.3. The Q-ILS jobs tend to occupy the best positions and to yield the best performances. The span over performances as can be read off from the column containing the average excesses are very similar for both multi-instance experiments for the TSP. The slight trend that Q-ILS jobs or at least the best Q-ILS jobs perform better in relation to the other jobs run for an instance, the larger the instance is, continues.

6.4.5 Discussion

The general conclusion to draw from the results of the first experiments for the TSP conducted with GAILS algorithms is that learning and transfer of what was learned across several problem instances, even of different sizes, is possible and can improve performance and therefore should be further pursued. Since this does not happens in all cases, further influences have to be identified and investigated in future experiments.

The Q-ILS algorithms were able to learn to choose the proper ILS-action during the local search procedure comparison experiment, in this case the 3-opt local search procedure. There, results for the normal ILS jobs and from Table 6.6 strongly suggest that the 3-opt ILS-action is always superior, so learning to choose this ILS-action all the time is also the best a Q-ILS algorithm could do. In fact, the thesis of the superiority of the 3-opt local search procedure has been further proven by the result of the experiments with GAILS algorithms conducted where running the automatically learning Q-ILS jobs justified the thesis and this way showed another profitable use case for the GAILS approach.

Learning a stable and beneficial action selection strategy (other than just using one ILS-action all the time as a normal ILS job does) is possible also and even occurs across several problem instances as was the case during the multi-instance experiments. There, action selection strategies changed for different instances as well as remained stable for others. The changes are not by chance, but systematic. What exactly influences them still remains to be identified, though. The opportunistic change of the action selection strategies induced by the transferred action-value functions from instance to instance indicates that the transfer is not restricted to find one single unchanged action selection strategy for any instance but that it in fact reacts to different feature values, most noticeably those of features indicating search progression. Only heuristic and state features were used but the built upon action selection strategies showed instance specific behavior. This is a remarkable effect. Apparently, some instance specific characteristic can be extracted from the heuristic and state features used which made learned action-value function to implement different policies for different instances. Together with the conjecture that the best ILS-action for each instance often is different, learning across instances and transfer of what was learned makes sense and seems to be a proper way to improve metaheuristics.

Learning did not happen in every case, though. The influence of the learning parameters thereby does not seem to be too high, so some other influences seem to be important and have to be identified and investigated. The not visible trend among the distribution of actual parameter values for the learning parameter otherwise is not too bad, since this also means that learning is to some extent independent from an actual parameter setting. One of the other influences certainly is the choice of the acceptance criterion employed whose influence proved to be crucial, potentially leveling out other learning influences. This possibility was already discussed in Subsection 4.3.5 and has been confirmed by the experiments described so far. Even though the learned action selection strategies probably could not be followed widely when using a better acceptance criterion in the application phase, learning has occurred nevertheless. Stabilizing the learning success and incorporating acceptance criterion influence beneficially are the next steps to do.

The results for the rerun of the single- and multi-instance experiments with regression tree function approximators basically showed no true differences. The performances as measured in average excess for example roughly are the same. If the choice of the function approximator has an influence on the performance and the learning success of Q-ILS algorithms, this at least is not discernible from the experimental results collected in this section. The conjecture is that the randomization and acceptance criterion choice effects prevail and cover any others.

6.5   Experiments for the FSP

This section presents the experiments conducted with GAILS algorithms on the FSP. The layout of the section is as before in Section 6.4: First, the FSP is formally described and the employed initial solution construction schemes, local search procedures, and perturbations are described in Subsection 6.5.1. Next, the features and their normalization are presented in Subsection 6.5.2.

Subsection 6.5.3 contains the description of the experiment results. Subsection 6.4.4 reruns some experiments using another function approximator. Finally, Subsection 6.5.4 wraps up this section and draws conclusions, in particular in comparison to the results for the experiments conducted with the TSP from the previous section.

6.5.1 Description

The formal problem formulation is given next. Subsequently, two initial solution construction schemes and several local search procedures and perturbations for the FSP are described.

Problem Formulation

In each problem instance of the Flow Shop Problem (FSP), a set $J := \{1, \ldots, n\}$ of n ($n \in \mathbb{N}^+$) *jobs* is given. Each job must be processed on m ($m \in \mathbb{N}^+$) *machines* from $M := \{1, \ldots, m\}$ in the order $1, \ldots, m$ yielding for each job j ($j \in J$) a set O_j ($O_j := \{o_{j1}, \ldots, o_{jm}\}$) of so-called *operations*. Each operation o_{jk} ($o_{jk} \in O_j, k \in \mathbb{N}^+$) has a processing time of p_{jk} ($p_{jk} \in \mathbb{R}_0^+$). A problem instance for the FSP with n jobs and m machines is denoted by $n \times m$ problem instance. The following constraints hold:

- Any machine can at any time only process one job.

- Any job can at any time only be processed by at most one machine.

- Once the processing of a job, i.e. an operation, on a machine has started, it cannot be interrupted.

- All jobs are available for processing at the beginning.

The experiments described in this section were conducted for a subclass of the FSP where the job order is the same on any machine. This subclass is called the *Permutation Flow Shop Problem* (PFSP), but will for ease of use be denoted by FSP also. The objective is to find a sequence or permutation π ($\pi: \{1, \ldots, n\} \to J$) of jobs (job $\pi(j)$ will be processed as j-th job) such that the completion time of the last operation processed on any machine over all jobs is minimal. This completion time is also called *makespan* since it determines the time span to process all jobs completely. A permutation π is a solution and the makespan represents the cost of a solution. Note that sign π is not to be mixed up with its usage in representing a policy (cf. Subsection 3.2.1). The set of all candidate and also feasible solutions Π is the set of all permutations of the n jobs: $\Pi := \{1, \ldots, n\} \to J$. The FSP and the PFSP are *NP*-hard [GJS76] so it is justified to apply heuristic methods, and local search methods in particular, to solve it. For each perturbation π of jobs a function p_π^* ($p_\pi^*: J \times M \to \mathbb{R}_0^+$) can be defined that assigns completion times $p_\pi^*(j, k)$ to operation o_{jk} according to ($l_a \in J, a \in \{0, \ldots, m\}$):

$$p_\pi^*(i, k) = \max_{1 = l_0 \leq l_1 \leq \ldots \leq l_{k-1} \leq i_k = j} \sum_{g=1}^{k} \sum_{h=l_{j-1}}^{l_j} p_\pi(\pi(h), g)$$

The makespan C_{max} ($C_{max} \in \mathbb{R}_0^+$) for a permutation π then is $C_{max} = p_\pi^*(n, m)$.

Each $n \times m$ FSP problem instance can be described as an $n \times m$ matrix which contains in cell in line j and column k the processing time for operation o_{jk}. The following matrix is an example for a 4×3 problem instance:

	Machines		
	1	2	3
1	7	5	4
2	4	2	7
3	10	3	4
4	2	5	4

Jobs

The makespan $C_{max} = p^*_\pi(4, 3)$ for a solution $id = (1, 2, 3, 4)$ for this 4×3 problem instance is 33. A solution for an FSP problem instance can be illustrated as so-called *Gantt-diagram*. As an example, a Gantt-diagram for the solution id for the example just presented is shown in Figure 6.17. The x-axis of such a diagram denotes time. Each block represents an operation o_{jk} and is labeled with "o_{jk}". Each line corresponds to one machine

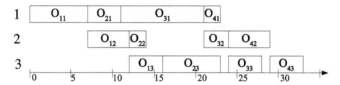

Figure 6.17: Solution for a 4×3 FSP problem instance

Initial solution

One of the two initial solution construction schemes that were used in the experiments described here is a construction algorithm called NEH from [NJH83]. This algorithm sorts the jobs according to their average processing times over the machines and constructs an initial solution by successively inserting the jobs in the resulting order into the respective partial solution. In each insertion, the insertion position which minimizes the resulting makespan of the resulting next and extended partial solution is chosen. The NEH algorithm is denoted by "neh" in plots and tables. Another initial solution construction scheme simply yields a random permutation of jobs. It is denoted by "rand" in plots and tables.

Neighborhood Structure

Five neighborhood structures that are the basis for local search procedures and perturbations for the FSP were implemented in the integration of the FSP into the GAILS framework in [Dot05]. Three of them were used in the experiments described here and will be explained next. In the *2-exchange* neighborhood structure two solutions are neighbors, if they can be transformed into each other by exchanging exactly two jobs. Figure 6.18 illustrates how to construct a new solution from a current one based on the 2-exchange neighborhood structure. The new solution in the lower part of this figure results from the current solution taken from Figure 6.17 in the upper part by exchanging jobs 3 and 4 indicated by operations o_{3k} and o_{4k}. The *insert* neighborhood structure relates two solutions as neighbors, if one solution can be constructed from the other by removing exactly one job of a solution at some position and inserting it before or after another job of the same solution. Figure 6.19 illustrates how to construct a new solution from a current one based on the insert neighborhood structure. The solution in the lower part of this figure results from the solution in the upper part (again taken from Figure 6.17) by removing job 4 (indicated by operations o_{4k}) and inserting it before job 1 (indicated by operations o_{1k}). The third neighborhood structure is called *partial NEH*. There, two solutions are neighbors, if one can construct one from the other by first removing a number of jobs from the first solution and reconstructing a complete solution using the NEH algorithm yielding the second solution. A *strength* parameter h ($h \in [0,1]$) indicates how many jobs are to be removed in percentage of all jobs of a solution. The jobs to be removed are picked randomly.

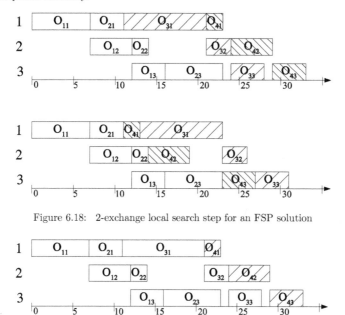

Figure 6.18: 2-exchange local search step for an FSP solution

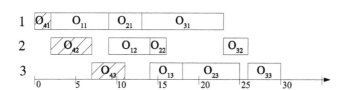

Figure 6.19: Insert local search step for an FSP solution

Local Search Procedures and Perturbations

Based on the neighborhood structures described just now, local search steps and procedures, and perturbations can be constructed. The local search steps and derived local search procedures are called *2-exchange* and *insert local search steps procedures* for the 2-exchange and insert neighborhood structures, respectively, and consist of one or more transitions from a solution to one of its neighbors. The neighbor of a transition in each step is chosen according to the first or best improvement neighbor selection strategy (cf. Section 2.3). The two resulting first and best improvement local search procedures for the 2-exchange neighborhood structure are denoted by "ex[first]" and "ex[best]" in plots and tables, respectively. The two resulting first and best improvement local search procedures for the insert neighborhood structure are denoted by "in[first]" and "in[best]" in plots and tables, respectively.

Two perturbations are used in the experiments for the FSP. The first perturbation simply is a transition to a randomly chosen neighbor according to the partial NEH neighborhood structure. This perturbation is called *partial NEH perturbation* and denoted by "partNeh[h]" in plots and tables with h indicating the percentage of jobs to be removed in order to insert them again according to the NEH algorithm. The other perturbation is based on a composition of two perturbations which both are based on the 2-exchange neighborhood structure. A perturbation for a given neighborhood structure in varying strength can be obtained by carrying out several transitions to always randomly chosen neighbors in a row. Additionally, several perturbations can be applied in succession yielding a composition of perturbations which itself can be regarded as a perturbation. The second perturbation used throughout the experiments for the FSP described here consists of carrying out g ($g \in \mathbb{N}^+$) random transitions to a neighbor in a row for a 2-exchange neighborhood structure variant as a first stage and next carrying out a random transition to a neighbor according to the standard 2-exchange neighborhood structure in a second stage. The 2-exchange neighborhood structure variant used in the first stage restricts exchanges to those between neighboring jobs only. This second, composite perturbation is called *composite exchange perturbation* and denoted by "compRandEx[g]" in plots and tables where g indicates the number of transitions for the first stage. The number g essentially serves a strength parameter here. A more detailed description of the initial solution construction schemes, local search procedures and perturbations for the FSP and their implementation and integration into the GAILS framework is given in [Dot05].

6.5.2   Features for the FSP

This subsection describes the features used for the FSP experiments. After their description, a short analysis based on results of an empirical bounds estimation for all problem instances used during the FSP experiments is carried out. Based on this analysis, a final set of features to be used was established and is described.

Features Description

Many features are based on the mean processing time of the operations of a job j over some part of the machines. This mean processing time is denoted by μ_j^v ($\mu_j^v \in \mathbb{R}_0^+, u, v \in M$) and can be computed as:

$$\mu_j^v := \frac{1}{n} \sum_{i=u}^{v} p(j,i)$$

where u and v denote machines. The *mean processing time* μ_j ($\mu_j \in \mathbb{R}_0^+$) of a job j (over all machines) then is $\mu_j := \mu_j^1_m$. The mean processing time of a job is an instance feature and so would be the sum of the mean processing times over all jobs. By considering only some jobs of a solution, though, feature values will change from solution to solution and hence will become state features describing some characteristic of a solution. Dividing the jobs into quarter, four so-called *accumulated mean jobs quarter processing time* features result. These are denoted by *SumMeanTimeJobQuart1*, $\cdots$, *SumMeanTimeJobQuart4* for the four job quarters of a solution, respectively. By furthermore considering for each such feature only machines from one of four machine quarters, $4 \times 4 = 16$ additional so-called *accumulated mean jobs and machine quarters processing time* features result. These are denoted by *MeanTimeJobQuart1MachineQuart1*, $\cdots$, *MeanTimeJobQuart1MachineQuart2*, *MeanTimeJobQuart2MachineQuart1*, $\cdots$, *MeanTimeJobQuart4MachineQuart4*.

Some further features based on mean processing times of jobs are obtained by using a discounting technique similar to reinforcement learning (cf. Subsection 3.1.2). The mean processing times of the jobs of a solution π are summed as for the mean jobs quarter processing time features, now, however, each with a weight or discount factor of $\gamma_j := (\frac{\pi(j)}{n})^2$ ($\gamma_j \in [0, 1]$). The resulting feature for a solution π is denoted by *DiscountedMeanTimeJobBack* and $\overleftarrow{\lambda}_\pi$ and is computed as:

$$\overleftarrow{\lambda}_\pi := \frac{1}{n} \sum_{i=1}^{n} \left(\frac{i}{n}\right)^2 \mu_{\pi(i)}$$

The more jobs with higher mean processing time cluster at the end of a solution, the higher will be the weighed sum. Analog, jobs with higher mean time clustering at the begin of the solution can be assigned a higher weight yielding a feature denoted by *DiscountedMeanTimeJobFront* and $\overrightarrow{\lambda}_\pi$ and computed as:

$$\overrightarrow{\lambda}_\pi := \frac{1}{n} \sum_{i=0}^{n-1} \left(\frac{n-i}{n}\right)^2 \mu_{\pi(i)}$$

Finally, clustering of jobs with lower mean processing at both ends of a solution can be made to yield higher feature values. The respective feature is denoted by *DiscountedMeanTimeJobEnds* and $\overleftrightarrow{\lambda}_\pi$ and computed as:

$$\overleftrightarrow{\lambda}_\pi := \frac{1}{n} \sum_{i=1}^{n} \left(\frac{n-|n-2i-1|}{n}\right)^2 \mu_{\pi(i)}$$

Each of the last three so called *discounted mean job processing time* features can be varied by taking into account jobs from only one of the four job quarters of a solution yielding features called *discounted quarter mean job processing time* and denoted by *DiscountedMeanTimeJobBack1*, $\ldots$, *DiscountedMeanTimeJobBack4*, *DiscountedMeanTimeJobFront1* , $\ldots$, *DiscountedMeanTimeJobFront4*, and *DiscountedMeanTimeJobEnds1*, $\ldots$, *DiscountedMeanTimeJobEnds4*, respectively.

The final group of features developed in [Dot05] is called *idle times features*. They are computed based on idle times of job operations when jobs wait to be processed on a machine which still is occupied by the respective operation of the predecessor job. The idle times correspond to gaps in a Gantt-diagram for a solution of a FSP instance. The idle time features comprise one feature denoted by *IdleTimesCount* counting the number of idle times that occur altogether in a solution, one denoted by *MeanJobIdleTime* computing the mean idle time per job over all the job's idle times, three denoted by *Quart1IdleTimes*, $\ldots$, *Quart3IdleTimes* for the first, second and third quartiles over all operation idle times, one for the sum of all operation idle times denoted by *SumIdleTimes*, and finally, one computing the variance over all operation idle times denoted by *VarIdleTimes*.

All the features for the FSP described have been developed during the integration of the FSP into the GAILS framework in [Dot05]. For each feature individually, bounds have been developed and are described in detail in [Dot05] also. As an example, lower and upper bounds $\overline{C}_{max}$ and $\underline{C}_{max}$ ($\overline{C}_{max}, \underline{C}_{max} \in \mathbb{R}_0^+$) for the makespan, i.e. solution cost, can be computed according to:

$$\overline{C}_{max} := \max_{D \in \{D' \subseteq J \times M \, | \, |D'| = n+m-1\}} \sum_{(j,k) \in D} p_{jk}$$

$$\underline{C}_{max} := \max_{j \in J} \sum_{k \in M} p_{jk}$$

The upper bound is justified by the fact that for each machine n operations, one for each job, must be executed in sequence and the fact that the last machine can only start its processing for the first job when $m-1$ sequentially started operations for the first job on the preceding machines have finished, yielding altogether $n+m-1$ operations that will be run sequentially. As concerns the justification of the lower bound, it certainly holds true that for all jobs at least all of their operations must be processed by all machines in sequence. For each machine, the sum over a job's operation processing times represents the minimum time a jobs needs processing time over all machines, if no idle time occurs. The minimal makespan then cannot be less than the maximum computed over these job-wise minimal processing times. The lower bound computation basically is the solution to a corresponding single machine problem. A shorter makespan than for the best solution for the corresponding single machine problem instance is not possible for the original problem instance. The cost in the form of the makespan is denoted by *Makespan* also, the theoretically normalized cost or makespan is denoted by *NormMakespan*.

Feature Bounds Estimation

The parameter settings for the bounds estimation experiment for the FSP features is shown in Table 6.31. It only differs from the bounds estimation experiment parameter settings for the TSP in the problem instances, termination criterion, local search procedures, and perturbations used (cf. Table 6.3). The perturbations were chosen in order to cover a stronger and a weaker strength for each perturbation used for the FSP experiments (cf. Subsection 6.5.1). The local search procedure based on the 2-exchange neighborhood structure was not used since it showed too long runtimes. The problem instances used for the bounds estimation and for the other FSP experiments were taken from the Taillard-repository [Tai05]. The problem instances are of different sizes and have been generated randomly. The problem instances used in the FSP experiments carried out for this thesis are listed in Table 6.31 together with the cost of their known global optima or with a lower bound for the cost of the global optima. For problem instances ta051, ..., ta060, and ta081, ..., ta090, only lower bounds are known. All other costs refer to known and proven global optima. The sizes for the problem instances are given in the last column named "Size" in the format #Jobs × #Machines.

Short	Actual Parameter Values
tc	3000
t_a	15
i_a	ta021, ta022, ta023, ta024, ta025, ta026, ta027, ta028, ta029, ta030
$pert$	partNeh[0.1], partNeh[0.2], compRandEx[1], compRandEx[2]
ls	in[first]
$init$	rand
ϵ_{a_a}	0.0, 0.4, 1.0

Table 6.31: Actual parameter values for the first feature bounds estimation for the FSP

All FSP experiments were run on a cluster of computers located at the Center for Scientific Computing at the University of Frankfurt [Url05a] on 1.8GHz 64 bit AMD Opteron 244 processors with 1024 KB cache and 4096 MB main memory.

The results of the empirical bounds estimation are shown in Table 6.35 and in Table 6.34. As before in Subsection 6.4.2, tables 6.34 and 6.35 present the feature bounds results. The former is sorted

Short	Actual Parameter Values
tc	15000
i_a	ta051, ta052, ta053, ta054, ta055, ta056, ta057, ta058, ta059, ta060, ta071, ta072, ta073, ta074, ta075, ta076, ta077, ta078, ta079, ta080, ta081, ta082, ta083, ta084, ta085, ta086, ta087, ta088, ta089, ta090

Table 6.32: Actual parameter values for the second feature bounds estimation for the FSP

	Problem Instances										Size
Name	ta021	ta022	ta023	ta024	ta025	ta026	ta027	ta028	ta029	ta030	20×20
Cost	2297	2099	2326	2223	2291	2226	2273	2200	2237	2178	
Name	ta051	ta052	ta053	ta054	ta055	ta056	ta057	ta058	ta059	ta060	50×20
Cost	3771	2668	3591	3635	3553	3667	3672	3627	3645	3696	
Name	ta071	ta072	ta073	ta074	ta075	ta076	ta077	ta078	ta079	ta080	100×10
Cost	5770	5349	5676	5781	5467	5303	5595	5617	5871	5845	
Name	ta081	ta082	ta083	ta084	ta085	ta086	ta087	ta088	ta089	ta090	100×20
Cost	6106	6183	6252	6254	6262	6302	6184	6315	6204	6404	

Table 6.33: Problem instances for the FSP and costs of or lower bounds on global optima

according to descending coverage except for the first two individual features named "Makespan" and "NormMakespan". Note that feature "Makespan" does not have entries in its last two columns since the criteria presented in these two columns do not apply. Table 6.35 is sorted according to increasing common coverage.

Two features of Table 6.34 have a coverage larger than 100% which means that their theoretical bounds are too narrow. They exceed 100% only slightly, though, so this does not pose a problem (cf. Subsection 6.3.2). All other features of Table 6.34 except for the theoretically normalized makespan have a very good coverage of at least 76.99%, so there is no need to employ empirical normalization for them. Even the coverage of 33.75% for the theoretically normalized makespan is considered high enough to do without empirical bounds normalization altogether. As a result, the experiments conducted for the FSP and described in this section were run without employing empirical normalization. In fact, during the experiments described next, no problems with too close feature or cost values were encountered. The common coverage of the FSP features from Table 6.34 is high compared to the results for the TSP features in Table 6.34. As a result, all features presented in Table 6.34 were used in the experiments for the FSP that are described in the next subsection.

In contrast, the coverage and common coverage for all idle times features presented in Table 6.35 is comparably poor. Some bounds are far too narrow and the common coverage is only in one case above zero. For these reasons, the idle times features were not used.

6.5.3   Experiments

The design of the experiments for the FSP is as before for the experiments conducted for the TSP. First, a good local search procedure is looked for in a local search procedure comparison experiment

Feature Name	Minimum	Maximum	∩ (in %)	∪ (in %)
Makespan	1908.0000	7244.0000	–	–
NormMakespan	0.2954	0.6329	33.75	10.47
DiscountedMeanTimeJobEnds	0.2421	1.3164	107.43	20.38
DiscountedMeanTimeJobEnds1	0.2007	1.2121	101.14	29.36
DiscountedMeanTimeJobEnds2	0.2338	1.1985	96.47	33.11
DiscountedMeanTimeJobEnds3	0.2185	1.1687	95.02	33.93
MeanTimeJobQuart4MachineQuart4	0.0745	1.0000	92.55	36.31
MeanTimeJobQuart1MachineQuart1	0.0822	1.0000	91.78	28.28
DiscountedMeanTimeJobEnds4	0.2053	1.1219	91.66	33.61
DiscountedMeanTimeJobFront4	0.0245	0.9410	91.64	35.43
DiscountedMeanTimeJobBack3	0.0552	0.9704	91.52	30.13
DiscountedMeanTimeJobFront	0.0395	0.9533	91.38	28.54
DiscountedMeanTimeJobFront3	0.0471	0.9542	90.71	29.04
DiscountedMeanTimeJobBack	0.0520	0.9581	90.61	35.05
DiscountedMeanTimeJobBack4	0.0726	0.9691	89.66	39.31
DiscountedMeanTimeJobBack2	0.0546	0.9463	89.17	30.57
DiscountedMeanTimeJobFront2	0.0575	0.9414	88.39	30.28
DiscountedMeanTimeJobBack1	0.0221	0.9036	88.14	38.86
DiscountedMeanTimeJobFront1	0.0915	0.9724	88.09	40.87
MeanTimeJobQuart2MachineQuart1	0.1322	1.0000	86.78	35.76
MeanTimeJobQuart2MachineQuart2	0.1337	1.0000	86.63	41.58
MeanTimeJobQuart3MachineQuart3	0.1378	1.0000	86.22	44.95
MeanTimeJobQuart2MachineQuart3	0.1431	1.0000	85.69	43.20
MeanTimeJobQuart3MachineQuart4	0.1482	1.0000	85.18	37.32
MeanTimeJobQuart4MachineQuart3	0.1483	1.0000	85.17	43.87
MeanTimeJobQuart3MachineQuart2	0.1490	1.0000	85.10	45.35
MeanTimeJobQuart3MachineQuart1	0.1490	1.0000	85.10	35.72
SumMeanTimeJobQuart1	0.1510	1.0000	84.90	27.81
MeanTimeJobQuart1MachineQuart2	0.1598	1.0000	84.02	38.72
MeanTimeJobQuart2MachineQuart4	0.1718	1.0000	82.82	38.33
MeanTimeJobQuart4MachineQuart1	0.1798	1.0000	82.02	41.99
MeanTimeJobQuart4MachineQuart2	0.1864	1.0000	81.36	44.95
MeanTimeJobQuart1MachineQuart3	0.1982	1.0000	80.18	45.86
SumMeanTimeJobQuart4	0.2003	1.0000	79.97	30.97
SumMeanTimeJobQuart2	0.2158	1.0000	78.42	41.79
MeanTimeJobQuart1MachineQuart4	0.2164	1.0000	78.36	42.56
SumMeanTimeJobQuart3	0.2301	1.0000	76.99	41.90

Table 6.34:   Feature bounds analysis for the FSP

Feature Name	Minimum	Maximum	∩ (in %)	∪ (in %)
VarIdleTimes	0.0000	0.4599	45.98	0.15
Quart1IdleTimes	0.0000	0.7987	79.87	-2.12
IdleTimesCount	0.0556	0.6789	62.34	-4.27
Quart3IdleTimes	0.0124	2.2864	227.40	-7.42
Quart2IdleTimes	0.0000	1.2291	122.91	-7.74
SumIdleTimes	0.0468	1.7289	168.20	-21.31
MeanJobIdleTime	0.0258	1.2942	126.85	-23.13

Table 6.35: Feature bounds analysis for idle times features of the FSP

on relatively fast to solve instances. Next, some good perturbations and parameterizations thereof are identified in a perturbation comparison experiment. Finally, based on the results of these experiments, some good performing ILS-actions with varying strengths are assembled and tested on some harder to solve instances in a single-instance and a multi-instance experiment. The results that are not displayed in this section directly can be viewed in Appendix B.

Local Search Procedure and Perturbation Comparison

The actual parameter settings and results for the local search procedure and perturbation comparison experiment are presented in Appendix B.1 for the local search procedure experiment and in Appendix B.2 for the perturbation comparison experiment. Only summarizing results are presented next.

The situation for the two local search procedures employed, one first improvement insert and one first improvement 2-exchange local search procedure is clear. According to the results of the normal ILS jobs (see Table B.3) also and the ranking results (see Table B.4), the insert local search procedure is performing substantially better. The second rank for the random ILS algorithm, which chooses randomly between the two local search procedures, also indicates that it is better to choose the insert local search procedure, even if this happens only every second move. As regards the results of the Q-ILS jobs (see Table B.6), the random ILS jobs do not perform well. Noticeable again is the influence of the acceptance criterion. Only ϵ-better acceptance criteria were used according to the actual parameter values presented in Table B.5 (the full table would contain 161 entries), but the Q-ILS jobs with lowest ϵ-value for the application phase performed best. This table does not show any consistent trend over all instances among the actual parameter settings for learning parameters. The action usage plots for the best Q-ILS job for each instance ta029 and ta030 in Figure B.1 also are very clear: Basically exclusively the ILS-action based on the insert local search procedure is used. All subsequently described experiments were conducted with the insert local search procedure only.

The perturbation comparison experiment is split into two parts. The first part investigates the behavior of the composite exchange perturbation from Subsection 6.5.1, the second part that of the partial NEH perturbation introduced in the same subsection. Both perturbations were combined in three different strengths with the insert local search procedure and run as normal and random ILS and as Q-ILS jobs on some selection of instances with different sizes in single- and multi-instance mode sub-experiments. The actual parameter values for the diverse sub-experiments and the results are presented in Appendix B.2 (the full Q-ILS presentation tables there would contain 225 entries, in the Q-ILS ranking tables one random ILS job algorithm is missing).

Summarizing, according to the single- and multi-instance sub-experiments run for small instances, the best Q-ILS jobs typically performed significantly better than the best normal ILS jobs, but no trend for learning parameters becomes apparent. The performances of all ILS-action variants for both parts were rather mixed. For example, the ranking values for a ranking procedure according to Subsection 6.1.2 carried out for three small instances in each experiment part were not severely different. What strikes for these experiments is that the average excess for many instances is very small. The values are rather close to each other for the jobs run, reducing the validity of these results. The only true trend that is discernible for both experiment parts is that the lower the strength of a perturbation of an ILS-action, the better the performance for the corresponding normal ILS. Another remarkable effect is the different influence of the choice of acceptance criterion compared to the TSP. Whereas for the TSP a true better acceptance criterion by far excelled any other choice, for the FSP, a small ϵ-value for an ϵ-better acceptance criterion seems to be better than a "pure" better acceptance criterion. Still the ϵ-values for the best jobs run are either 0.0 or 0.1 (none used 0.5) during the application phase, but the 0.1 values are predominant and a definite trend towards using the ϵ-better acceptance criterion with a small ϵ-value was visible. Together with the fact that at least on larger instances ILS-actions with lower strength perturbations performed better, the conjecture is that the solutions of the FSP rather have to be improved by small steps and gradually. Exploration should better be provided by sometimes accepting a worsening move instead of using stronger ILS-actions.

Short	Actual Parameter Values
tc	15000
i_{tr}	ta054, ta074, ta084
i_a	ta054, ta074, ta084
$strat$	Q(0), Q(λ)
$pert$	partNeh[0.01], partNeh[0.05], partNeh[0.1], compRandEx[1], compRandEx[2], compRandEx[3]
ls	in[first]
$feat$	FSP-Features-AllButIdletime.feat
$init$	neh
ϵ_{a_t}	0.1, 0.5
ϵ_{a_a}	0.0, 0.1, 0.3
ϵ_{π_a}	0.0, 0.2
γ	0.3, 0.8
r	c_Δ, c_i
λ	0.4, 0.8
fa	svr[0.0035]
lf	2000

Table 6.36: Actual parameter values for the FSP single-instance experiment

Since it was not possible to truly exclude any perturbation strength for either perturbation tested after the perturbation comparison experiment as inferior, the subsequently described experiments provided six ILS-actions, one for each combination of the insert local search procedure and one of the two perturbations in one of the three strengths. The resulting ILS-actions are denoted by *strength g partial NEH ILS-action* or *partial NEH ILS-action with strength g* ($g \in [0,1]$) for the

ILS-actions using the partial NEH perturbation with strength g, and *strength h composite exchange ILS-Action* or *composite exchange ILS-action with strength h* ($h \in \mathbb{N}^+$) for the ILS-actions using a composite exchange perturbation with strength h. The ILS-actions are denoted by "partNeh[g]" and "compEx[h]", respectively, in plots and tables.

Single-Instance

The actual parameter values for the single-instance experiment for the FSP is displayed in Table 6.36. The results for the normal ILS jobs are shown in Table 6.37. These results show a clear trend. The best ILS-action according to the performance of the corresponding normal ILS jobs is the strength 0.01 partial NEH ILS-action followed by the strength 1 composite exchange ILS-action. The difference to the next best normal ILS jobs according to average excess are substantial. The worst performing ILS-actions are the partial NEH ILS-actions with strengths 0.05 and 0.1. Assuming that these two ILS-actions are rather strong, the trend that became visible for the perturbation comparison experiment already is continued: For larger instances, ILS-actions with low strength perturbations perform better.

| Instance | Perturbation | LS | Statistics | | | | |
			avg	σ	avg.excess	min	max
ta054	partNeh[0.01]	in[first]	3750.28	7.99	0.5168 %	3731	3767
(3635)	compRandEx[1]	in[first]	3753.44	6.49	0.6014 %	3742	3767
	partNeh[0.05]	in[first]	3755.52	5.21	0.6572 %	3748	3768
	compRandEx[2]	in[first]	3759.40	6.87	0.7612 %	3739	3767
	compRandEx[3]	in[first]	3764.28	7.83	0.8920 %	3745	3778
	partNeh[0.1]	in[first]	3773.56	6.44	1.1407 %	3756	3782
ta074	partNeh[0.01]	in[first]	5794.80	8.78	0.2387 %	5781	5809
(5781)	compRandEx[1]	in[first]	5795.40	15.77	0.2491 %	5781	5826
	compRandEx[3]	in[first]	5809.00	12.43	0.4843 %	5791	5826
	compRandEx[2]	in[first]	5809.40	14.38	0.4913 %	5782	5826
	partNeh[0.05]	in[first]	5828.08	3.79	0.8144 %	5821	5838
	partNeh[0.1]	in[first]	5834.12	5.43	0.9189 %	5822	5842
ta084	partNeh[0.01]	in[first]	6316.72	12.47	0.7612 %	6303	6350
(6254)	compRandEx[1]	in[first]	6324.12	12.13	0.8792 %	6303	6353
	compRandEx[2]	in[first]	6336.04	17.73	1.0694 %	6303	6366
	compRandEx[3]	in[first]	6339.88	22.84	1.1306 %	6304	6366
	partNeh[0.05]	in[first]	6363.72	8.92	1.5109 %	6342	6372
	partNeh[0.1]	in[first]	6378.28	9.97	1.7432 %	6364	6397

Table 6.37: Normal ILS job results in the FSP single-instance experiment

Looking at Table 6.38 displaying the results for the Q-ILS jobs of the single-instance experiment for the FSP, one can see that the choice of the better acceptance criterion for larger instance for the FSP also is crucial. Among the best Q-ILS jobs shown for each instance in Table 6.38, no one employs an ϵ-better acceptance criterion even with a small ϵ-value for 0.1. For the training phase, only ϵ-better acceptance criterion with varying ϵ-values were used. The trend here is also to better choose a small ϵ-value. Only for the largest instance, ta084, this trend is not confirmed. Another trend that strikes is that the inverse cost reward signal is mostly used by the best Q-ILS jobs and only values of 0.8 are used by the best Q-ILS jobs during their training phase for the γ parameter. Additional regularities comprise a slight clustering of some Q-ILS jobs with the same actual parameter setting on the best positions. The actual parameter setting for the Q-ILS job at position 1 for instance ta054 is the same as for position 5 Q-ILS on instance ta074 and position 1 Q-ILS job on instance ta084. The actual parameter setting for position 2 Q-ILS job on instance ta074 is same as for position 3 Q-ILS job on instance ta084. This kind of concentration was not observed for the TSP. What can also be seen is that the larger an instance is, the more elaborated are the differences in performance between the best Q-ILS jobs and the best normal ILS job for this instance. In case of instance ta084, the difference becomes significant according to the t-test at least. The Wilcoxon test shows no reaction, though. On instance ta084, two training phase jobs are in the top six positioned Q-ILS jobs which is remarkable also.

Action usage plots for position 1 Q-ILS job on instance ta054 (subfigure (a)), position 1, 2, and 5 Q-ILS jobs on instance ta074 (subfigures (c), (b), and (d), respectively), and positions 1 and 3 Q-ILS jobs on instance ta084 (subfigures (e) and (f)) are shown in Figure 6.20. Note that only position 1 Q-ILS job on instance ta074 employs an ϵ-greedy policy during the application phase. Most of the action usage plots provide a clear picture: The partial NEH ILS-action with strength 0.01 is basically chosen all the time or most frequently in combination with the strength 1 composite exchange ILS-action. This makes sense, since the corresponding normal ILS jobs for strength 0.01 partial NEH ILS-action on all three instances performs best, too, followed by the strength 1 composite exchange normal ILS jobs. Only position 1 and 2 Q-ILS jobs on instance ta074 (subfigures (c) and (b), respectively) show different action selection strategies. They are similar to each other in that first the partial NEH ILS-action with strength 0.01 is used, but after some 4000 iterations, all of a sudden the action usage is changed to using two other ILS-actions, one of which, however, differs for the position 1 and 2 Q-ILS jobs.

The common ILS-action, used between 20 and 50 % of the time by both Q-ILS jobs , is the strength 1 composite exchange ILS-action. The differing ILS-action is the strength 2 composite exchange ILS-action for position 1 Q-ILS job and the strength 3 composite exchange ILS-action for position 2 Q-ILS job. Note also that in the very end of the search, position 1 Q-ILS job on instance ta074 also increasingly uses the strength 3 composite exchange ILS-action. Since both corresponding normal ILS jobs perform third and fourth best on instance ta074, this is not really surprising. Comparing the action usage for the Q-ILS jobs with identical actual parameter settings (on the one hand position 1 Q-ILS job for instance ta054, position 5 Q-ILS on instance ta074, and position 1 Q-ILS job on instance ta084, on the other hand position 2 Q-ILS job on instance ta074 and position 3 Q-ILS job on instance ta084), in one case it can be seen that the Q-ILS-jobs with actual parameter setting identical to the position 1 Q-ILS job on instance t054 basically all learned the same action selection strategy. The action selection strategy for the other Q-ILS jobs with identical actual parameter setting (see position 2 Q-ILS on instance ta074, subfigure (b) of Figure 6.20 and position 3 Q-ILS job on instance ta084, subfigure (b) of Figure 6.20) differ, though. The action usage of the Q-ILS jobs run on instance ta074 differ also (see subfigures (b), (c), and (d)). No trend was established in this direction.

The single-instance experiment for the FSP shows that learning is possible, but also shows the need to identify yet invisible other influences that lead to partly confusing results. The distinct superiority of two ILS-actions that were indeed most of the time learned to be used make it difficult to clearly show improvements due to learning. As regards learning nothing else than to show that the obvious was learned can be achieved in this case. In this context it comes as a surprise that other action selection strategies that do not use the two best ILS-actions all of the time yield also good performances which is worth further investigation.

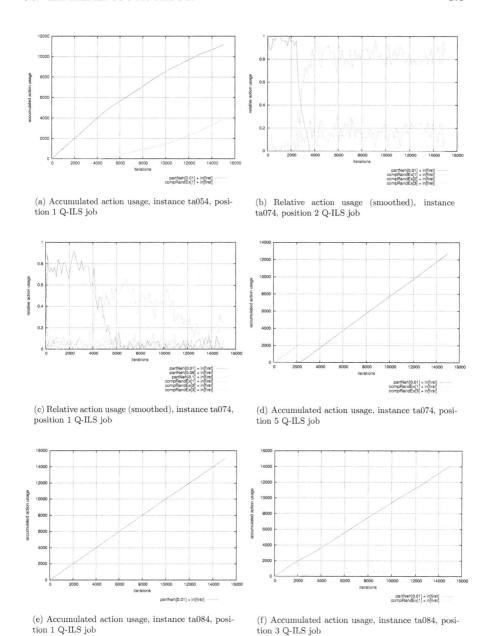

(a) Accumulated action usage, instance ta054, position 1 Q-ILS job

(b) Relative action usage (smoothed), instance ta074, position 2 Q-ILS job

(c) Relative action usage (smoothed), instance ta074, position 1 Q-ILS job

(d) Accumulated action usage, instance ta074, position 5 Q-ILS job

(e) Accumulated action usage, instance ta084, position 1 Q-ILS job

(f) Accumulated action usage, instance ta084, position 3 Q-ILS job

Figure 6.20: Action usage plots for several Q-ILS jobs run on instance ta054, ta074, and ta084 in the FSP single-instance experiment

| Instance | Job | Type | Parameters | | | | | | | H_0: equal | | H_0: not better | | T | Statistics | | | | |
			ϵ_{π_a}	ϵ_{α_a}	ϵ_{α_t}	γ	λ	strat	r	0.05	0.1	0.05	0.1		avg	σ	avg. exc.	min	max
ta054 (3635)	1	apply	0.0	0.0	0.1	0.8	0.8	$Q(\lambda)$	c_i	=	=	=	=	0.4216	3749.36	7.43	0.4921 %	3739	3768
	2	apply	0.2	0.0	0.1	0.8		$Q(0)$	c_i	=	=	=	=	0.0909	3750.08	7.56	0.5114 %	3736	3763
	3	apply	0.2	0.0	0.1	0.8	0.8	$Q(\lambda)$	c_i	=	=	=	=	0.0717	3750.12	7.79	0.5125 %	3735	3766
	4	apply	0.0	0.0	0.1	0.3		$Q(0)$	c_i	=	=	=	=	-0.4183	3751.24	8.23	0.5425 %	3739	3768
	⋯	⋯	⋯	⋯	⋯	⋯	⋯	⋯	⋯	⋯	⋯	⋯	⋯	⋯	⋯	⋯	⋯	⋯	⋯
	45	rand		0.0						=	–	=	=	-1.7127	3754.20	8.19	0.6218 %	3739	3773
ta074 (5781)	1	apply	0.2	0.0	0.1	0.8	0.4	$Q(\lambda)$	c_i	=	=	=	=	1.0280	5792.24	8.83	0.1944 %	5781	5810
	2	apply	0.0	0.0	0.1	0.8		$Q(0)$	c_i	=	=	=	=	0.7951	5792.56	11.02	0.2000 %	5781	5826
	3	apply	0.0	0.0	0.1	0.3		$Q(0)$	c_i	=	=	=	=	0.8099	5792.76	9.03	0.2034 %	5781	5805
	4	apply	0.2	0.0	0.1	0.3		$Q(0)$	c_i	=	=	=	=	0.5840	5793.36	8.65	0.2138 %	5781	5811
	5	apply	0.0	0.0	0.1	0.8	0.8	$Q(\lambda)$	c_i	=	=	=	=	0.4507	5793.72	8.15	0.2200 %	5781	5810
	⋯	⋯	⋯	⋯	⋯	⋯	⋯	⋯	⋯	⋯	⋯	⋯	⋯	⋯	⋯	⋯	⋯	⋯	⋯
	43	rand		0.0						=	–	=	=	-1.5773	5799.56	12.27	0.3211 %	5781	5826
ta084 (6254)	1	apply	0.0	0.0	0.1	0.8	0.8	$Q(\lambda)$	c_i	=	=	+	+	1.8471	6310.84	9.89	0.6674 %	6293	6326
	2	train			0.5	0.8		$Q(0)$	c_Δ	+	=	+	+	1.5847	6311.60	10.27	0.6795 %	6303	6326
	3	apply	0.0	0.0	0.1	0.8		$Q(0)$	c_i	=	=	+	+	1.6017	6311.76	9.18	0.6821 %	6303	6325
	4	apply	0.2	0.0	0.5	0.8		$Q(0)$	c_i	=	=	=	+	1.3802	6312.08	11.27	0.6872 %	6303	6336
	5	apply	0.0	0.0	0.5	0.8		$Q(0)$	c_i	=	=	=	=	1.1913	6312.52	12.46	0.6942 %	6292	6344
	6	train			0.1	0.8	0.8	$Q(\lambda)$	c_i	=	=	=	=	1.0622	6313.05	11.96	0.7027 %	6303	6347
	⋯	⋯	⋯	⋯	⋯	⋯	⋯	⋯	⋯	⋯	⋯	⋯	⋯	⋯	⋯	⋯	⋯	⋯	⋯
	45	rand		0.0						–	–	=	=	-1.5357	6321.04	6.51	0.8301 %	6303	6326

Table 6.38: Q-ILS job results in the FSP single-instance experiment

The suspicion, however, remains that some results simply are gross outliers that can happen for highly randomized algorithms. A reason why the results might not be overly representative can be the fact that all jobs were not run long enough. Coming back to the problem of "saturation" discussed during the discourse for the perturbation comparison experiment for the TSP in Subsection 6.5.3, Figure 6.20 reveals that the search process might not have been stabilized after 15000 iteration for the larger instances and hence might have penalized some ILS-actions with the effect that they even at the end of the search could not unfold their full power. On the other hand, the lines reveal a small and steady improving which rather seems to favor the application of low strength ILS-actions that are able to gradually improve a solution for a long period of time.

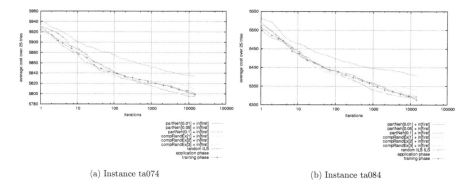

     (a) Instance ta074   (b) Instance ta084

Figure 6.21: Runtime development plot for Q-ILS job 1 and normal ILS job 1 run on instances ta074 and ta084 in the FSP single-instance experiment

Multi-Instance

Finalizing the experiment presentation for the FSP, the multi-instance experiment for the FSP is presented. It is also divided into two parts as the multi-instance experiments for the (cf. Subsection 6.5.3). The first part only uses instances of size 50×20 for training, application and test phase, the second part varies instances over sizes 50×20, 100×10, and 100×20. The other parameters are as for the single-instance experiment for the FSP, except for an adjustment of the ν parameter for the function approximator as discussed in Subsection 6.3.1. The actual parameter values for the two parts are shown in tables 6.39 and 6.40.

The number of iterations was set to 20000 in order to ensure a proper stabilization of the search process. Running 15000 iterations was only used for the training and application phase in order to save some computation time. The ranking tables for the normal ILS algorithms for the two parts of the multi-instance experiment are shown as tables 6.41 and 6.42. They again underline the superior performance of the supposedly two least strength ILS-actions (partial NEH ILS-action with strength 0.01 and the composite exchange ILS-action with strength 1). The third place of the random ILS algorithm and the difference in the ranking values for the then following ranks indicate that even only occasionally using the two best ILS-actions is better than never using them. The worst ILS-actions, with huge difference, is the strength 1 partial NEH ILS-action. Perhaps the perturbation of this ILS-action simply is too strong and not useful for the FSP.

Short	Actual Parameter Values
tc	15000
tc_{te}	20000
i_{tr}	ta057:ta058:ta059:ta060
i_a	ta057:ta058:ta059:ta060
i_{te}	ta052, ta053, ta056
$strat$	$Q(0)$, $Q(\lambda)$
$pert$	partNeh[0.01], partNeh[0.05], partNeh[0.1], compRandEx[1], compRandEx[2], compRandEx[3]
ls	in[first]
$feat$	FSP-Features-AllButIdletime.feat
$init$	neh
ϵ_{a_t}	0.1, 0.5
ϵ_{a_a}	0.0, 0.1, 0.5
ϵ_{π_a}	0.0, 0.2
γ	0.3, 0.8
r	c_Δ, c_i
λ	0.4, 0.8
fa	svr[0.0015]
lf	2000

Table 6.39: Actual parameter values for the first part of the FSP multi-instance experiment

Short	Actual Parameter Values
i_{tr}	ta052:ta056:ta072:ta076:ta086
i_a	ta052:ta056:ta072:ta076:ta086
i_{te}	ta055, ta075, ta084
fa	svr[0.001]

Table 6.40: Changed actual parameter values for the second part of the FSP multi-instance experiment

The ranks for the Q-ILS jobs are shown in tables 6.43 and 6.44. In total each of the tables would comprise 102 Q-ILS jobs. In contrast to the experience for the single-instance experiment, there does not seem to be a regularity underlying the distribution of good actual parameter settings for Q-ILS jobs other than the choice of the acceptance criterion. Only for the second experiment part, the inverse cost reward signal and the $Q(0)$-ILS learning strategy are used by the Q-ILS jobs that occupy the first ranks. Whether this is by chance or not is not evident. The ranking values, except for the one for the rank 1 Q-ILS of the second experiment part, are relatively close, for example.

As was done for the multi-instance experiment for the TSP, for each combination of actual parameter values for parameters learning strategy, reward signal, and ϵ-policy (during) application, the top ranked Q-ILS algorithms were chosen to be run in the test phase, but only, if they were not ranked too bad. For the first experiment part, rank 1, 2, 3, 4, 5, 8, and 12 Q-ILS algorithms were run, for the second part rank 1, 4, 5, 6, 8, 9, 13, 14 Q-ILS jobs. The test instances were chosen randomly

Rank	Perturbation	LS	Value
1	compRandEx[1]	in[first]	5330
2	partNeh[0.01]	in[first]	5386
3	random ILS		6863
4	partNeh[0.05]	in[first]	7905
5	compRandEx[2]	in[first]	8100
6	compRandEx[3]	in[first]	10904
7	partNeh[0.1]	in[first]	15757

Table 6.41:   Ranks of normal and random ILS algorithms in the first part of the FSP multi-instance experiment

Rank	Perturbation	LS	Value
1	partNeh[0.01]	in[first]	4154
2	compRandEx[1]	in[first]	5521
3	random ILS		5983
4	compRandEx[2]	in[first]	7642
5	partNeh[0.05]	in[first]	8819
6	compRandEx[3]	in[first]	9769
7	partNeh[0.1]	in[first]	15339

Table 6.42:   Ranks of normal and random ILS algorithms in the second part of the FSP multi-instance experiment

Rank	r	γ	λ	ϵ_{a_t}	ϵ_{π_a}	ϵ_{a_a}	strat	Value
				Parameters				
1	c_Δ	0.3		0.1	0.2	0.0	Q(0)	35336
2	c_i	0.8	0.8	0.5	0.0	0.0	Q(λ)	36166
3	c_i	0.8		0.5	0.2	0.0	Q(0)	38144
4	c_Δ	0.3		0.1	0.0	0.0	Q(0)	38934
5	c_i	0.8		0.5	0.0	0.0	Q(0)	39084
6	c_i	0.8	0.4	0.1	0.2	0.0	Q(λ)	40784
7	c_i	0.8	0.8	0.5	0.2	0.0	Q(λ)	41049
8	c_Δ	0.8	0.4	0.1	0.2	0.0	Q(λ)	41178
9	c_i	0.8		0.1	0.2	0.0	Q(0)	41429
10	c_i	0.8	0.4	0.5	0.0	0.0	Q(λ)	41484
11	c_Δ	0.8		0.1	0.2	0.0	Q(0)	41500
12	c_Δ	0.8	0.8	0.5	0.0	0.0	Q(λ)	41932
	$\vdots$	$\vdots$	$\vdots$	$\vdots$	$\vdots$	$\vdots$	$\vdots$	$\vdots$

Table 6.43:   Ranks of Q-ILS algorithms in the first part of the FSP multi-instance experiment

Rank	Parameters							Value
	r	γ	λ	ϵ_{a_t}	ϵ_{π_a}	ϵ_{d_a}	strat	
1	c_i	0.8		0.5	0.0	0.0	Q(0)	25473
2	c_i	0.3		0.1	0.0	0.0	Q(0)	32109
3	c_i	0.8		0.5	0.0	0.0	Q(0)	36454
4	c_i	0.8		0.5	0.2	0.0	Q(0)	37524
5	c_Δ	0.3		0.5	0.0	0.0	Q(0)	38872
6	c_i	0.8		0.1	0.2	0.0	Q(0)	39343
7	c_i	0.8		0.1	0.0	0.0	Q(0)	40109
8	c_i	0.8	0.4	0.5	0.0	0.0	Q(λ)	40624
9	c_Δ	0.3	0.8	0.1	0.0	0.0	Q(λ)	40664
10	c_i	0.8		0.5	0.2	0.0	Q(0)	41159
11	c_Δ	0.3	0.8	0.5	0.0	0.0	Q(λ)	43057
12	c_Δ	0.3		0.5	0.2	0.0	Q(0)	44625
13	c_Δ	0.3	0.8	0.5	0.2	0.0	Q(λ)	44827
14	c_i	0.8	0.4	0.5	0.2	0.0	Q(λ)	45250
$\vdots$	$\vdots$	$\vdots$	$\vdots$	$\vdots$	$\vdots$	$\vdots$	$\vdots$	$\vdots$

Table 6.44:   Ranks of Q-ILS algorithms in the second part of the FSP multi-instance experiment

as to comprise instances not yet in the respective parts. For the second part, each of the three available instance sizes had to be included. The results of the test phase of the multi-instance experiment for the FSP are shown in Table 6.45 for the first experiment part and in Table 6.46 for the second. The average excesses shown in these tables do not fall below 6 % for one instance and are otherwise not lower than 30 % which is an indication that the runtime was not too high leveling out any performance differences again. Looking at the runtime development plot for all jobs run on the two largest instances for both experiment parts in Figure 6.23, the runtime perhaps was still not long enough, since the lines still show some slope at the end of the search and hence probably the represented jobs still occasionally were able to find new overall best solutions (although it has to be taken into account that the plots are logarithmically scaled for the x-axis).

Except for the two best ranked ILS algorithms, the Q-ILS algorithms in tendency did better than the ILS algorithms (including the random ILS jobs during the first part of the multi-instance experiments for the FSP). The ranking trend for the ILS algorithms is affirmed by the test phase results, whereas the positions of the Q-ILS algorithms are rather scattered. The second part of the multi-instance experiment confirms the good ranking result for the best ranked Q-ILS algorithm from the application phase and the ranking for the normal ILS algorithms also. The positions of the Q-ILS jobs over the test instances are more consistent than for the first part, though. It is seldom the case that a Q-ILS job is significantly worse than the best job and only for an instance of size 50×20 a normal ILS job is best. Looking at the runtime development plots for instances ta056 and ta083, it becomes clear that the worst normal ILS jobs are far behind as regards performance. Especially for the second part of the multi-instance experiment for the FSP, the Q-ILS jobs are performing generally better than the normal ILS jobs, for the two largest instance ta075 and ta084 the differences seem to be rather significant. There, the rank 1 Q-ILS algorithm performs significantly better than almost any other algorithm.

Instance	Rank	Type	H_0: equal		H_0: not better			Statistics				
			0.05	0.1	0.05	0.1	T	avg	σ	avg. exc.	min	max
ta052	1	ILS(1)						3719.65	6.75	0.3412 %	3709	3740
(2668)	3	Q-ILS	=	=	=	=	0.3396	3720.12	5.72	0.3541 %	3708	3735
	4	Q-ILS	=	=	=	=	0.3978	3720.30	7.83	0.3588 %	3708	3741
	2	Q-ILS	=	=	=	=	0.4693	3720.38	7.06	0.3608 %	3709	3742
	12	Q-ILS	=	–	=	=	1.1131	3721.38	7.11	0.3878 %	3714	3744
	5	Q-ILS	=	=	=	=	1.0631	3721.53	8.88	0.3918 %	3708	3749
	8	Q-ILS	=	–	=	–	1.4498	3722.10	8.29	0.4073 %	3710	3741
	1	Q-ILS	–	–	–	–	1.6879	3722.62	8.87	0.4215 %	3708	3744
	2	ILS(0.01)	–	–	–	–	1.8719	3722.80	8.23	0.4262 %	3709	3744
	6	Q-ILS	–	–	–	–	1.9981	3722.97	8.07	0.4309 %	3714	3751
	5	ILS(2)	–	–	–	–	2.4365	3723.43	7.10	0.4431 %	3714	3744
	3	ILS(R)	–	–	–	–	2.5796	3724.62	10.16	0.4755 %	3713	3751
	4	ILS(0.05)	–	–	–	–	4.5417	3727.28	8.20	0.5469 %	3714	3745
	6	ILS(3)	–	–	–	–	4.4472	3727.65	9.16	0.5571 %	3715	3751
	7	ILS(0.1)	–	–	–	–	15.9982	3751.10	10.44	1.1896 %	3727	3775
ta053	6	Q-ILS						3675.25	8.93	0.8853 %	3661	3699
(3591)	2	ILS(0.01)	=	=	=	=	0.4501	3676.15	8.95	0.9100 %	3655	3696
	4	Q-ILS	=	=	=	=	1.0264	3677.43	9.99	0.9450 %	3658	3696
	3	Q-ILS	=	=	=	=	1.0598	3677.55	10.42	0.9484 %	3662	3706
	1	Q-ILS	=	=	=	=	1.1476	3677.62	9.57	0.9505 %	3659	3695
	8	Q-ILS	–	–	–	–	2.1924	3679.47	8.30	1.0012 %	3663	3700
	2	Q-ILS	–	–	–	–	2.3034	3679.78	8.64	1.0095 %	3661	3697
	12	Q-ILS	–	–	–	–	2.2613	3679.88	9.36	1.0122 %	3659	3696
	1	ILS(1)	–	–	–	–	2.5097	3680.55	9.93	1.0307 %	3659	3701
	4	ILS(0.05)	–	–	–	–	2.6971	3680.78	9.39	1.0369 %	3658	3708
	5	Q-ILS	–	–	–	–	2.4976	3681.18	12.06	1.0479 %	3659	3713
	3	ILS(R)	–	–	–	–	3.5016	3683.32	11.53	1.1069 %	3661	3706
	5	ILS(2)	–	–	–	–	3.9968	3684.50	11.60	1.1392 %	3659	3711
	6	ILS(3)	–	–	–	–	8.9238	3693.07	8.94	1.3746 %	3676	3716
	7	ILS(0.1)	–	–	–	–	19.1682	3710.80	7.61	1.8611 %	3698	3727
ta056	6	Q-ILS						3706.82	8.20	0.5377 %	3689	3722
(3667)	5	Q-ILS	=	=	=	=	0.4026	3707.53	7.33	0.5567 %	3692	3723
	2	Q-ILS	=	=	=	=	0.5274	3707.90	9.95	0.5669 %	3691	3734
	1	Q-ILS	=	=	=	=	0.9358	3708.53	8.05	0.5838 %	3693	3729
	3	Q-ILS	=	=	=	=	0.9101	3708.60	9.21	0.5858 %	3686	3730
	2	ILS(0.01)	=	=	=	=	1.2351	3709.22	9.16	0.6028 %	3691	3729
	1	ILS(1)	=	–	–	–	1.8949	3710.50	9.12	0.6374 %	3697	3730
	4	Q-ILS	–	–	–	–	2.0288	3710.65	8.66	0.6414 %	3695	3729
	8	Q-ILS	–	–	–	–	1.9770	3711.03	10.64	0.6516 %	3689	3742
	12	Q-ILS	–	–	–	–	2.5914	3711.88	9.20	0.6747 %	3701	3739
	5	ILS(2)	–	–	–	–	3.1537	3713.15	9.68	0.7092 %	3690	3738
	3	ILS(R)	–	–	–	–	3.3650	3713.78	10.17	0.7262 %	3693	3736
	4	ILS(0.05)	–	–	–	–	3.9214	3714.05	8.28	0.7337 %	3699	3731
	6	ILS(3)	–	–	–	–	4.3536	3715.32	9.23	0.7682 %	3693	3731
	7	ILS(0.1)	–	–	–	–	19.8364	3741.40	7.37	1.4755 %	3726	3758

Table 6.45: Results of the test phase jobs in the first part of the FSP multi-instance experiment

Instance	Rank	Type	H_0: equal		H_0: not better			Statistics				
			0.05	0.1	0.05	0.1	T	**avg**	σ	avg. exc.	min	max
ta055	2	ILS(1)						3638.85	10.51	0.5485 %	3621	3668
(3553)	14	Q-ILS	=	=	=	=	0.4982	3640.03	10.59	0.5810 %	3620	3673
	6	Q-ILS	=	=	=	=	0.6673	3640.30	8.86	0.5886 %	3616	3658
	1	ILS(0.01)	=	=	=	=	0.7585	3640.55	9.52	0.5955 %	3623	3663
	5	Q-ILS	=	=	=	=	0.8472	3640.78	9.80	0.6017 %	3623	3664
	4	Q-ILS	=	=	=	=	1.0372	3641.22	9.97	0.6141 %	3623	3663
	1	Q-ILS	=	=	=	=	1.2124	3641.50	8.98	0.6217 %	3623	3662
	9	Q-ILS	=	–	=	–	1.3434	3641.55	7.15	0.6231 %	3625	3655
	8	Q-ILS	=	=	=	=	1.2440	3641.70	9.98	0.6272 %	3624	3666
	13	Q-ILS	–	–	–	–	2.2985	3644.25	10.51	0.6977 %	3627	3668
	5	ILS(0.05)	–	–	–	–	2.1841	3644.35	11.97	0.7005 %	3622	3674
	3	ILS(R)	–	–	–	–	2.5018	3645.03	11.55	0.7191 %	3627	3672
	4	ILS(2)	–	–	–	–	2.7362	3645.62	11.61	0.7357 %	3620	3676
	6	ILS(3)	–	–	–	–	5.2497	3650.80	9.84	0.8787 %	3632	3668
	7	ILS(0.1)	–	–	–	–	15.2460	3670.57	7.93	1.4251 %	3653	3686
ta075	1	Q-ILS						5470.70	5.94	0.0677 %	5467	5485
(5467)	4	Q-ILS	–	–	=	=	0.7197	5471.73	6.77	0.0864 %	5467	5491
	1	ILS(0.01)	–	–	=	–	1.3739	5472.82	7.78	0.1065 %	5467	5491
	13	Q-ILS	–	–	=	–	1.3836	5472.88	7.97	0.1075 %	5467	5491
	9	Q-ILS	–	–	=	–	1.5460	5473.25	8.58	0.1143 %	5467	5491
	8	Q-ILS	–	–	–	–	2.3658	5474.62	8.65	0.1395 %	5467	5491
	6	Q-ILS	–	–	–	–	2.5286	5474.75	8.21	0.1418 %	5467	5491
	14	Q-ILS	–	–	–	–	3.1848	5476.70	10.33	0.1774 %	5467	5497
	5	Q-ILS	–	–	–	–	3.7309	5476.90	8.67	0.1811 %	5467	5496
	3	ILS(R)	–	–	–	–	3.6060	5477.20	9.73	0.1866 %	5467	5498
	2	ILS(1)	–	–	–	–	4.5531	5479.48	10.65	0.2282 %	5467	5498
	4	ILS(2)	–	–	–	–	5.3388	5481.00	10.66	0.2561 %	5467	5498
	6	ILS(3)	–	–	–	–	8.2483	5486.20	10.30	0.3512 %	5467	5501
	5	ILS(0.05)	–	–	–	–	23.8904	5502.05	5.80	0.6411 %	5486	5512
	7	ILS(0.1)	–	–	–	–	32.8696	5508.65	4.25	0.7618 %	5498	5513
ta083	1	Q-ILS						6312.32	10.42	0.6590 %	6286	6331
(6252)	4	Q-ILS	=	=	=	=	0.8168	6314.20	10.12	0.6889 %	6295	6337
	1	ILS(0.01)	=	=	=	=	1.1210	6315.15	12.07	0.7040 %	6296	6342
	9	Q-ILS	–	–	–	–	2.5292	6318.65	11.90	0.7598 %	6285	6345
	13	Q-ILS	–	–	–	–	2.7451	6318.82	10.76	0.7626 %	6301	6344
	6	Q-ILS	–	–	–	–	2.9147	6319.12	10.45	0.7674 %	6289	6341
	14	Q-ILS	–	–	–	–	3.0038	6320.32	13.24	0.7866 %	6300	6348
	8	Q-ILS	–	–	–	–	3.4955	6320.82	11.32	0.7945 %	6300	6344
	5	Q-ILS	–	–	–	–	3.9440	6321.32	9.99	0.8025 %	6304	6346
	2	ILS(1)	–	–	–	–	4.6636	6324.00	11.92	0.8452 %	6300	6354
	3	ILS(R)	–	–	–	–	6.7033	6328.05	10.57	0.9097 %	6306	6348
	4	ILS(2)	–	–	–	–	10.3581	6340.38	13.60	1.1063 %	6311	6363
	6	ILS(3)	–	–	–	–	13.7679	6345.35	11.03	1.1856 %	6324	6368
	5	ILS(0.05)	–	–	–	–	29.5752	6380.65	10.25	1.7485 %	6343	6402
	7	ILS(0.1)	–	–	–	–	42.5660	6405.00	9.01	2.1368 %	6380	6421

Table 6.46: Results of the test phase jobs in the second part of the FSP multi-instance experiment

Looking at the action usage plots from Figure 6.22, the thesis that there is a best ILS-action per instance size or per instance (as was conjectured for the TSP also) can be established. The action usage plots presented in Figure 6.22 are representative for all action usage plots, of the respective Q-ILS algorithms, since the plots look almost exactly the same for the other instances for each Q-IL job. This is true for both experiment parts and for all other Q-ILS jobs run in the test phase also (which are not shown, though). All learned action selection strategies basically only use the two best previously identified ILS-actions – the strength 0.01 partial NEH and the strength 1 composite exchange ILS-actions – in varying frequencies. Dependent on which of the two ILS-action is predominantly used, a Q-ILS job performed differently well on different instances. The same is true for the corresponding normal ILS jobs.

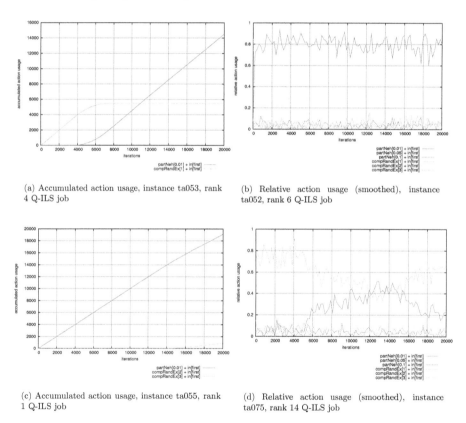

(a) Accumulated action usage, instance ta053, rank 4 Q-ILS job

(b) Relative action usage (smoothed), instance ta052, rank 6 Q-ILS job

(c) Accumulated action usage, instance ta055, rank 1 Q-ILS job

(d) Relative action usage (smoothed), instance ta075, rank 14 Q-ILS job

Figure 6.22: Action usage plots for several Q-ILS jobs run on test instances ta052, ta053, ta055, and ta075 in the FSP multi-instance experiment

Note that the action usage plot for rank 4 Q-ILS job on instance ta053 (see subfigure (a) of Figure 6.22) for the first part of the multi-instance experiment looks identical to the one displayed in subfigure (d) of Figure 6.20 for position 5 Q-ILS job run on instance ta075. This is an indication that the learning effort does not produce action selection strategies on a random basis. Finally, it can be seen from the (representative) action plots presented in Figure 6.22 that the Q-ILS

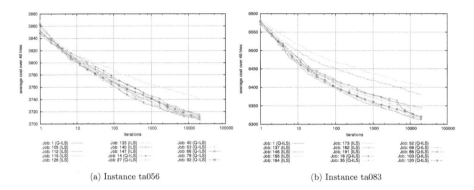

(a) Instance ta056   (b) Instance ta083

Figure 6.23:   Runtime development plots for Q-ILS jobs and normal ILS jobs run on test instances ta056 and ta083 in the FSP multi-instance experiment

algorithms learned to exclude the bad performing ILS-actions almost completely.

6.5.4   Discussion

The results for the single-instance experiment for the FSP just described show far more regularities as concerns successful actual parameter settings than were extractable for the respective experiment results for the TSP experiments. Future work might identify a superior actual parameter setting for learning, at least easier than it would be for the TSP. The regularities found for the single-instance experiment unfortunately were not repeated for the multi-instance experiment. This experiment showed some further evidence that there typically is an instance individually one ILS-action to use which simply is performing best for the instance. Perhaps there is even a best ILS-action for all instances of the same size. This phenomenon of instance best ILS-action relationship was observed for the TSP also. There, however, the Q-ILS algorithms were able to learn at least sometimes to vary the predominantly chosen ILS-action instance dependent. This did not happen for the FSP experiments which showed very stable action selection strategies for each test instance in the multi-instance experiment. As was the case for the TSP experiment, the influence of randomization should not be underestimated. For the FSP experiment some changing results were obtained, also, especially when comparing the regularities found for good performing actual parameter settings between the single- and the multi-instance experiments.

One thing that is remarkable is the insight that for the FSP it seems better to use very low strength perturbations and ILS-actions and allow them to run for a long runtime, since they tend to gradually and continuously improve solutions. Even for very long runtimes of 20000 iterations for the test instances in the multi-instance experiment runtime development plots indicated that the search has not been saturated.

Both Q-ILS algorithms applications to the FSP and the TSP need further examination with Q-ILS algorithms. In particular, it might be necessary to provide the Q-ILS algorithms with more closely related ILS-actions with respect to performance, in order to enable them to prove their usefulness ultimately. In other words, the actions provided did not allow to learn anything else than simply to use the best action all the time. Further development effort should be spend on inventing more useful actions for learning. Until then, the situation for Q-ILS algorithms applied to the FSP is similar to the situation for the Q-ILS algorithms applied to the TSP: Learning happens and showed to be useful. In particular it was learned to exclude bad performing ILS-actions altogether. Nevertheless, further influences on the learning success must be identified.

Chapter 7

Summary and Perspective

Concluding this theses, this chapter will briefly summarize in Section 7.1 what was presented so far, will draw final and summarizing conclusions in Section 7.2, will point out the scientific contributions made in and by this thesis in Section 7.3 and will, based on the conclusions, give diverse directions of future work in Section 7.4. Finally, Section 7.5 will put the approach described here into a broader context and will give an outlook.

7.1 Summary

Local search methods are useful tools for tackling hard problems such as many combinatorial optimization problems (COP). Experience has shown that utilizing regularities in problem solving pays off. In fact, the field of mathematics is concerned with identifying, describing and subsequently utilizing regularities. Consequently, identifying and exploiting regularities in the context of local search methods is desirable, too. Due to the complexity of the COPs tackled, regularities better might be detected and learned about automatically. Hence, it seems to be a good idea to extend existing local search methods with machine learning capability. One way to incorporate existing machine learning methods into local search methods is by means of the concept of a local search agent (LSA). The proceeding of a trajectory-based local search method can be regarded as the proceeding of a virtual agent whose states basically consist of solutions and whose actions are composed from local search operators, altogether yielding the same setting as a Markov decision process (MDP). In order to learn, a learning LSA needs feedback. Although instructive feedback is not available for a learning LSA – global optima are not known in advance – evaluative feedback can be derived from the cost function of COPs. Combined with the MDP problem setting it follows that reinforcement learning techniques can be applied directly yielding the concept of a learning LSA and resulting in the invention of the GAILS method using a learning LSA and a hierarchical action structure for building actions for learning LSAs.

There are manifold possibilities to build actions according to the hierarchical action structure as well as there are several possibilities to devise and assign rewards yielding many conceivable learning scenarios. Prominent among these scenarios is the GAILS standard learning scenario which is based on ILS-actions. ILS-actions coincide to the application of one iteration of the well-known Iterated Local Search (ILS) metaheuristic. The advantage of this learning scenario is that a learning LSA only moves in the subset S^* of local optima and thus introduces a search space abstraction which in turn can improve performance.

In order to evaluate the resulting new local search algorithms according to the GAILS method and the concept of learning LSAs, empirical experiments have to be conducted in manifold combinations of the three central parts of a learning LSA. These are:

- the problem-specific part with its solution and problem instance encodings and basic local search operators,

- the strategy part with its search and learning strategy implementing action-hierarchies and its heuristic states, and

- the learning-part with its function approximators needed by the reinforcement learning techniques employed.

Supporting the empirical experiments can be done by providing an implementation framework that exactly can decouple these three main parts in any GAILS algorithm program instantiation thus allowing for an arbitrary combination and reuse of components with the aim to enable rapid prototyping. The GAILS (implementation) framework is such an implementation framework which is designed to rapidly implement learning LSAs reflecting the separation of a learning LSA into its three main parts. It provides generic interfaces between components of the three parts and allows to map actions built according to the hierarchical action structure to object instantiation-based actions-hierarchies. It additionally not only separates problem type specific states from search control, but also provides a separation of search control from its state in the form of actions, action-hierarchies, and heuristic states. This way, learning LSAs can be build easily for any newly integrated problem type, reusing any newly integrated function approximator also. The framework supports all components needed for reinforcement learning and implements two reinforcement learning algorithms, namely $Q(0)$ and $Q(\lambda)$, for the GAILS standard learning scenario. These so-called Q-ILS Q-learning algorithms were tested for two problem types using different function approximators. The results showed that learning and transfer of what was learned across multiple problem instance, even of different sizes, is possible and useful.

The whole GAILS framework with three problem types and three types of function approximators integrated comprises approximately 300 C++ classes and approximately 150000 lines of code. A jobserver was implemented specifically for GAILS algorithms in [Ban04] and extended during the experiments conducted for this thesis. The extended jobserver in total comprises approximately 25000 lines of code, written in Ruby [TFH05].

7.2 Conclusions

Practical experience with the GAILS framework showed the usefulness of such a framework. Six people were integrating two function approximators and three problem types at a time simultaneously without any major conflicts. Their integration efforts were completely independent of each other and did not interfere. The integrated function approximators and problem types could be combined smoothly and arbitrarily, too. Also, the people involved in integrating new components did not need to know anything about the details of the implementation of other parts they were not concerned with directly which also emphasizes a successful decoupling of the several parts of GAILS algorithms by the GAILS framework. Once function approximators and problem type specific components were integrated, three exemplary possible combinations could be tested in the experiments described in Chapter 6 without any further implementation effort other than changing some names in a main file. Other combinations could be tested immediately, too. In particular, no copy and paste of problem type, function approximator, or reinforcement learning technique specific code is necessary. Although the one-time effort designing and implementing the GAILS framework was substantial, it paid off when finally starting experiments. Many people involved benefited from it [Kor04, Ban04, Dot05, Gim05].

Also, any changes that were made due to insights stemming from the first preparing experiments could be made easily. The integration of new functionality required only local changes and was mostly transparent for problem type and completely transparent for function approximator specific parts. For example, the integration of a procedure for empirical feature bounds estimation did only require changes to class **Features** (cf. Subsection 5.2.9) and some minor modification to problem type specific classes in the form of adjustments of method declarations. Also, the additional implementation of one-time rewards had no impact on any other parts of the framework other than the reward and move cost hierarchies (cf. Subsection 5.2.11).

The lessons learned in the context of the GAILS framework clearly are that the strict separation of states of search strategies from their control structures (actions here) is very beneficial. The benefits comprise conceptual simplicity for users of the framework and, as regards implementation, reduced effort due to reuse. The building blocks principle enabled by means of action-hierarchies leads to easy utilization and reuse. As a final conclusion concerning the GAILS framework, it can be said that it, as intended, enables rapid prototyping of general learning enhanced local search methods.

For the GAILS method itself, the experiments conducted and described in Chapter 6 yielded some promising and some confusing results. The general conclusion is that learning is possible, also the transfer of what was learned. It became clear that the transfer of learned action-value function can improve performance and that further attempts and improvements should be attempted. Since learning does not happen in all cases, still work has to be invested to stabilize the learning success. The reasons that this has not happened already cannot be identified conclusively.

One observation is that the features used were not as predictive as they were hoped to be. The only feature a learned action selection strategy showed to be dependent on so far was that of the search progression. Of course, all action usage plots presented and analyzed were only in dependence of search progression. It would be useful to have action usage plots in dependence of feature values for other features also. Nevertheless, if one feature pivotally had influenced an action selection strategy, the action usage plots should have showed more randomization. Instead, often one single ILS-action seemed to be performing best for an individual instance or only the timely change between two or more best performing ILS-actions showed to be useful. This behavior rather is an indication that state and heuristic features except for search progression are not predictive enough or useful for learning. If there is only exactly one best ILS-action to apply, the best a learning algorithm then can do is to learn to choose exactly this best ILS-action all the time, regardless what kinds of features are available. In such a case, even small variations during the learning process due to the inherent randomization can lead to noisy training examples and hence to some small variations in the action selection strategies which in turn necessarily will lead to inferior results. Then, there simply is nothing to learn on the basis of state and heuristic features, and it would be better to learn on the basis of instance features to choose an ILS-action or even a complete metaheuristic a priori. As a further conclusion, the balance between intensification and exploration during the search is not as important as assumed, at least on a state by state level. Perhaps, the assumption is wrong altogether, and nothing can be learned in this direction. Instead, it might be better to have a balancing on an instance by instance basis by means of instance features. Developing and computing instance features thereby is nothing else than an a priori search space analysis. Also, maybe there is nothing to learn at all. Maybe large search spaces such as those of COPs from a certain point on practically are distributed randomly. No predictive features then will be available and no learning is sensible. Randomized metaheuristics will do best then.

Other possible reasons why learning not always was successful could be that the function approximators are not powerful enough for the huge amount of training examples. Or, conversely, that the number of training examples was not enough or not representative enough to allow for transfer of what was learned, since the search spaces tackled simply are too large. Furthermore, the ILS-actions provided might not be suitable for learning. If they are equally well performing when used exclusively, nothing else than a random selection can be learned. This, for example, is supported by the fact that the random ILS algorithm showed very robust performance for all instance. Perhaps it is better not to concentrate on finding *the* "best" ILS-actions of instances and an LSA state dependent action selection strategy based on them, but to find some decently performing ILS-actions and to alternate between them more or less randomly or perhaps in the direction of a variable neighborhood search [HM99, HM01, HM02] (cf. Subsection 4.1.1). In contrast, if ILS-actions are significantly different in performance, learning will become trivial and does not need an overhead as is introduced by the GAILS algorithms.

Another related obvious challenge for empirical experiments with highly randomized algorithms is the problem of randomization and outliers. The GAILS algorithms that were run are highly randomized and in fact showed many outliers during the experiments. Together with the observation that the results for normal ILS and Q-ILS algorithms due to the high quality of the best solutions found (average excesses over the globally optimal cost typically between 0.2 and 0.8 % !) are very close to each other, it is difficult to do a reliable generalization of results other than the conclusions already presented. This also means that neither intended nor unintended results should not be overestimated. One cannot closing say whether the Q-ILS algorithms and even less the more general GAILS method is more useful than the standard ILS algorithm. Instead, future work trying to decrease the randomization influences by means of more elaborate experimental setup is needed. Further investigation is also needed to handle the enormous impact the choice of the acceptance criterion had. The better acceptance criterion clearly was performing best for all problem types and experiments conducted. The short term goal of improving the cost of successor LSA states yields better results than trying to maximize the long-term reward according to a learned action-value function.

Summarizing, it has to be said that the GAILS method is very new and very general. The experiments and hence results presented here can only cover a small aspect and leave plenty of room for variation, extension and improvements. As is the case for any new method invented, good performance is only achieved after a long engineering and optimization process. The first results confirm that the method can do what it is expected for, namely learn profitably, but also indicate that future work is still necessary.

7.3   Contributions

The scientific contributions of this thesis will be briefly summarized next. This thesis for the first time in this clarity has justified the need to incorporate automated and as such machine learning into existing search technology to further improve it, in particular aimed at local search methods. It provided criteria that must be fulfilled in order to consider a local search method to be adaptive and even learning. Such a distinction has not been made before. Next, the theoretically very fruitful concept of an LSA and a learning LSA was introduced and elaborated. During this elaboration the ILS metaheuristic was identified as a role model for any successful local search method in that they all have to visit local optima repeatedly. The realization that the ILS metaheuristic is *the* prototypical metaheuristic and generic for any local search method has not been made in this abstraction before.

This thesis emphasized a necessary abstraction of local search methods to the space of local optima,

S^*, and consequently pursued and transfered this abstraction to LSAs yielding the notion of ILS-actions. The idea of ILS-actions was carried further and based on it the concept of a hierarchical action structure for building arbitrary actions for LSAs was developed. This way, practical means for abstraction were introduced by enabling a building blocks principle. The integration of machine learning then was smoothly and straightforwardly possible with the help of the concept of an LSA and action hierarchies. As a side effect, a comprehensive theoretical framework for applying machine learning and reinforcement learning techniques in particular to solve COPs was provided. Learning LSAs were developed as a rigorous reinforcement learning application. The integration of reinforcement learning techniques was elaborated theoretically by presenting and discussing effects of action design for learning LSAs, how to devise different reward signals, by identifying learning opportunities and scenarios, and by presenting possible design choices and their likely consequences with respect to performance and learning success. Additionally, potentials, drawbacks, and hazards of the resulting GAILS method were discussed. The theoretical analysis resulted practically in the augmentation and adaptation of the $Q(\lambda)$ reinforcement learning technique to work with ILS-actions yielding the Q-ILS algorithms. In particular the proper integration of the acceptance criterion part of ILS-actions and enabling Q-ILS algorithms to work with function approximators and features was achieved. Summarizing, the GAILS method is the first method (at least trajectory-based one) that incorporates the full concepts of reinforcement learning into local search methods by explicitly viewing the search process as an interaction of a learning LSA which explicitly learns to choose between explicitly given actions. As such, it subsumes many existing and also conceivable combination of local search methods with machine learning techniques (such as the STAGE method, for example, cf. Subsection 4.4.3).

On the more practical side, the GAILS implementation framework was designed and implemented. The GAILS (implementation) framework complements the newly invented GAILS method with an implementation framework to enable rapid prototyping and comprehensive study of GAILS algorithms. It is the first framework in the context of local search methods that incorporates learning facilities. It realized the LSA and hierarchical action structure concepts in practice by a new "typing" of local search methods according to their heuristic and search states. These two concepts enabled a new approach for building a framework for local search methods that is different from existing ones. Whereas the conventional local search method frameworks concentrate on separating and decoupling problem type specific aspects, mainly in the form of solution encodings, from the search control strategies, the GAILS framework in addition also separates and decouples the search control from its state. It also for the first time provided for a smooth integration of all reinforcement learning related components such as policies, rewards, learning methods and lots of utilities. All components are practically decoupled and can be combined flexibly and easily which was confirmed in practice.

Besides the theoretical invention and analysis of the GAILS method and the design and implementation of the complementing GAILS framework, practical experiments with several then rapidly prototyped GAILS algorithms were conducted. The experiments showed that learning according to one GAILS method learning scenario using ILS-actions works in principle and also often yielded improving performances compared to the closely related ILS metaheuristic. As a side-effect, several insights into the working of local search methods and the ILS metaheuristic were complemented such as the importance of the choice of acceptance criterion and the true impact of balancing intensification and exploration.

Altogether, a "A Practical Framework for Adaptive Metaheuristics" has been invented (where the notion framework both denotes an implementation framework and the conceptual framework of a methodology).

7.4 Future Work

As is often the case especially for newly invented methods, the analysis conducted in Chapter 6 has posed more open questions than were answered. As a consequence, future work is required to investigate the GAILS method and the GAILS framework further. Some possible extensions of the GAILS method, the GAILS framework and the experiments conducted for GAILS algorithms are proposed next.

Beginning with the experiments, it would be beneficial to find means to reduce the variety of contrary experimental results by reducing the influence of randomization. This can be done, for example, by tackling harder and larger problem instances where differences in performance hopefully are more elaborated. Next, further investigations how influential and predictive features really are, for example leave-one-out tests can be made together with the development of more features, in particular instance features, and an a priori search space analysis. More features then can be used to insert an a priori features selection process which for example could be based on feature scores computed by regression tress (cf. Subsection 6.3.1) or look like the APA method described in [Ban04]. Perhaps it is possible to extract more information from a feature bounds analysis and to investigate whether there can be made any predictions based on how homogenous coverage and common coverage for features for several problem instances are. Comparing feature bounds for the Traveling Salesman Problem (TSP) and the Flow Shop Problem (FSP) reveals that bounds differ greatly. In this context it certainly also would be useful to have action usage plots in dependence of feature values for features other than those obviously indicating search progression.

Since the acceptance criterion influence showed to be substantial, either new kinds of acceptance criteria especially adjusted to ILS-actions could be developed, or, for the beginning, ILS-actions with different and individual acceptance criteria could be used. Also, the suggestion from Subsection 4.5.2 to only concentrate on search trajectories consisting of improving moves could be realized.

Although there were no differences discernible for multi-instance experiments that were trained and tested for supposedly differently coherent sets of training, application, and test phase problem instances, the influence of coherence among problem instance on the learning success could be further investigated. In order to do so, one could attempt to identify clusters of problem instances or use already known subclasses of problem types and learn to exploit their specialties more. In this respect, instance features that might be able to identify and distinguish subclasses biasing what was learned for each subclass individually can be beneficial. This will also increase insights into the distribution and characteristic of problem instances for different problem types which certainly will improve the capability to build better performing local search methods.

Many action selection strategies or policies found during the experiments of Chapter 6 showed a behavior which depends on runtime only. It is easy to reimplement and hard-code these policies and investigate and compare different such policies explicitly. Also, more policy implementations in addition to greedy or ϵ-greedy ones could be used. These could also be extended to incorporate knowledge from other kinds of resources such as insights learned by human researchers. This argument carries over to trying many more actual parameter values, especially for the learning parameter, since up to now only very few actual values for each parameter have been tested due to lack of time.

Other possible sources for improvements are to develop better theoretical or empirical bounds for normalization or complete other means of normalization, to develop and vary for more kinds of move costs, and to improve and specially adjust function approximators. Complementary, improvements for experiment evaluation is desirable. Additional and perhaps sometimes more appropriate statistical tests could be used, new kinds of plots can be invented to find out more about search state characteristics, and other kinds of information produced can be looked at such as the Q-value updates and action-values produced, feature values, and so on. This basically means to use the GAILS method for search space analysis in order to not only improve GAILS algorithms but other local search methods this way also. Finally, other learning scenarios than the standard GAILS learning scenario using ILS-actions and other reinforcement learning techniques such as average reward reinforcement learning or an actor-critic architecture could be experimented with. Other learning scenarios would comprise using different kinds of actions up to complete metaheuristics for an LSA. These could perhaps also be learned automatically by identifying proper and useful subgoals and learn policies to achieve them as is attempted in hierarchical reinforcement learning (cf. Subsection 4.4.5). Results from there could be transferred.

As concerns the GAILS framework, it could be extended manifold besides further improving the decoupling of the various parts and aspects and enhancing it towards a black box framework. First, other trajectory-based local search methods such as Tabu Search (TS) could be implemented as action-hierarchies and extended with machine learning techniques to make them more adaptive and learning also. Next, construction- and population-based local search methods and multi-agent support could be integrated to repeat this extensions for other construction- and population-based metaheuristics. The other metaheuristics then can be used to compare their performance to this GAILS algorithms. Also, the building blocks principle for constructing learning LSAs and hence local search methods could be enhanced. Currently, a new main program for each new problem type has to be instantiated which only differs in the classes used for problem and search states. One could unify this procedure and invent a meta- or LSA specification language for easy assembly of learning LSAs. This specification language could be made such that it is more human readable or even a graphical user interface (GUI) could be provided for it. Carrying these contemplations a little bit further, additional support for conducting experiments with GAILS algorithms can be attempted. This partly has already been done by the jobserver that was implemented specifically for GAILS algorithms in [Ban04] and a testbed for algorithms in [VHE03] which is GUI-based and basically only has to be extended to work together with the jobserver from [Ban04] and with the specialties of GAILS algorithms (such as that they require several input files for problem instance specification, feature bounds, function approximator models, and so on). This could even result in a complete experiment specification language automating the tedious process of experimentation with its parameter settings definition, data management, and statistical analysis as is proposed in [VHE03].

7.5 Perspective

A final direction of future work or rather perspective of the whole GAILS approach is so fundamental that it is covered in detail concluding this thesis. Recall that a central trait of intelligence is the ability to learn (cf. Section 2.6). However, learning can only become useful, if the learned, e.g. in the form of knowledge (about regularities), can be represented *and* used in new situations. As presupposition for usage in new situations, the knowledge must be adapted to these, which requires the ability to logically combine knowledge with other knowledge. This way, effectively a deduction process (as humans do also) producing new knowledge is initiated. This deduction process is another central trait of intelligence exhibited by humans. Belief in the advancements achieved by science is elementarily connected with making knowledge explicit and with combining knowledge to deduce new knowledge. In essence, the knowledge that is made explicit is about regularities and their underlying laws or rules (cf. Section 2.6). Consequently, it is desirable to make experience and knowledge as explicit as possible, for example by representing it in some high-level form such as logic (as humans typically do), in order to further work with it. Looked at it from another point of view, the process of extracting and making knowledge explicit by representing it in some form is nothing else than the principle of abstraction. In both cases, effectively models of how "things" or "the world" work, i.e. regularities, are built. This abstraction process is central to science and research, and intelligence.

In the context of learning LSAs and using reinforcement learning by means of value functions, knowledge is not made explicit and hence chances for further improvements are missed. For example, Q-learning does not learn explicit rules or underlying laws, but learns to somehow exploit effects of these. Learning becomes really useful, if reinforcement learning can be combined with extraction of knowledge, either by human researchers that manually analyze learned behavior (such as polices) yielding computer aided human learning, or automatically by a machine learning based post-processing. The second possibility allows to interprete and analyze resulting action value function or rather their representing function approximators directly with the aim to extract explicit rules from the implicit representation in the form of the function approximators. For example, as has already been mentioned, recurring good policies could be hard-coded and compared directly or even collected in a repository for good working policies. Or action-value functions could be analyzed with the aim to find out that some parts of a search space cannot contain reasonably good local optima and based on these insights establish constraints that can speed-up searches. This value-function independent knowledge then is even better suitable for transfer, maybe not only across problem instances of the same type, but also as generic laws across many or all problem types.

In any case, a persistent and explicit representation of knowledge is needed. The field of knowledge representation exactly is concerned with means to represent knowledge and also has elaborated deduction and inference mechanisms [Bib93a, Bib93b]. Examples of knowledge representation means are all mathematical notions such as for logic or constraint programming languages and logic programming such as Prolog. In fact, constraint programming has already been applied very successfully in the context of optimization [Hen99, VDLL01, MH00, VD02, DV04]. For the learning LSAs, knowledge for example could be extracted in the form of constraints and integrated back into them by means of constraint programming. Incorporation of explicit knowledge in the form of constraints also additionally can aid the search from a more abstract perspective and this way would combine abstract and explicit knowledge with very specific and implicit knowledge (in the form of value functions). First attempts aimed at this combination have already been undertaken for local search methods [Hen99, VDLL01, MH00, VD02, DV04] and could get a new spin when doing so for learning LSAs.

Closing, the GAILS method and implementation framework are further steps on the way towards empowering humans beings with intelligent problem solving tools.

Appendix A

Experiment results for the TSP

A.1 Second Part of Perturbation Comparison Experiment

Short	Actual Parameter Values
i_{tr}	rbx711, rl1304
i_a	rbx711, rl1304
$pert$	randDB[4], randDB[10], randDB[15]
ls	3opt[first]
ϵ_{a_a}	0.0, 0.1, 0.2

Table A.1: Changed actual parameter values for the second part of the single-instance mode sub-experiment of the TSP perturbation comparison experiment

Short	Actual Parameter Values
i_{tr}	rbx711:rl1304:vm1748
i_a	rbx711:rl1304:vm1748
i_{te}	d657, lim963, dka1376, u1817
$pert$	randDB[4], randDB[7], randDB[10], randDB[15]
ϵ_{a_a}	0.0, 0.1, 0.2

Table A.2: Changed actual parameter values for the second part of the multi-instance mode sub-experiment of the TSP perturbation comparison experiment

Instance	Job	Type	Parameters						H_0: equal		H_0: not better					Statistics		
			ϵ_{a_a}	ϵ_{a_t}	γ	λ	strat	r	0.05	0.1	0.05	0.1	T	avg	σ	avg, exc.	min	max
rbx711 (3115)	1	apply	0.0	0.0	0.8	0.4	Q(λ)	c_Δ	=	=	=	=	1.2575	3119.36	3.55	0.1400 %	3115	3128
	2	apply	0.0	0.5	0.3	0.4	Q(λ)	c_Δ	=	=	=	=	0.6694	3119.92	5.02	0.1579 %	3115	3138
	3	apply	0.0	0.5	0.3	0.8	Q(λ)	c_Δ	=	=	=	=	0.5727	3120.20	3.03	0.1669 %	3115	3128
	4	apply	0.0	0.5	0.3	0.8	Q(λ)	c_i	=	=	=	=	0.4719	3120.24	4.28	0.1682 %	3115	3129
	5	apply	0.0	0.0	0.8	0.4	Q(λ)	c_i	=	=	=	=	0.4302	3120.32	3.80	0.1708 %	3115	3128
	6	apply	0.0	0.0	0.8	0.8	Q(0)	c_i	=	=	=	=	0.0902	3120.72	4.71	0.1836 %	3116	3132
	7	train		0.0	0.8	0.8	Q(0)	c_i	=	=	=	=	-0.0090	3120.85	3.01	0.1878 %	3117	3129
	8	apply	0.0	0.5	0.8	0.4	Q(λ)	c_i	=	=	=	=	-0.1143	3121.00	5.19	0.1926 %	3115	3136
	9	apply	0.0	0.0	0.3		Q(0)	c_i	=	=	=	=	-0.2770	3121.20	4.49	0.1990 %	3116	3136
	...	...	...		...	...	...	...	...	...	...	...	...	...	...	...	...	...
	39	rand	0.0						−	−	=	=	-2.7688	3125.88	7.80	0.3493 %	3117	3147
rl1304 (252948)	1	apply	0.0	1.0	0.8	0.8	Q(λ)	c_i	+	+	+	+	2.9589	253277.56	276.02	0.1303 %	252948	253900
	2	apply	0.0	1.0	0.8		Q(0)	c_Δ	=	+	+	+	1.8994	253353.80	335.89	0.1604 %	252948	254102
	3	apply	0.0	0.0	0.8	0.4	Q(λ)	c_Δ	=	+	+	+	1.8454	253354.60	352.11	0.1607 %	252948	254125
	4	apply	0.0	1.0	0.8	0.4	Q(λ)	c_Δ	=	=	+	+	1.7649	253366.32	336.67	0.1654 %	252948	254050
	5	apply	0.0	0.0	0.3	0.8	Q(λ)	c_i	=	=	=	+	1.3423	253395.48	391.44	0.1769 %	252948	254394
	6	apply	0.0	1.0	0.3	0.8	Q(λ)	c_i	=	=	=	+	1.5191	253399.24	290.33	0.1784 %	252948	253900
	7	apply	0.0	0.0	0.3	0.8	Q(λ)	c_Δ	=	=	=	+	1.4539	253403.36	298.83	0.1800 %	252948	254173
	8	apply	0.0	0.5	0.3		Q(λ)	c_i	=	=	=	=	1.0028	253432.92	374.55	0.1917 %	252948	254203
	9	apply	0.0	1.0	0.8	0.8	Q(0)	c_Δ	=	=	=	=	0.8492	253448.16	375.76	0.1977 %	252948	254480
	10	train		0.0	0.8		Q(0)	c_Δ	=	=	=	=	0.8924	253452.55	307.38	0.1995 %	252948	254050
	11	train		0.0	0.8	0.8	Q(λ)	c_i	=	=	=	=	0.7842	253461.20	318.17	0.2029 %	252948	253909
	...	...	...		...	...	...	...	...	...	...	...	...	...	...	...	...	...
	24	rand	0.0						=	=	=	=	0.1799	253516.80	319.46	0.2249 %	252948	254096

Table A.3:   Q-ILS job results in the second part of the single-instance mode sub-experiment of the TSP perturbation comparison experiment

Rank	Perturbation	LS	Value
1	randDB[4]	3opt[first]	2516
2	randDB[7]	3opt[first]	3420
3	random ILS		3721
4	randDB[10]	3opt[first]	5587
5	randDB[15]	3opt[first]	8045

Table A.4: Ranks of normal and random ILS algorithms in the second part of the multi-instance mode sub-experiment of the TSP perturbation comparison experiment

Rank	Parameters						Value
	r	γ	λ	ϵ_{a_t}	ϵ_{a_a}	strat	
1	c_i	0.3	0.4	0.5	0.0	$Q(\lambda)$	19872
2	c_i	0.3	0.8	0.5	0.0	$Q(\lambda)$	20357
3	c_i	0.8	0.4	1.0	0.0	$Q(\lambda)$	20719
4	c_i	0.3	0.8	1.0	0.0	$Q(\lambda)$	21273
5	c_Δ	0.8	0.4	1.0	0.0	$Q(\lambda)$	23020
6	c_Δ	0.3		1.0	0.0	$Q(0)$	23457
7	c_i	0.8	0.8	1.0	0.0	$Q(\lambda)$	23840
8	c_Δ	0.3		0.5	0.0	$Q(0)$	24105
9	c_i	0.3		0.0	0.0	$Q(0)$	24983
	⋮	⋮	⋮	⋮	⋮	⋮	⋮

Table A.5: Ranks of Q-ILS algorithms in the second part of the multi-instance mode sub-experiment of the TSP perturbation comparison experiment

Instance	Rank	Type	H_0: equal		H_0: not better			Statistics				
			0.05	0.1	0.05	0.1	T	avg	σ	avg. exc.	min	max
d657	5	Q-ILS						49009.93	34.42	0.2002 %	48927	49087
(48912)	1	ILS(4)	=	=	=	=	0.0504	49010.40	48.64	0.2012 %	48938	49171
	1	Q-ILS	=	=	=	–	1.3497	49022.93	50.26	0.2268 %	48939	49141
	9	Q-ILS	–	–	–	–	2.0798	49029.35	48.01	0.2399 %	48931	49117
	3	ILS(R)	–	–	–	–	7.3506	49095.32	64.92	0.3748 %	48988	49285
	2	ILS(7)	–	–	–	–	12.4505	49163.60	70.06	0.5144 %	48989	49337
	6	Q-ILS	–	–	–	–	22.9464	49288.50	68.63	0.7697 %	49181	49414
	4	ILS(10)	–	–	–	–	25.1913	49299.45	64.02	0.7921 %	49146	49420
	5	ILS(15)	–	–	–	–	39.3302	49417.55	55.78	1.0336 %	49293	49500
lim963	1	Q-ILS						2799.78	5.81	0.3863 %	2789	2809
(2789)	1	ILS(4)	=	=	=	=	0.2462	2800.07	5.06	0.3971 %	2790	2813
	9	Q-ILS	=	=	=	=	0.5560	2800.45	5.02	0.4105 %	2789	2812
	5	Q-ILS	=	=	=	=	0.7593	2800.68	4.74	0.4186 %	2789	2808
	3	ILS(R)	–	–	–	–	1.7455	2802.15	6.35	0.4715 %	2791	2814
	2	ILS(7)	–	–	–	–	3.2427	2804.22	6.45	0.5459 %	2790	2815
	6	Q-ILS	–	–	–	–	6.2467	2808.15	6.18	0.6866 %	2796	2823
	4	ILS(10)	–	–	–	–	7.8919	2811.65	7.54	0.8121 %	2799	2829
	5	ILS(15)	–	–	–	–	19.3450	2825.78	6.20	1.3186 %	2807	2837
dka1376	5	Q-ILS						4686.62	11.57	0.4420 %	4666	4705
(4666)	1	ILS(4)	=	=	=	=	0.7539	4688.32	8.34	0.4785 %	4669	4704
	3	ILS(R)	=	=	=	–	1.4331	4690.18	10.57	0.5181 %	4669	4715
	1	ILS(4)	–	–	–	–	2.6009	4692.02	6.21	0.5578 %	4680	4708
	2	ILS(7)	–	–	–	–	2.1419	4692.10	11.29	0.5594 %	4672	4711
	9	Q-ILS	–	–	–	–	3.1089	4694.30	10.49	0.6065 %	4669	4712
	4	ILS(10)	–	–	–	–	4.9354	4699.98	12.60	0.7281 %	4679	4725
	6	Q-ILS	–	–	–	–	5.5351	4702.55	14.05	0.7833 %	4676	4724
	5	ILS(15)	–	–	–	–	14.8021	4727.20	12.91	1.3116 %	4697	4752
u1817	1	Q-ILS						57490.10	98.20	0.5054 %	57310	57748
(57201)	1	ILS(4)	=	=	=	=	0.4491	57501.30	123.41	0.5250 %	57297	57768
	5	Q-ILS	=	=	=	–	1.5779	57523.28	89.66	0.5634 %	57349	57709
	6	Q-ILS	=	–	–	–	1.8687	57534.53	113.86	0.5831 %	57292	57726
	3	ILS(R)	–	–	–	–	2.6561	57552.68	112.06	0.6148 %	57377	57937
	9	Q-ILS	–	–	–	–	2.7768	57560.28	126.11	0.6281 %	57316	57798
	2	ILS(7)	–	–	–	–	2.7723	57560.35	126.65	0.6282 %	57330	57808
	4	ILS(10)	–	–	–	–	6.2476	57638.95	114.29	0.7656 %	57415	57950
	5	ILS(15)	–	–	–	–	14.8924	57979.28	183.07	1.3606 %	57587	58456

Table A.6: Results of test phase jobs in the second part of the multi-instance mode sub-experiment of the TSP perturbation comparison experiment

A.2 Further Results for Multi-Instance Experiment

Instance	Rank	Type	H_0: equal		H_0: not better		T	avg	σ	avg. exc.	min	max
			0.05	0.1	0.05	0.1				Statistics		
fra1488	4	ILS(W)						4279.60	6.76	0.3659 %	4267	4291
(4264)	16	Q-ILS	+	+	+	+	0.2335	4279.98	7.58	0.3746 %	4266	4298
	2	Q-ILS	+	+	+	+	0.9163	4281.00	6.91	0.3987 %	4266	4300
	1	ILS(R)	−	−	−	−	1.8436	4282.48	7.18	0.4333 %	4266	4299
	1	Q-ILS	−	−	−	−	2.3119	4282.95	6.19	0.4444 %	4270	4300
	9	Q-ILS	−	−	−	−	2.3457	4283.18	6.87	0.4497 %	4268	4299
	4	Q-ILS	−	−	−	−	2.3522	4283.52	8.11	0.4579 %	4266	4301
	5	Q-ILS	−	−	−	−	3.1000	4284.62	7.71	0.4837 %	4267	4301
	3	Q-ILS	−	−	−	−	3.1994	4284.65	7.35	0.4843 %	4269	4299
	3	ILS(2)	−	−	−	−	3.5273	4285.20	7.43	0.4972 %	4269	4304
	6	ILS(G)	−	−	−	−	4.8209	4286.73	6.46	0.5330 %	4270	4296
	6	Q-ILS	−	−	−	−	5.4122	4287.35	6.03	0.5476 %	4274	4301
	2	ILS(4)	−	−	−	−	4.6782	4287.73	8.66	0.5564 %	4268	4308
	5	ILS(1)	−	−	−	−	5.7018	4288.48	7.16	0.5740 %	4277	4304
	7	ILS(7)	−	−	−	−	12.1489	4306.07	12.01	0.9867 %	4283	4337
u1817	2	Q-ILS						57438.65	91.55	0.4155 %	57251	57622
(57201)	5	Q-ILS	+	+	+	+	0.0875	57440.45	92.50	0.4186 %	57262	57660
	6	Q-ILS	+	+	+	+	1.1544	57465.88	117.75	0.4631 %	57293	57690
	3	ILS(2)	+	+	+		1.3106	57467.85	107.11	0.4665 %	57295	57742
	16	Q-ILS	+	−	−	−	1.9030	57476.12	84.44	0.4810 %	57292	57636
	1	Q-ILS	+	+	+	−	1.5462	57476.72	126.00	0.4820 %	57277	57902
	4	Q-ILS	+	−	−	−	1.8810	57482.88	117.17	0.4928 %	57254	57761
	9	Q-ILS	−	−	−	−	2.0806	57484.65	105.69	0.4959 %	57306	57718
	5	ILS(1)	−	−	−	−	2.5351	57491.88	96.18	0.5085 %	57311	57693
	3	Q-ILS	+	−	−	−	2.2997	57492.40	116.06	0.5094 %	57338	57866
	1	ILS(R)	−	−	−	−	2.5390	57493.05	99.90	0.5106 %	57301	57727
	2	ILS(4)	−	−	−	−	2.6347	57494.07	96.54	0.5124 %	57303	57728
	4	ILS(W)	−	−	−	−	3.6752	57520.10	106.14	0.5579 %	57325	57712
	7	ILS(7)	−	−	−	−	5.2034	57558.68	113.58	0.6253 %	57361	57834
	6	ILS(G)	−	−	−	−	7.0369	57616.20	130.70	0.7259 %	57425	58169

Table A.7: Results of test phase jobs in the first part of the TSP multi-instance experiment

Instance	Rank	Type	H_0: equal 0.05	0.1	H_0: not better 0.05	0.1	T	avg	σ	avg. exc.	min	max
dja1436	7	Q-ILS						5267.57	6.81	0.2012 %	5257	5283
(5257)	6	Q-ILS	=	=	=	=	0.6861	5268.82	9.29	0.2249 %	5257	5290
	9	Q-ILS	=	=	=	=	1.0460	5269.43	8.87	0.2364 %	5257	5293
	1	Q-ILS	=	=	=	=	1.1259	5269.52	8.58	0.2383 %	5258	5292
	1	ILS(W)	=	=	=	=	1.2458	5269.75	8.69	0.2425 %	5258	5298
	4	ILS(4)	=	–	–	–	1.7140	5270.55	8.61	0.2578 %	5259	5289
	2	Q-ILS	–	–	–	–	2.0228	5270.75	7.22	0.2616 %	5259	5285
	3	Q-ILS	–	–	–	–	2.3736	5271.15	6.66	0.2692 %	5258	5284
	2	ILS(R)	–	–	–	–	2.3586	5271.95	9.55	0.2844 %	5258	5300
	6	ILS(G)	–	–	–	–	3.4931	5273.73	8.81	0.3181 %	5259	5296
	3	ILS(2)	–	–	–	–	3.4944	5274.00	9.42	0.3234 %	5260	5298
	5	Q-ILS(1)	–	–	–	–	4.6956	5275.48	8.17	0.3514 %	5262	5293
	7	ILS(7)	–	–	–	–	6.2596	5278.60	8.81	0.4109 %	5264	5305
	5	ILS(1)	–	–	–	–	6.8709	5280.35	9.59	0.4442 %	5260	5305
dbj2924	2	Q-ILS						10181.17	13.07	0.5250 %	10161	10227
(10128)	2	ILS(R)	=	=	=	=	0.0457	10181.33	16.14	0.5265 %	10154	10226
	1	Q-ILS	=	=	=	=	0.1633	10181.62	11.53	0.5295 %	10159	10209
	9	Q-ILS	=	=	=	=	0.2648	10181.92	12.25	0.5324 %	10157	10208
	6	Q-ILS	=	=	=	=	0.6658	10183.08	12.44	0.5438 %	10157	10212
	1	ILS(W)	=	=	=	=	1.0526	10184.25	13.06	0.5554 %	10158	10211
	3	ILS(2)	=	–	=	=	1.2480	10185.10	14.99	0.5638 %	10142	10215
	7	Q-ILS	–	–	=	–	1.4041	10185.48	14.29	0.5675 %	10154	10215
	5	Q-ILS	–	–	–	–	1.9608	10186.75	12.35	0.5801 %	10158	10226
	5	ILS(1)	–	–	–	–	2.5630	10188.50	12.48	0.5974 %	10166	10216
	4	ILS(4)	–	–	–	–	2.8340	10189.23	12.32	0.6045 %	10169	10226
	3	Q-ILS	–	–	–	–	3.4052	10191.17	13.19	0.6238 %	10168	10223
	6	ILS(G)	–	–	–	–	4.4646	10194.30	13.22	0.6546 %	10170	10229
	7	ILS(7)	–	–	–	–	8.7929	10210.10	16.18	0.8106 %	10173	10251

Table A.8: Results of test phase jobs in the second part of the TSP multi-instance experiment

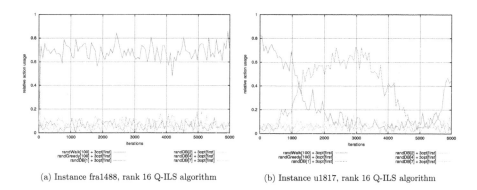

(a) Instance fra1488, rank 16 Q-ILS algorithm   (b) Instance u1817, rank 16 Q-ILS algorithm

Figure A.1:   Relative action usage plots (smoothed) for rank 16 Q-ILS algorithm run on test instances fra1488 and u1817 in the first part of the TSP multi-instance experiment

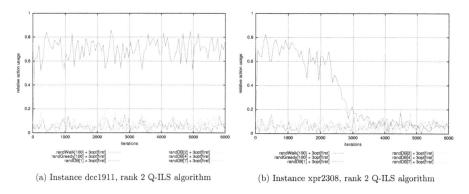

(a) Instance dcc1911, rank 2 Q-ILS algorithm   (b) Instance xpr2308, rank 2 Q-ILS algorithm

Figure A.2:   Relative action usage plots (smoothed) for rank 2 Q-ILS algorithm run on test instances dja1436 and dbj2924 in the second part of the TSP multi-instance experiment

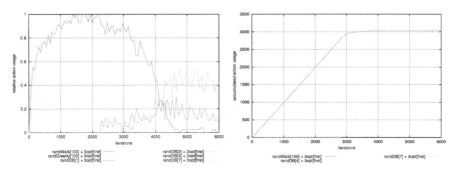

(a) Instance dcc1911, rank 9 Q-ILS algorithm

(b) Instance dlb3694, rank 9 Q-ILS algorithm, accumulated action usage

Figure A.3:   Action usage plots for rank 9 Q-ILS algorithm run on test instances dja1436 and dbj2924 in the second part of the TSP multi-instance experiment

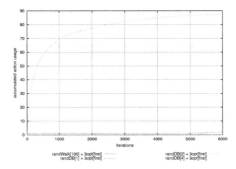

Figure A.4:   Accumulated improving action usage plot for rank 6 Q-ILS algorithm run on instance u2152 in the first part of the TSP multi-instance experiment

A.3 Results for the Experiments with Regression Trees

A.3.1 Single-Instance Experiments

Instance	Perturbation	LS	Statistics				
			avg	σ	avg.excess	min	max
pbd984	randDB[2]	3opt[first]	2803.24	6.23	0.2231 %	2797	2815
(2797)	randDB[4]	3opt[first]	2804.00	6.70	0.2503 %	2797	2817
	randWalk[100]	3opt[first]	2808.12	6.46	0.3976 %	2799	2818
	randDB[1]	3opt[first]	2809.00	5.37	0.4290 %	2797	2817
	randDB[7]	3opt[first]	2811.36	8.42	0.5134 %	2801	2833
	randGreedy[100]	3opt[first]	2811.88	5.90	0.5320 %	2800	2822
nrw1379	randDB[1]	3opt[first]	56821.44	45.25	0.3239 %	56720	56905
(56638)	randDB[2]	3opt[first]	56836.12	50.06	0.3498 %	56728	56944
	randGreedy[100]	3opt[first]	56888.88	47.72	0.4430 %	56783	56969
	randWalk[100]	3opt[first]	56949.80	40.74	0.5505 %	56854	57045
	randDB[4]	3opt[first]	56985.64	67.81	0.6138 %	56840	57115
	randDB[7]	3opt[first]	57238.28	79.43	1.0599 %	57128	57477
rl1889	randDB[4]	3opt[first]	317649.08	545.18	0.3516 %	316788	318637
(316536)	randDB[2]	3opt[first]	317835.68	940.17	0.4106 %	316842	320860
	randDB[7]	3opt[first]	317976.16	976.21	0.4550 %	316627	321152
	randDB[1]	3opt[first]	318276.96	822.93	0.5500 %	316842	320911
	randWalk[100]	3opt[first]	318822.24	1466.26	0.7223 %	316761	321767
	randGreedy[100]	3opt[first]	319614.76	1228.56	0.9726 %	317619	321443
dcb2086	randWalk[100]	3opt[first]	6639.60	13.96	0.6000 %	6612	6674
(6600)	randDB[4]	3opt[first]	6640.84	14.54	0.6188 %	6616	6671
	randDB[2]	3opt[first]	6650.44	19.72	0.7642 %	6612	6696
	randDB[7]	3opt[first]	6652.32	14.36	0.7927 %	6627	6686
	randGreedy[100]	3opt[first]	6652.44	15.16	0.7945 %	6623	6682
	randDB[1]	3opt[first]	6654.92	14.47	0.8321 %	6625	6690
pds2566	randWalk[100]	3opt[first]	7680.68	8.86	0.4930 %	7668	7701
(7643)	randGreedy[100]	3opt[first]	7684.60	9.06	0.5443 %	7667	7704
	randDB[2]	3opt[first]	7685.84	8.48	0.5605 %	7665	7704
	randDB[4]	3opt[first]	7687.24	9.98	0.5788 %	7667	7709
	randDB[1]	3opt[first]	7690.16	9.77	0.6170 %	7667	7708
	randDB[7]	3opt[first]	7702.84	16.63	0.7829 %	7669	7734

Table A.9: Normal ILS job results in the TSP single-instance experiment with regression trees

Instance	Rank	Type	ϵ_{π_a}	ϵ_{a_t}	γ	λ	strat	r	H_0: equal 0.05	H_0: equal 0.1	H_0: not better 0.05	H_0: not better 0.1	T	avg	σ	avg. exc.	min	max
pbd984 (2797)	1	apply	0.0	0.3	0.8		$Q(0)$	c_i	=	=	=	=	0.8067	2801.88	5.68	0.1745 %	2797	2815
	2	apply	0.3	0.0	0.8		$Q(0)$	c_i	=	=	=	=	0.3150	2802.68	6.34	0.2031 %	2797	2821
	3	train		0.0	0.8	0.4	$Q(\lambda)$	c_Δ	=	=	=	=	0.2144	2802.85	6.63	0.2092 %	2797	2816
	…	…	…	…	…	…	…	…	…	…	…	…	…	…	…	…	…	…
	20	rand							=	=	=	=	-0.9062	2804.88	6.57	0.2817 %	2797	2820
nrw1379 (56638)	1	apply	0.0	0.3	0.8		$Q(0)$	c_Δ	+	+	+	+	1.9073	56799.80	34.22	0.2857 %	56723	56874
	2	apply	0.0	0.3	0.3	0.4	$Q(\lambda)$	c_i	+	+	=	+	1.6005	56801.68	41.99	0.2890 %	56722	56915
	3	train		0.0	0.3		$Q(0)$	c_Δ	=	=	=	=	1.0735	56808.60	39.10	0.3012 %	56732	56884
	…	…	…	…	…	…	…	…	…	…	…	…	…	…	…	…	…	…
	42	rand							-	-	=	=	-2.1025	56853.08	60.12	0.3797 %	56729	56939
rl1889 (316536)	1	apply	0.0		0.8	0.8	$Q(\lambda)$	c_i	=	=	=	=	0.1820	317615.68	737.91	0.3411 %	316638	320394
	2	train		0.0	0.3	0.4	$Q(\lambda)$	c_Δ	=	=	=	=	0.0956	317630.45	807.39	0.3458 %	316638	319137
	3	train		0.0	0.8		$Q(0)$	c_Δ	=	=	=	=	-0.0334	317654.55	609.56	0.3534 %	316823	319335
	…	…	…	…	…	…	…	…	…	…	…	…	…	…	…	…	…	…
dcb2086 (7643)	1	apply	0.3	0.3	0.3		$Q(0)$	c_Δ	=	=	=	=	0.4817	6637.36	18.60	0.5661 %	6608	6673
	2	apply	0.3	0.3	0.3	0.8	$Q(\lambda)$	c_Δ	=	=	=	=	0.2237	6638.68	15.11	0.5861 %	6611	6669
	3	apply	0.3	0.0	0.8	0.8	$Q(\lambda)$	c_i	=	=	=	=	0.2027	6638.76	15.32	0.5873 %	6613	6665
	4	apply	0.3	0.0	0.8	0.8	$Q(\lambda)$	c_Δ	=	=	=	=	-0.0197	6639.68	14.70	0.6012 %	6611	6673
	…	…	…	…	…	…	…	…	…	…	…	…	…	…	…	…	…	…
pds2566 (7643)	1	train		0.0	0.3	0.8	$Q(\lambda)$	c_Δ	=	=	=	=	-0.0970	7681.00	13.92	0.4972 %	7655	7713
	2	apply	0.0	0.3	0.8	0.4	$Q(\lambda)$	c_i	=	=	=	=	-0.1957	7681.16	8.48	0.4993 %	7663	7700
	…	…	…	…	…	…	…	…	…	…	…	…	…	…	…	…	…	…
	72	rand							-	-	=	=	-3.1059	7689.00	10.04	0.6019 %	7672	7712

Table A.10:   Q-ILS job results in the TSP single-instance experiment with regression trees

A.3.2   Multi-Instance Experiments

Rank	Perturbation	LS	Value
1	randDB[2]	3opt[first]	5578
2	random ILS		5659
3	randWalk[100]	3opt[first]	6120
4	randDB[4]	3opt[first]	6352
5	randDB[1]	3opt[first]	6755
6	randGreedy[100]	3opt[first]	7620
7	randDB[7]	3opt[first]	10094

Table A.11:   Ranks of normal and random ILS algorithms in the first part of the TSP multi-instance experiment with regression trees

Rank	\multicolumn{7}{c}{Parameters}							Value
	r	γ	λ	ϵ_{a_t}	ϵ_{π_a}	ϵ_{a_a}	strat	Value
1	c_i	0.8		0.3	0.3	0.0	Q(0)	57778
2	c_Δ	0.3	0.8	0.3	0.0	0.0	Q(λ)	58436
3	c_Δ	0.3	0.4	0.0	0.3	0.0	Q(λ)	60869
4	c_Δ	0.3	0.8	0.3	0.3	0.0	Q(λ)	60880
5	c_i	0.3	0.8	0.0	0.0	0.0	Q(λ)	61751
6	c_Δ	0.8	0.8	0.0	0.0	0.0	Q(λ)	62163
7	c_i	0.3	0.4	0.3	0.3	0.0	Q(λ)	63380
8	c_i	0.8		0.0	0.3	0.0	Q(0)	64472
9	c_i	0.3		0.3	0.3	0.0	Q(0)	64669
10	c_Δ	0.8	0.4	0.0	0.3	0.0	Q(λ)	65337
11	c_i	0.3		0.3	0.0	0.0	Q(0)	65417
⋮	⋮	⋮	⋮	⋮	⋮	⋮	⋮	⋮

Table A.12:   Ranks of Q-ILS algorithms in the first part of the TSP multi-instance experiment with regression trees

Rank		Parameters						Value
	r	γ	λ	ϵ_{a_t}	ϵ_{π_a}	ϵ_{a_a}	strat	
1	c_i	0.8		0.3	0.0	0.0	Q(0)	38612
2	c_Δ	0.3	0.4	0.3	0.0	0.0	Q(λ)	48258
3	c_i	0.8		0.3	0.3	0.0	Q(0)	49205
4	c_i	0.3		0.3	0.3	0.0	Q(0)	50299
5	c_Δ	0.8	0.4	0.0	0.3	0.0	Q(λ)	53797
6	c_i	0.3	0.8	0.3	0.0	0.0	Q(λ)	53949
7	c_i	0.8	0.8	0.3	0.3	0.0	Q(λ)	54400
8	c_i	0.3		0.0	0.3	0.0	Q(0)	54773
9	c_i	0.8		0.3	0.3	0.0	Q(0)	54961
10	c_i	0.3		0.0	0.3	0.0	Q(0)	55091
11	c_i	0.3	0.8	0.3	0.3	0.0	Q(λ)	55444
12	c_i	0.8	0.8	0.3	0.0	0.0	Q(λ)	55963
13	c_i	0.8	0.4	0.3	0.3	0.0	Q(λ)	56181
14	c_Δ	0.3		0.0	0.3	0.0	Q(0)	56792
⋮	⋮	⋮	⋮	⋮	⋮	⋮	⋮	⋮

Table A.13: Ranks of Q-ILS algorithms in the second part of the TSP multi-instance experiment with regression trees

Rank	Perturbation	LS	Value
1	randWalk[100]	3opt[first]	4855
2	random ILS		5609
3	randDB[4]	3opt[first]	6067
4	randDB[2]	3opt[first]	6301
5	randGreedy[100]	3opt[first]	7131
6	randDB[1]	3opt[first]	7507
7	randDB[7]	3opt[first]	10563

Table A.14: Ranks of normal and random ILS algorithms in the second part of the TSP multi-instance experiment with regression trees

| Instance | Rank | Type | H_0: equal | | H_0: not better | | | Statistics | | | | |
			0.05	0.1	0.05	0.1	T	avg	σ	avg. exc.	min	max
fnb1615	2	ILS(R)						4973.98	8.32	0.3627 %	4960	4994
(4956)	1	Q-ILS	=	=	=	=	0.1305	4974.23	8.80	0.3677 %	4961	4993
	3	Q-ILS	=	=	=	=	0.2828	4974.45	6.60	0.3723 %	4961	4987
	2	Q-ILS	=	=	=	=	0.3727	4974.60	6.57	0.3753 %	4962	4993
	7	Q-ILS	=	=	=	=	0.7995	4975.48	8.46	0.3930 %	4961	4996
	11	Q-ILS	=	=	=	=	0.8061	4975.50	8.60	0.3935 %	4961	4998
	1	ILS(2)	=	=	=	=	0.9029	4975.55	7.24	0.3945 %	4960	4994
	3	ILS(W)	=	=	=	−	1.3126	4976.48	8.71	0.4131 %	4959	4996
	5	ILS(1)	=	−	=	−	1.6024	4976.95	8.28	0.4227 %	4964	5000
	4	Q-ILS	=	−	=	−	1.5937	4977.38	10.62	0.4313 %	4960	5000
	4	ILS(4)	=	−	−	−	1.7833	4977.48	9.21	0.4333 %	4961	5001
	6	ILS(G)	−	−	−	−	2.3225	4978.82	10.26	0.4606 %	4961	5004
	7	ILS(7)	−	−	−	−	6.9493	4988.25	9.98	0.6507 %	4968	5014
u2152	3	Q-ILS						64600.07	101.66	0.5402 %	64397	64800
(64253)	1	Q-ILS	=	=	=	=	0.5122	64612.45	114.08	0.5594 %	64337	64827
	4	ILS(4)	=	=	=	=	0.8964	64621.85	115.19	0.5741 %	64404	64873
	2	ILS(R)	=	=	=	=	1.0123	64624.60	114.65	0.5783 %	64410	64864
	7	Q-ILS	=	=	=	=	1.2793	64631.62	118.29	0.5893 %	64392	64886
	3	ILS(W)	=	=	=	−	1.3748	64635.65	128.26	0.5955 %	64370	64939
	2	Q-ILS	=	−	−	−	1.7985	64641.85	106.05	0.6052 %	64405	64842
	1	ILS(2)	=	−	−	−	1.9751	64650.03	123.49	0.6179 %	64437	64996
	11	Q-ILS	−	−	−	−	2.2611	64654.62	113.79	0.6251 %	64447	64927
	5	ILS(1)	−	−	−	−	2.4365	64659.90	117.39	0.6333 %	64383	64935
	4	Q-ILS	−	−	−	−	3.0969	64670.10	100.58	0.6492 %	64408	64832
	7	ILS(7)	−	−	−	−	3.7374	64699.32	133.70	0.6946 %	64479	64947
	6	ILS(G)	−	−	−	−	4.6244	64717.57	124.46	0.7230 %	64469	64950
ley2323	3	ILS(W)						8371.33	13.35	0.2314 %	8353	8407
(8352)	1	ILS(2)	=	=	=	=	0.1334	8371.73	13.47	0.2362 %	8357	8420
	2	ILS(R)	=	=	=	=	0.4153	8372.65	15.13	0.2472 %	8353	8427
	3	Q-ILS	=	=	=	=	0.6908	8373.60	15.99	0.2586 %	8352	8413
	2	Q-ILS	=	=	=	=	0.8273	8374.45	19.81	0.2688 %	8356	8453
	7	Q-ILS	=	=	=	=	0.9951	8374.55	15.56	0.2700 %	8352	8409
	7	ILS(7)	=	−	=	=	1.2854	8374.98	12.01	0.2751 %	8354	8402
	4	ILS(4)	−	−	−	−	1.7357	8376.40	12.80	0.2921 %	8355	8403
	1	Q-ILS	=	=	=	−	1.5762	8377.35	20.16	0.3035 %	8353	8431
	11	Q-ILS	−	−	−	−	3.3077	8382.83	17.47	0.3691 %	8361	8450
	6	ILS(G)	−	−	−	−	3.3096	8384.25	20.78	0.3861 %	8361	8445
	5	ILS(1)	−	−	−	−	3.7946	8385.50	19.49	0.4011 %	8358	8440
	4	Q-ILS	−	−	−	−	3.3630	8385.77	23.67	0.4044 %	8358	8445

Table A.15: Results of the test phase jobs in the first part of the TSP multi-instance experiment with regression trees

Instance	Rank	Type	H_0: equal		H_0: not better			Statistics				
			0.05	0.1	0.05	0.1	T	avg	σ	avg. exc.	min	max
fra1488	2	Q-ILS						4282.35	5.60	0.4303 %	4273	4293
(4264)	7	Q-ILS	=	=	=	=	0.2034	4282.65	7.46	0.4374 %	4267	4302
	3	Q-ILS	=	=	=	=	0.4370	4282.95	6.64	0.4444 %	4269	4296
	3	ILS(W)	=	=	=	=	0.5306	4283.12	7.35	0.4485 %	4267	4297
	2	ILS(R)	=	=	=	=	0.7231	4283.32	6.43	0.4532 %	4269	4304
	1	ILS(2)	=	=	=	=	1.1776	4284.02	7.04	0.4696 %	4269	4298
	11	Q-ILS	=	=	=	=	1.2370	4284.40	8.86	0.4784 %	4266	4306
	1	Q-ILS	–	–	–	–	2.1990	4285.30	6.38	0.4995 %	4270	4300
	4	ILS(4)	–	–	–	–	2.4607	4286.05	7.69	0.5171 %	4270	4300
	6	ILS(G)	–	–	–	–	2.8984	4286.38	6.77	0.5247 %	4275	4303
	5	ILS(1)	–	–	–	–	3.9140	4287.98	7.16	0.5623 %	4276	4308
	4	Q-ILS	–	–	–	–	3.7709	4288.43	8.52	0.5728 %	4272	4309
	7	ILS(7)	–	–	–	–	10.2278	4300.70	9.87	0.8607 %	4281	4324
u1817	2	ILS(R)						57451.60	84.09	0.4381 %	57309	57646
(57201)	4	ILS(4)	=	=	=	=	0.4694	57462.05	112.92	0.4564 %	57286	57865
	4	Q-ILS	=	=	=	=	0.6063	57464.20	101.01	0.4601 %	57244	57633
	1	ILS(2)	=	=	=	=	0.7290	57465.53	86.74	0.4624 %	57344	57707
	2	Q-ILS	=	=	=	=	0.8134	57470.80	123.35	0.4717 %	57283	57746
	3	Q-ILS	=	=	=	=	1.0820	57473.82	99.02	0.4770 %	57294	57713
	3	ILS(W)	=	=	=	–	1.5868	57485.93	107.92	0.4981 %	57332	57711
	11	Q-ILS	=	=	=	–	1.5717	57493.25	144.97	0.5109 %	57244	57888
	5	ILS(1)	–	–	–	–	2.2856	57501.15	108.30	0.5247 %	57329	57825
	7	Q-ILS	–	–	–	–	2.6673	57506.28	98.67	0.5337 %	57322	57702
	1	Q-ILS	–	–	–	–	2.7234	57506.70	96.45	0.5344 %	57295	57708
	7	ILS(7)	–	–	–	–	4.0560	57533.47	96.06	0.5812 %	57362	57760
	6	ILS(G)	–	–	–	–	6.1755	57591.22	115.66	0.6822 %	57335	57867

Table A.16: Results test phase jobs in the first part of the TSP multi-instance experiment with regression trees

Instance	Rank	Type	H_0: equal		H_0: not better			Statistics				
			0.05	0.1	0.05	0.1	T	avg	σ	avg. exc.	min	max
dcc1911	1	Q-ILS						6420.95	9.55	0.3901 %	6409	6450
(6396)	2	ILS(R)	=	=	=	=	0.3224	6421.62	9.18	0.4006 %	6405	6444
	5	Q-ILS	=	=	=	=	0.5215	6422.05	9.32	0.4073 %	6403	6442
	2	Q-ILS	=	=	=	=	0.7054	6422.48	9.79	0.4139 %	6400	6457
	6	Q-ILS	=	=	=	=	0.8209	6422.57	8.10	0.4155 %	6407	6443
	3	Q-ILS	=	=	=	=	0.7123	6422.65	11.69	0.4167 %	6401	6451
	7	Q-ILS	=	–	=	=	0.9664	6422.98	9.19	0.4217 %	6403	6440
	1	ILS(W)	=	–	=	=	1.1297	6423.18	8.00	0.4249 %	6406	6438
	14	Q-ILS	=	=	=	=	1.1505	6423.30	8.70	0.4268 %	6405	6445
	5	ILS(G)	–	–	–	–	2.0216	6425.27	9.59	0.4577 %	6405	6445
	3	ILS(4)	–	–	–	–	2.4340	6426.68	11.41	0.4796 %	6408	6448
	4	ILS(2)	–	–	–	–	2.4519	6426.73	11.43	0.4804 %	6404	6448
	6	ILS(1)	–	–	–	–	3.0228	6427.35	9.39	0.4902 %	6404	6461
	7	ILS(7)	–	–	–	–	6.8348	6439.00	13.70	0.6723 %	6419	6475
xpr2308	6	Q-ILS						7250.88	9.77	0.4415 %	7233	7272
(7219)	2	Q-ILS	=	=	=	=	0.8897	7252.75	9.06	0.4675 %	7225	7267
	7	Q-ILS	=	=	=	=	0.8544	7252.77	10.11	0.4679 %	7233	7273
	1	ILS(W)	=	=	=	=	0.9756	7253.10	10.61	0.4724 %	7234	7276
	3	Q-ILS	=	=	=	–	1.4036	7254.15	11.06	0.4869 %	7234	7288
	3	ILS(4)	=	–	=	–	1.6216	7254.43	9.81	0.4907 %	7236	7277
	2	ILS(R)	=	–	=	–	1.6073	7254.80	11.96	0.4959 %	7235	7279
	1	Q-ILS	–	–	–	–	2.5313	7256.25	9.21	0.5160 %	7238	7276
	5	Q-ILS	–	–	–	–	2.5841	7256.93	11.12	0.5253 %	7233	7283
	4	ILS(2)	–	–	–	–	3.1300	7257.77	9.94	0.5371 %	7234	7280
	14	Q-ILS	–	–	–	–	2.9949	7257.98	11.37	0.5399 %	7228	7281
	6	ILS(1)	–	–	–	–	4.9501	7261.43	9.28	0.5877 %	7242	7287
	7	ILS(7)	–	–	–	–	4.5888	7262.38	12.48	0.6008 %	7233	7286
	5	ILS(G)	–	–	–	–	6.0685	7263.20	8.33	0.6123 %	7238	7281
dlb3694	2	Q-ILS						11017.75	10.78	0.5361 %	10994	11040
(10959)	1	Q-ILS	=	=	=	=	1.2672	11021.15	13.10	0.5671 %	10991	11046
	6	Q-ILS	–	–	–	–	1.7616	11022.23	11.91	0.5769 %	10998	11047
	1	ILS(W)	–	–	–	–	1.9511	11022.23	9.70	0.5769 %	11006	11040
	3	Q-ILS	=	–	–	–	1.8659	11022.42	11.61	0.5787 %	11002	11045
	14	Q-ILS	–	–	–	–	3.6971	11027.20	12.05	0.6223 %	11003	11057
	2	ILS(R)	–	–	–	–	4.1827	11029.52	14.17	0.6435 %	10999	11057
	7	Q-ILS	–	–	–	–	4.7923	11029.70	11.51	0.6451 %	11000	11059
	4	ILS(2)	–	–	–	–	6.1014	11033.00	11.56	0.6752 %	11013	11051
	5	Q-ILS	–	–	–	–	6.8647	11034.55	11.11	0.6894 %	11011	11060
	6	ILS(1)	–	–	–	–	7.4798	11038.55	13.90	0.7259 %	11004	11061
	3	ILS(4)	–	–	–	–	7.2960	11039.02	14.96	0.7302 %	11017	11075
	5	ILS(G)	–	–	–	–	10.7656	11044.52	11.45	0.7804 %	11021	11061
	7	ILS(7)	–	–	–	–	13.9644	11058.42	14.94	0.9072 %	11030	11095

Table A.17: Results of test phase jobs in the second part of the TSP multi-instance experiment with regression trees

Instance	Rank	Type	H_0: equal		H_0: not better			Statistics				
			0.05	0.1	0.05	0.1	T	**avg**	σ	avg. exc.	min	max
dja1436	2	ILS(R)						5269.10	8.62	0.2302 %	5257	5291
(5257)	3	Q-ILS	=	=	=	=	0.3994	5269.85	8.17	0.2444 %	5259	5286
	6	Q-ILS	=	=	=	=	0.4714	5269.95	7.47	0.2463 %	5258	5286
	1	Q-ILS	=	=	=	=	0.5825	5270.15	7.46	0.2501 %	5257	5284
	4	ILS(2)	=	=	=	=	0.7205	5270.55	9.37	0.2578 %	5259	5293
	2	Q-ILS	=	=	=	=	0.7666	5270.60	8.88	0.2587 %	5257	5298
	1	ILS(W)	=	–	=	=	1.0625	5270.95	6.86	0.2654 %	5258	5284
	7	Q-ILS	–	–	=	–	1.6199	5272.05	7.64	0.2863 %	5259	5286
	5	Q-ILS	–	–	–	–	1.7405	5272.45	8.60	0.2939 %	5259	5295
	3	ILS(4)	=	–	–	–	1.9287	5273.62	12.08	0.3162 %	5258	5301
	14	Q-ILS	–	–	–	–	2.4794	5274.45	10.58	0.3319 %	5259	5300
	5	ILS(G)	–	–	–	–	4.3018	5276.52	6.70	0.3714 %	5263	5290
	7	ILS(7)	–	–	–	–	5.8965	5278.82	5.88	0.4152 %	5263	5288
	6	ILS(1)	–	–	–	–	5.2172	5279.27	8.83	0.4237 %	5262	5298
dbj2924	2	Q-ILS						10178.50	11.52	0.4986 %	10155	10203
(10128)	1	Q-ILS	=	=	=	=	0.7649	10180.38	10.38	0.5171 %	10154	10200
	6	Q-ILS	=	=	=	=	0.6762	10180.48	14.44	0.5181 %	10150	10214
	3	Q-ILS	=	=	=	=	0.7142	10180.48	13.16	0.5181 %	10145	10203
	7	Q-ILS	=	=	=	=	0.9152	10181.20	14.68	0.5253 %	10141	10214
	1	ILS(W)	=	–	=	–	1.4442	10182.45	12.91	0.5376 %	10154	10208
	2	ILS(R)	–	–	–	–	2.0658	10184.05	12.49	0.5534 %	10154	10214
	5	Q-ILS	–	–	–	–	2.6217	10185.20	11.34	0.5648 %	10166	10207
	14	Q-ILS	–	–	–	–	2.7027	10186.08	13.47	0.5734 %	10162	10213
	3	ILS(4)	–	–	–	–	3.0648	10187.60	14.83	0.5885 %	10159	10226
	4	ILS(2)	–	–	–	–	4.3808	10190.05	12.06	0.6127 %	10164	10212
	5	ILS(G)	–	–	–	–	4.7858	10191.88	13.41	0.6307 %	10169	10231
	6	ILS(1)	–	–	–	–	4.9192	10191.92	12.85	0.6312 %	10168	10222
	7	ILS(7)	–	–	–	–	9.5660	10208.02	15.76	0.7901 %	10181	10251

Table A.18: Results of test phase jobs in the second part of the TSP multi-instance experiment with regression trees

Appendix B

Experiment results for the FSP

B.1 Results for Local Search Procedure Comparison Experiment

Short	Actual Parameter Values
tc	2000
i_a	ta022, ta029, ta030, ta023:ta024:ta025
$pert$	compRandEx[2]
ls	in[first], ex[first]
$init$	neh
acc_a	better

Table B.1: Actual parameter values for normal ILS jobs of the FSP local search procedure comparison experiment

Short	Actual Parameter Values
tc	15000
i_a	ta051
$pert$	compRandEx[2]
ls	in[first], ex[first]
$init$	neh
acc_a	better

Table B.2: Changed actual parameter values for normal ILS jobs of the FSP local search procedure comparison experiment on instance ta051

Instance	Perturbation	LS	Statistics				
			avg	σ	avg.excess	min	max
ta022	compRandEx[2]	in[first]	2099.28	0.46	0.0133 %	2099	2100
(2099)	compRandEx[2]	ex[first]	2106.48	5.33	0.3564 %	2099	2116
ta029	compRandEx[2]	in[first]	2238.20	2.18	0.0536 %	2237	2242
(2237)	compRandEx[2]	ex[first]	2242.32	2.72	0.2378 %	2237	2252
ta030	compRandEx[2]	in[first]	2178.88	1.17	0.0404 %	2178	2183
(2178)	compRandEx[2]	ex[first]	2186.12	4.59	0.3728 %	2178	2195
ta051	compRandEx[2]	in[first]	3895.04	7.68	1.0124 %	3875	3911
(3771)	compRandEx[2]	ex[first]	3906.12	12.39	1.2998 %	3871	3932

Table B.3: Normal ILS job results in the FSP local search procedure comparison experiment

Rank	Perturbation	LS	Value
1	compRandEx[2]	in[first]	1323
2	random ILS		2082
3	compRandEx[2]	ex[first]	3898

Table B.4: Ranks of normal and random ILS algorithms for the FSP local search comparison experiment

Short	Actual Parameter Values
tc	2000
i_{tr}	ta022, ta029, ta030
i_a	ta022, ta029, ta030
$strat$	Q(0), Q(λ)
$feat$	FSP-Features-AllButIdletime.feat
$init$	neh
acc_a	ϵ-better
ϵ_{a_t}	0.1, 0.5
ϵ_{a_a}	0.1, 0.5
γ	0.3, 0.8
r	c_Δ, c_i
λ	0.4, 0.8
fa	svr[0.025]
lf	200

Table B.5: Changed actual parameter values for Q-ILS jobs of the FSP local search procedure comparison experiment

Instance	Rank	Type	Parameters						H_0: equal		H_0: not better		T	Statistics				
			ϵ_{α_a}	ϵ_{α_t}	γ	λ	strat	r	0.05	0.1	0.05	0.1		avg	σ	avg. exc.	min	max
ta022 (2099)	1	apply	0.1	0.5	0.3		Q(0)	c_Δ	+	+	+	+	2.4000	2099.04	0.20	0.0019 %	2099	2100
	2	apply	0.1	0.5	0.3	0.8	Q(λ)	c_Δ	+	+	+	+	2.4000	2099.04	0.20	0.0019 %	2099	2100
	3	train		0.1	0.8		Q(0)	c_Δ	+	+	+	+	2.2553	2099.05	0.22	0.0024 %	2099	2100
	4	apply	0.1	0.5	0.3	0.4	Q(λ)	c_Δ	+	+	+	+	1.8677	2099.08	0.28	0.0038 %	2099	2100
	5	apply	0.1	0.5	0.8		Q(0)	c_Δ	+	+	+	+	1.8677	2099.08	0.28	0.0038 %	2099	2100
	6	apply	0.1	0.1	0.3		Q(0)	$c_{\tilde{q}}$	+	+	+	+	1.8677	2099.08	0.28	0.0038 %	2099	2100
	...								...	...	...	...	...	...	...	...	...	...
	41	rand	0.0						+	+	=	=	0.2877	2099.24	0.52	0.0114 %	2099	2101
ta029 (2237)	1	apply	0.1	0.5	0.3		Q(0)	$c_{\tilde{q}}$	+	+	=	=	0.5232	2237.92	1.55	0.0411 %	2237	2242
	2	apply	0.1	0.5	0.3		Q(0)	c_Δ	+	+	=	=	0.4168	2237.96	1.88	0.0429 %	2237	2242
	3	apply	0.1	0.1	0.3		Q(0)	c_Δ	+	+	=	=	0.2069	2238.08	1.91	0.0483 %	2237	2242
	4	apply	0.5	0.5	0.3		Q(0)	c_Δ	−	−	=	=	−0.4444	2238.48	2.28	0.0662 %	2237	2242
	5	apply	0.1	0.1	0.3		Q(0)	c_Δ	−	−	=	=	−0.5567	2238.52	1.87	0.0679 %	2237	2242
	6	apply	0.1	0.1	0.8	0.4	Q(λ)	c_Δ	−	−	=	=	−0.6606	2238.60	2.10	0.0715 %	2237	2242
	...								...	...	...	...	...	...	...	...	...	...
	49	rand	0.0						−	−	=	=	−2.0870	2239.60	2.55	0.1162 %	2237	2242
ta030 (2178)	1	apply	0.1	0.5	0.8	0.4	Q(λ)	c_Δ	=	=	+	+	2.7213	2178.16	0.62	0.0073 %	2178	2181
	2	apply	0.1	0.5	0.3		Q(0)	$c_{\tilde{q}}$	=	=	+	+	2.5039	2178.24	0.52	0.0110 %	2178	2180
	3	apply	0.1	0.5	0.3		Q(0)	c_Δ	=	=	+	+	1.7596	2178.36	0.91	0.0165 %	2178	2181
	4	apply	0.1	0.1	0.3		Q(0)	$c_{\tilde{q}}$	=	=	+	+	1.4559	2178.44	0.96	0.0202 %	2178	2181
	...								...	...	...	...	...	...	...	...	...	...
	125	rand	0.0						−	−	=	=	−1.8849	2179.68	1.77	0.0771 %	2178	2185

Table B.6: Q-ILS job results in the FSP local search procedure comparison experiment

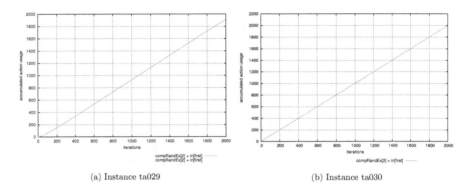

(a) Instance ta029   (b) Instance ta030

Figure B.1:   Accumulated action usage plots for best Q-ILS job run on instances ta029 and ta030 in the FSP local search procedure comparison experiment

B.2   Results for Perturbation Comparison Experiment

Short	Actual Parameter Values
tc	2000
i_a	ta022, ta023, ta024
$pert$	compRandEx[1], compRandEx[2], compRandEx[3]
ls	in[first]
$init$	neh
acc_a	better

Table B.7:   Actual parameter values for normal ILS jobs of the first part of the FSP perturbation comparison experiment

Short	Actual Parameter Values
tc	10000
i_a	ta055, ta052, ta072, ta075, ta081

Table B.8:   Changed actual parameter values for normal ILS jobs of the first part of the FSP perturbation comparison experiment on instances ta055, ta052, ta072, ta075, and ta081

Instance	Perturbation	LS	Statistics				
			avg	σ	avg.excess	min	max
ta022	compRandEx[3]	in[first]	2099.16	0.37	0.0076 %	2099	2100
(2099)	compRandEx[2]	in[first]	2099.16	0.37	0.0076 %	2099	2100
	compRandEx[1]	in[first]	2099.28	0.46	0.0133 %	2099	2100
ta023	compRandEx[3]	in[first]	2329.04	2.54	0.1307 %	2326	2334
(2326)	compRandEx[2]	in[first]	2330.68	3.57	0.2012 %	2326	2337
	compRandEx[1]	in[first]	2330.72	3.01	0.2029 %	2326	2335
ta024	compRandEx[1]	in[first]	2224.36	1.89	0.0612 %	2223	2229
(2223)	compRandEx[3]	in[first]	2224.56	2.10	0.0702 %	2223	2229
	compRandEx[2]	in[first]	2224.96	2.21	0.0882 %	2223	2229
ta052	compRandEx[1]	in[first]	3723.56	8.57	0.4467 %	3714	3749
(2668)	compRandEx[2]	in[first]	3730.56	10.76	0.6356 %	3715	3755
	compRandEx[3]	in[first]	3734.56	8.68	0.7435 %	3718	3750
ta055	compRandEx[1]	in[first]	3642.20	9.82	0.6411 %	3625	3662
(3553)	compRandEx[2]	in[first]	3649.80	10.99	0.8511 %	3628	3668
	compRandEx[3]	in[first]	3654.72	8.93	0.9870 %	3640	3672
ta072	compRandEx[2]	in[first]	5358.36	5.96	0.1750 %	5349	5362
(5349)	compRandEx[3]	in[first]	5360.12	4.48	0.2079 %	5349	5362
	compRandEx[1]	in[first]	5360.44	4.31	0.2139 %	5349	5362
ta075	compRandEx[1]	in[first]	5483.40	10.57	0.3000 %	5467	5498
(5467)	compRandEx[2]	in[first]	5484.12	12.73	0.3132 %	5467	5501
	compRandEx[3]	in[first]	5489.16	10.23	0.4053 %	5467	5498
ta081	compRandEx[1]	in[first]	6288.08	13.35	0.9647 %	6260	6310
(6106)	compRandEx[2]	in[first]	6300.44	13.07	1.1631 %	6282	6324
	compRandEx[3]	in[first]	6308.52	11.10	1.2929 %	6289	6332

Table B.9: Normal ILS job results in the first part of the FSP perturbation comparison experiment

Instance	Job	Type	Parameters						H_0: equal		H_0: not better			Statistics				
			ϵ_{a_o}	ϵ_{a_t}	γ	λ	strat	r	0.05	0.1	0.05	0.1	T	avg	σ	avg. exc.	min	max
ta022	1	apply	0.1	0.1	0.3		Q(0)	c_Δ	+	+	+	+	2.1381	2099.00	0.00	0.0000 %	2099	2099
(2099)	2	apply	0.1	0.5	0.3	0.8	Q(λ)	c_Δ	+	+	+	+	2.1381	2099.00	0.00	0.0000 %	2099	2099
	3	apply	0.1	0.5	0.3	0.4	Q(λ)	c_i	+	+	+	+	2.1381	2099.00	0.00	0.0000 %	2099	2099
	4	apply	0.1	0.5	0.8	0.8	Q(λ)	c_i	+	+	+	+	2.1381	2099.00	0.00	0.0000 %	2099	2099
	5	apply	0.1	0.1	0.8	0.4	Q(λ)	c_i	+	+	+	+	2.1381	2099.00	0.00	0.0000 %	2099	2099
	6	apply	0.1	0.1	0.3	0.4	Q(λ)	c_Δ	+	+	=	+	2.1381	2099.00	0.00	0.0000 %	2099	2099
	7	apply	0.1	0.5	0.8		Q(0)	c_i	+	+	=	+	1.4142	2099.04	0.20	0.0019 %	2099	2100
	…	…	…	…	…	…	…	…	…	…	…	…	…	…	…	…	…	…
	29	rand	0.0						+	+	=	=	0.4000	2099.12	0.33	0.0057 %	2099	2100
ta023	1	apply	0.1	0.5	0.8	0.4	Q(λ)	c_Δ	=	+	+	+	1.5754	2328.08	1.68	0.0894 %	2326	2334
(2326)	2	apply	0.1	0.1	0.8	0.8	Q(λ)	c_Δ	=	=	=	+	1.4022	2328.16	1.84	0.0929 %	2326	2334
	3	apply	0.1	0.1	0.8		Q(0)	c_i	=	=	=	=	1.2217	2328.32	1.49	0.0997 %	2326	2331
	4	apply	0.1	0.5	0.8	0.4	Q(λ)	c_i	=	=	=	=	1.2045	2328.32	1.57	0.0997 %	2326	2332
	5	apply	0.1	0.1	0.3		Q(0)	c_Δ	=	=	=	=	1.0237	2328.36	2.14	0.1015 %	2326	2333
	6	apply	0.1	0.1	0.3		Q(0)	c_Δ	=	=	=	=	1.2255	2328.36	1.11	0.1015 %	2326	2332
	…	…	…	…	…	…	…	…	…	…	…	…	…	…	…	…	…	…
	103	rand	0.0						−	=	=	=	-1.5574	2330.16	2.54	0.1788 %	2326	2334
ta024	1	apply	0.1	0.1	0.8	0.8	Q(λ)	c_Δ	=	=	+	+	2.5605	2223.36	0.49	0.0162 %	2223	2224
(2223)	2	apply	0.1	0.5	0.3		Q(0)	c_Δ	=	=	+	+	2.2233	2223.48	0.59	0.0216 %	2223	2225
	3	apply	0.1	0.1	0.3		Q(0)	c_i	=	=	+	+	2.1000	2223.52	0.65	0.0234 %	2223	2225
	4	apply	0.0	0.5	0.8		Q(0)	c_Δ	=	=	+	+	1.8478	2223.52	1.26	0.0234 %	2223	2229
	5	apply	0.1	0.5	0.8	0.4	Q(λ)	c_Δ	=	=	+	+	1.8045	2223.64	0.64	0.0288 %	2223	2225
	…	…	…	…	…	…	…	…	…	…	…	…	…	…	…	…	…	…
	126	rand	0.0						−	=	=	=	-1.9056	2225.56	2.52	0.1152 %	2223	2229

Table B.10: Q-ILS job results in the first part of the FSP perturbation comparison experiment

Short	Actual Parameter Values
tc	2000
i_{tr}	ta022, ta023, ta024
i_a	ta022, ta023, ta024
$strat$	Q(0), Q(λ)
$pert$	compRandEx[1], compRandEx[2], compRandEx[3]
ls	in[first]
$feat$	FSP-Features-AllButIdletime.feat
$init$	neh
ϵ_{a_t}	0.1, 0.5
ϵ_{a_a}	0.0, 0.1, 0.5
γ	0.3, 0.8
r	c_Δ, c_i
λ	0.4, 0.8
fa	svr[0.025]
lf	200

Table B.11: Actual parameter values for the single-mode sub-experiment of the first part of the FSP perturbation comparison experiment

Rank	Perturbation	LS	Value
1	compRandEx[2]	in[first]	1349
2	random ILS		1603
3	compRandEx[1]	in[first]	1840
4	compRandEx[3]	in[first]	2007

Table B.12: Ranks of normal and random ILS algorithms in the multi-instance mode sub-experiment of the first part of the FSP perturbation comparison experiment

Short	Actual Parameter Values
t_{tr}	15
t_a	20
tc	2000
i_{tr}	ta026:ta027:ta028
i_a	ta026:ta027:ta028
i_{te}	ta021, ta029, ta030
$strat$	Q(0), Q(λ)
$pert$	compRandEx[1], compRandEx[2], compRandEx[3]
ls	in[first]
$feat$	FSP-Features-AllButIdletime.feat
$init$	neh
ϵ_{a_t}	0.1, 0.5
ϵ_{a_a}	0.0, 0.1, 0.5
γ	0.3, 0.8
r	c_Δ, c_i
λ	0.4, 0.8
fa	svr[0.01]
lf	200

Table B.13:   Actual parameter values for the multi-instance mode sub-experiment of the first part of the FSP perturbation comparison experiment

Rank	Parameters						Value
	r	γ	λ	ϵ_{a_t}	ϵ_{a_a}	strat	
1	c_i	0.8		0.1	0.0	Q(0)	32514
2	c_i	0.8	0.8	0.1	0.0	Q(λ)	32967
3	c_i	0.3	0.8	0.5	0.0	Q(λ)	35698
4	c_Δ	0.8	0.8	0.5	0.1	Q(λ)	36137
5	c_i	0.8		0.5	0.1	Q(0)	37717
6	c_Δ	0.8		0.5	0.0	Q(0)	41229
7	c_Δ	0.8		0.1	0.1	Q(0)	41526
8	c_i	0.3		0.5	0.1	Q(0)	42004
8	c_i	0.3		0.5	0.1	Q(0)	42004
10	c_Δ	0.8	0.8	0.1	0.1	Q(λ)	43026
	⋮	⋮	⋮	⋮	⋮	⋮	⋮

Table B.14:   Ranks of Q-ILS algorithms in the multi-instance mode sub-experiment of the first part of the FSP perturbation comparison experiment

| Instance | Rank | Type | H_0: equal | | H_0: not better | | | Statistics | | | | |
			0.05	0.1	0.05	0.1	T	avg	σ	avg. exc.	min	max
ta021	2	ILS(R)						2297.32	0.73	0.0141 %	2297	2300
(2297)	6	Q-ILS						2297.32	0.73	0.0141 %	2297	2300
	2	Q-ILS	–	–	=	=	0.3945	2297.40	0.96	0.0174 %	2297	2302
	1	ILS(2)	–	–	=	=	0.6677	2297.45	0.93	0.0196 %	2297	2301
	3	ILS(1)	–	–	=	=	0.9657	2297.57	1.47	0.0250 %	2297	2305
	4	ILS(3)	–	–	–	–	1.8615	2297.82	1.53	0.0359 %	2297	2305
	1	Q-ILS	–	–	–	–	1.7399	2297.90	1.96	0.0392 %	2297	2305
	4	Q-ILS	–	–	–	–	2.9947	2298.20	1.70	0.0522 %	2297	2303
ta029	2	ILS(R)						2238.00	2.03	0.0447 %	2237	2242
(2237)	6	Q-ILS	–	–	=	=	0.5297	2238.25	2.19	0.0559 %	2237	2242
	4	Q-ILS	–	–	=	–	1.6221	2238.75	2.11	0.0782 %	2237	2242
	4	ILS(3)	–	–	–	–	1.8482	2238.93	2.43	0.0861 %	2237	2242
	2	Q-ILS	–	–	–	–	1.9748	2239.00	2.48	0.0894 %	2237	2242
	1	ILS(2)	–	–	–	–	2.2096	2239.12	2.50	0.0950 %	2237	2242
	3	ILS(1)	–	–	–	–	2.4457	2239.25	2.52	0.1006 %	2237	2242
	1	Q-ILS	–	–	–	–	2.6841	2239.38	2.53	0.1062 %	2237	2242
ta030	4	Q-ILS						2178.68	1.05	0.0310 %	2178	2181
(2178)	1	ILS(2)	–	–	=	=	0.1139	2178.70	0.91	0.0321 %	2178	2181
	2	ILS(R)	–	–	=	=	0.2315	2178.72	0.88	0.0333 %	2178	2180
	2	Q-ILS	–	–	=	=	0.2873	2178.75	1.28	0.0344 %	2178	2183
	4	ILS(3)	–	–	=	=	0.5620	2178.80	0.94	0.0367 %	2178	2180
	1	Q-ILS	–	–	=	=	0.9148	2178.90	1.15	0.0413 %	2178	2183
	3	ILS(1)	–	–	=	=	1.1534	2178.95	1.08	0.0436 %	2178	2183
	6	Q-ILS	–	–	=	–	1.4592	2179.03	1.10	0.0471 %	2178	2183

Table B.15: Results of test phase jobs in the first part of the FSP perturbation comparison experiment

Short	Actual Parameter Values
tc	2000
i_a	ta025, ta026, ta027
$pert$	partNeh[0.01], partNeh[0.05], partNeh[0.1]
ls	in[first]
$init$	neh
acc_a	better

Table B.16: Actual parameter values normal ILS job of the second part of FSP perturbation comparison experiment

Short	Actual Parameter Values
tc	10000
i_a	ta053, ta056, ta073, ta076, ta082

Table B.17: Changed actual parameter values for normal ILS jobs of the second part of the FSP perturbation comparison experiment on instances ta053, ta056, ta073, ta076, and ta082

Instance	Perturbation	LS	Statistics				
			avg	σ	avg.excess	min	max
ta025	partNeh[0.1]	in[first]	2294.40	1.71	0.1484 %	2291	2298
(2291)	partNeh[0.01]	in[first]	2294.68	2.19	0.1606 %	2291	2300
	partNeh[0.05]	in[first]	2294.96	2.35	0.1729 %	2291	2299
ta026	partNeh[0.1]	in[first]	2228.40	1.04	0.1078 %	2226	2230
(2226)	partNeh[0.01]	in[first]	2228.84	2.25	0.1276 %	2226	2235
	partNeh[0.05]	in[first]	2229.68	2.79	0.1653 %	2226	2238
ta027	partNeh[0.1]	in[first]	2274.08	2.04	0.0475 %	2273	2279
(2273)	partNeh[0.05]	in[first]	2274.32	2.25	0.0581 %	2273	2279
	partNeh[0.01]	in[first]	2275.04	3.06	0.0897 %	2273	2282
ta053	partNeh[0.01]	in[first]	3676.68	8.70	0.9245 %	3663	3697
(3591)	partNeh[0.05]	in[first]	3683.84	9.71	1.1211 %	3663	3697
	partNeh[0.1]	in[first]	3717.40	8.48	2.0423 %	3694	3732
ta056	partNeh[0.01]	in[first]	3713.24	10.07	0.7117 %	3698	3730
(3667)	partNeh[0.05]	in[first]	3718.04	11.56	0.8419 %	3695	3738
	partNeh[0.1]	in[first]	3746.92	7.52	1.6252 %	3730	3761
ta073	partNeh[0.01]	in[first]	5678.76	0.83	0.0486 %	5676	5679
(5676)	partNeh[0.05]	in[first]	5679.00	0.00	0.0529 %	5679	5679
	partNeh[0.1]	in[first]	5679.00	0.00	0.0529 %	5679	5679
ta076	partNeh[0.01]	in[first]	5305.60	2.55	0.0490 %	5303	5308
(5303)	partNeh[0.05]	in[first]	5308.00	0.00	0.0943 %	5308	5308
	partNeh[0.1]	in[first]	5308.04	0.20	0.0950 %	5308	5309
ta082	partNeh[0.01]	in[first]	6251.68	12.19	0.6712 %	6221	6272
(6183)	partNeh[0.05]	in[first]	6309.00	7.01	1.5942 %	6294	6327
	partNeh[0.1]	in[first]	6326.72	8.44	1.8795 %	6312	6340

Table B.18:   Normal ILS job results in the second part of the FSP perturbation comparison experiment

Instance	Job	Type	Parameters						H_0: equal		H_0: not better		T	avg	σ	Statistics		
			ϵ_{a_a}	ϵ_{a_t}	γ	λ	strat	r	0.05	0.1	0.05	0.1				avg. exc.	min	max
ta025 (2291)	1	apply	0.1	0.5	0.8		Q(0)	c_Δ	=	=	+	+	2.3842	2293.20	1.85	0.0960 %	2291	2297
	2	apply	0.1	0.1	0.8	0.8	Q(λ)	c_i	=	=	+	+	2.5392	2293.20	1.63	0.0960 %	2291	2296
	3	apply	0.1	0.5	0.3		Q(0)	c_i	=	=	+	+	2.5392	2293.20	1.63	0.0960 %	2291	2296
	4	apply	0.1	0.5	0.8		Q(0)	c_Δ	=	=	+	+	1.7802	2293.48	1.94	0.1082 %	2291	2297
	5	apply	0.1	0.1	0.8		Q(0)	c_Δ	=	=	+	+	1.7028	2293.52	1.94	0.1100 %	2291	2296
	6	apply	0.1	0.1	0.3		Q(0)	c_Δ	=	=	+	+	1.6904	2293.56	1.80	0.1117 %	2291	2296
	...	...	...	...	...	...	...	...	...	...	...	...	...	...	...	...	...	...
	116	rand	0.0						-	=	=	=	-1.3258	2295.08	1.91	0.1781 %	2291	2298
ta026 (2226)	1	apply	0.1	0.5	0.3		Q(0)	c_Δ	=	=	+	+	2.5688	2227.56	1.26	0.0701 %	2226	2229
	2	apply	0.1	0.5	0.8	0.4	Q(λ)	c_Δ	=	=	+	+	2.1164	2227.68	1.35	0.0755 %	2226	2230
	3	apply	0.5	0.1	0.3		Q(0)	c_i	=	=	+	+	2.1802	2227.68	1.28	0.0755 %	2226	2231
	4	apply	0.1	0.1	0.3		Q(0)	c_i	=	=	+	+	2.1224	2227.76	1.09	0.0791 %	2226	2229
	5	apply	0.1	0.1	0.8		Q(0)	c_i	=	=	+	+	2.0845	2227.76	1.13	0.0791 %	2226	2230
	6	apply	0.1	0.5	0.3	0.8	Q(λ)	c_Δ	=	=	+	+	1.8933	2227.76	1.33	0.0791 %	2226	2231
	...	...	...	...	...	...	...	...	...	...	...	...	...	...	...	...	...	...
	105	rand	0.0						-	=	=	=	-1.3149	2228.92	1.68	0.1312 %	2226	2234
ta027 (2273)	1	apply	0.1	0.1	0.8		Q(0)	c_i	+	+	+	+	2.2577	2273.12	0.60	0.0053 %	2273	2276
	2	apply	0.1	0.5	0.8	0.8	Q(λ)	c_Δ	+	+	+	+	2.2577	2273.12	0.60	0.0053 %	2273	2276
	3	apply	0.1	0.1	0.3	0.8	Q(λ)	c_Δ	+	+	+	+	2.2577	2273.12	0.60	0.0053 %	2273	2276
	4	apply	0.1	0.1	0.8		Q(0)	c_Δ	+	+	+	+	2.0996	2273.16	0.80	0.0070 %	2273	2277
	5	apply	0.1	0.5	0.8	0.8	Q(λ)	c_i	+	+	+	+	2.0996	2273.16	0.80	0.0070 %	2273	2277
	6	apply	0.1	0.1	0.3	0.4	Q(λ)	c_i	+	+	+	+	1.9370	2273.20	1.00	0.0088 %	2273	2278
	...	...	...	...	...	...	...	...	...	...	...	...	...	...	...	...	...	...
	98	rand	0.0						-	=	=	=	-0.6545	2274.48	2.28	0.0651 %	2273	2279

Table B.19: Q-ILS job results in the second part of the FSP perturbation comparison experiment

Short	Actual Parameter Values
tc	2000
i_{tr}	ta025, ta026, ta027
i_a	ta025, ta026, ta027
$strat$	Q(0), Q(λ)
$pert$	partNeh[0.01], partNeh[0.05], partNeh[0.1]
ls	in[first]
$feat$	FSP-Features-AllButIdletime.feat
$init$	neh
ϵ_{a_t}	0.1, 0.5
ϵ_{a_a}	0.0, 0.1, 0.5
γ	0.3, 0.8
r	c_Δ, c_i
λ	0.4, 0.8
fa	svr[0.025]
lf	200

Table B.20: Actual parameter values for the single-mode sub-experiment of the second part of the FSP perturbation comparison experiment

Short	Actual Parameter Values
i_{tr}	ta028:ta029:ta030
i_a	ta028:ta029:ta030
i_{te}	ta021, ta022, ta023
$strat$	Q(0), Q(λ)
$pert$	partNeh[0.01], partNeh[0.05], partNeh[0.1]
ls	in[first]
$feat$	FSP-Features-AllButIdletime.feat
$init$	neh
ϵ_{a_t}	0.1, 0.5
ϵ_{a_a}	0.0, 0.1, 0.5
γ	0.3, 0.8
r	c_Δ, c_i
λ	0.4, 0.8
fa	svr[0.01]
lf	200

Table B.21: Actual parameter values for the multi-instance mode sub-experiment of the second part of the FSP perturbation comparison experiment

Rank	Perturbation	LS	Value
1	partNeh[0.1]	in[first]	1732
2	random ILS		2107
3	partNeh[0.05]	in[first]	2511
4	partNeh[0.01]	in[first]	2513

Table B.22: Ranks of normal and random ILS algorithms in the multi-instance mode sub-experiment of the second part of the FSP perturbation comparison experiment

Rank	Parameters						Value
	r	γ	λ	ϵ_{a_t}	ϵ_{a_a}	strat	
1	c_Δ	0.3		0.1	0.1	$Q(0)$	25022
2	c_i	0.3		0.5	0.1	$Q(0)$	25543
3	c_i	0.8	0.4	0.1	0.1	$Q(\lambda)$	26596
4	c_i	0.8		0.5	0.1	$Q(0)$	27040
4	c_i	0.8		0.5	0.1	$Q(0)$	27040
6	c_i	0.3	0.4	0.1	0.1	$Q(\lambda)$	27072
7	c_i	0.8		0.1	0.1	$Q(0)$	27304
8	c_i	0.8	0.8	0.1	0.1	$Q(\lambda)$	28502
9	c_Δ	0.3		0.1	0.1	$Q(0)$	29629
10	c_Δ	0.8	0.8	0.1	0.1	$Q(\lambda)$	30043
	⋮	⋮	⋮	⋮	⋮	⋮	⋮

Table B.23: Ranks of Q-ILS algorithms in the multi-instance mode sub-experiment of the second part of the FSP perturbation comparison experiment

Instance	Rank	Type	H_0: equal 0.05	0.1	H_0: not better 0.05	0.1	T	avg	σ	avg. exc.	min	max
ta021	2	ILS(R)						2297.50	1.36	0.0218 %	2297	2304
(2297)	3	ILS(0.05)						2297.50	1.60	0.0218 %	2297	2304
	1	ILS(0.1)						2297.50	1.60	0.0218 %	2297	2304
	4	ILS(0.01)	–	–	=	=	0.0657	2297.53	1.80	0.0229 %	2297	2308
	2	Q-ILS	–	–	=	=	0.7355	2297.72	1.09	0.0316 %	2297	2302
	1	Q-ILS	–	–	=	=	0.9021	2297.78	1.07	0.0337 %	2297	2302
	4	Q-ILS	–	–	=	=	0.9298	2297.80	1.26	0.0348 %	2297	2303
	5	Q-ILS	–	–	=	–	1.3969	2297.95	1.26	0.0414 %	2297	2303
ta022	5	Q-ILS						2099.07	0.35	0.0036 %	2099	2101
(2099)	1	Q-ILS	–	–	=	=	0.9427	2099.15	0.36	0.0071 %	2099	2100
	2	Q-ILS	–	–	=	=	0.8596	2099.15	0.43	0.0071 %	2099	2101
	4	Q-ILS	–	–	–	–	1.7284	2099.22	0.42	0.0107 %	2099	2100
	4	ILS(0.01)	–	–	–	–	3.1500	2099.38	0.49	0.0179 %	2099	2100
	2	ILS(R)	–	–	–	–	3.5174	2099.50	0.68	0.0238 %	2099	2101
	3	ILS(0.05)	–	–	–	–	4.6157	2099.62	0.67	0.0298 %	2099	2101
	1	ILS(0.1)	–	–	–	–	4.6157	2099.62	0.67	0.0298 %	2099	2101
ta023	1	Q-ILS						2328.12	1.60	0.0914 %	2326	2332
(2326)	5	Q-ILS	–	–	=	=	0.9052	2328.40	1.06	0.1032 %	2326	2331
	4	Q-ILS	–	–	–	–	1.8821	2328.85	1.83	0.1225 %	2326	2333
	2	Q-ILS	–	–	–	–	3.3096	2329.28	1.50	0.1408 %	2326	2333
	4	ILS(0.01)	–	–	–	–	2.9542	2329.53	2.53	0.1515 %	2326	2336
	2	ILS(R)	–	–	–	–	3.3199	2329.78	2.70	0.1623 %	2326	2335
	3	ILS(0.05)	–	–	–	–	4.6216	2330.38	2.63	0.1881 %	2326	2334
	1	ILS(0.1)	–	–	–	–	4.6216	2330.38	2.63	0.1881 %	2326	2334

Table B.24: Results of test phase jobs in the second part of the FSP perturbation comparison experiment

Bibliography

[AC04]   E. Alba and J. F. Chicano. Training neural networks with GA hybrid algorithms. In
           K. Deb, R. Poli, W. Banzhaf, H.-G. Beyer, E. K. Burke, P. J. Darwen, D. Dasgupta,
           D. Floreano, J. A. Foster, M. Harman, O. Holland, P. Luca Lanzi, L. Spector, A. Tet-
           tamanzi, D. Thierens, and A. M. Tyrrell, editors, *GECCO 2004: Proceedings of the
           Genetic and Evolutionary Computation Conference*, volume 3102 of *Lecture Notes in
           Computer Science*, pages 852 – 863. Springer Verlag, Berlin, Germany, 2004.

[ACG⁺99]   G. Ausiello, P. Crescenzi, G. Gambosi, V. Kann, A. Marchetti-Spaccamela, and
           M. Protasi. *Complexity and Approximation: Combinatorial Optimization Problems
           and Their Approximability Properties*. Springer Verlag, Berlin, Germany, 1999.

[ACR98]   A. A. Andreatta, S. E. R. Carvalho, and C. C. Ribeiro. An object-oriented framework
           for local search heuristics. In *Proceedings of the 26th Conference on Technology of
           Object-Oriented Languages and Systems*, pages 33 – 45. IEEE Computer Society, 1998.

[ADL02]   B. Adenso-Diaz and M. Laguna. Fine-tuning of algorithms us-
           ing fractional experimental designs and local search. Available at:
           http://leeds.colorado.edu/Faculty/Laguna/articles/finetune.html, 2002.
           Version visited last on 14 February 2005, submitted to publication.

[AKv97]   E. H. L. Aarts, J. H. M. Korst, and P. J. M. van Laarhoven. Simulated annealing. In
           E. H. L. Aarts and J. K. Lenstra, editors, *Local Search in Combinatorial Optimization*,
           pages 91 – 120. John Wiley & Sons, Chichester, UK, 1997.

[AL97]   E. H. L. Aarts and J. K. Lenstra, editors. *Local Search in Combinatorial Optimization*.
           John Wiley & Sons, Chichester, UK, 1997.

[AMS97]   C. G. Atkeson, A. W. Moore, and S. Schaal. Locally weighted learning. *Artificial
           Intelligence Review*, 11(1 – 5):11 – 73, 1997.

[Bäc96]   T. Bäck. *Evolutionary Algorithms in Theory and Practice*. Oxford University Press,
           New York, NY, USA, 1996.

[Bai95]   L. C. Baird. Residual algorithms: Reinforcement learning with function approxima-
           tion. In *Proceedings of the Twelfth International Conference on Machine Learning*,
           pages 33 – 37. Morgan Kaufmann Publishers, San Francisco, CA, USA, 1995.

[Bak74]   K. R. Baker. *Introduction to Sequencing and Scheduling*. John Wiley & Sons, New
           York, NY, USA, 1974.

[Bal94]   S. Baluja. Population-based incremental learning: A method for integrating genetic
           search based function optimization and competitive learning. Technical Report CMU-
           CS-94-163, Carnegie Mellon University, Pittsburgh, PA, USA, 1994.

[Ban04]   C. Bang. Inkrementelle Support-Vektor Maschinen im GAILS Framework: Integra-
            tion und Evaluierung — Anwendung von Techniken des Maschinellen Lernens zur
            Steuerung von iterierter lokaler Suche. Master's thesis, Fachgebiet Intellektik, Fach-
            bereich Informatik, Technische Universität Darmstadt, Darmstadt, Germany, 2004.

[Bat96]   R. Battiti. Reactive search: Toward self-tuning heuristics. In V. J. Rayward-Smith,
            editor, *Modern Heuristic Search Methods*, pages 61 – 83. John Wiley & Sons, 1996.

[BC95]   S. Baluja and R. Caruana. Removing the genetics from the standard genetic algo-
            rithm. In A. Prieditis and S. Russell, editors, *Proceedings of the Twelfth International
            Conference on Machine Learning*, pages 38 – 46. Morgan Kaufmann Publishers, Palo
            Alto, CA, USA, 1995.

[BC99]   D. Bertsekas and D. Castanon. Rollout algorithms for stochastic scheduling problems.
            Journal of Heuristics, pages 89 – 108, 1999.

[BÇPP98]   R. E. Burkard, E. Çela, P. M. Pardalos, and L. S. Pitsoulis. The quadratic assignment
            problem. In P. M. Pardalos and D.-Z. Du, editors, *Handbook of Combinatorial Opti-
            mization*, pages 241 – 338. Kluwer Academic Publishers, Dordrecht, The Netherlands,
            1998.

[BCT94]   P. Briggs, K. D. Cooper, and L. Torczon. Improvements to graph coloring register
            allocation. *ACM Transactions on Programming Languages and Systems*, 16(3):428 –
            455, 1994.

[BD97]   S. Baluja and S. Davies. Using optimal dependency-trees for combinatorial optimiza-
            tion: Learning the structure of the search space. Technical Report CMU-CS97-107,
            Departement of Computer Science, Carnegie Mellon University, Pittsburgh, USA,
            1997.

[BD98]   S. Baluja and S. Davies. Fast probabilistic modeling for combinatorial optimization.
            In *Proceedings of the Fifteenth National Conference on Artificial Intelligence*, pages
            469 – 476. AAAI Press / The MIT Press, Menlo Park, CA, USA, 1998.

[Bel57]   R. E. Bellman. *Dynamic Programming*. Princeton University Press, Princeton, NJ,
            USA, 1957.

[Ben92]   J. L. Bentley. Fast algorithms for geometric traveling salesman problems. *ORSA
            Journal on Computing*, 4:347 – 411, 1992.

[Ber82]   D. P. Bertsekas. Distributed dynamic programming. *IEEE Transactions on Evolu-
            tionary Computation*, 27:610 – 616, 1982.

[Ber83]   D. P. Bertsekas. Distributed asynchronous computation of fixed points. *Mathematical
            Programming*, 27:107 – 120, 1983.

[Ber87]   D. P. Bertsekas. *Dynamic Programming: Deterministic and Stochastic Models*. Pren-
            tice Hall, Englewood Cliffs, NJ, USA, 1987.

[Ber95]   D. P. Bertsekas. *Dynamic Programming and Optimal Control*. Athena Scientific,
            Belmont, MA, USA, 1995.

[Ber00]   A. Berny. Selection and reinforcement learning for combinatorial optimization. In
            M. Schoenauer, K. Deb, G. Rudolph, X. Yao, E. Lutton, J. J. Merelo, and H.-P.
            Schwefel, editors, *Proceedings of PPSN-VI, Sixth International Conference on Parallel
            Problem Solving from Nature*, pages 601 – 610. Springer Verlag, Berlin, Germany, 2000.

[Bey95]   Hans-Georg Beyer. Towards a theory of evolution strategies: Self-adaptation. *Evolutionary Computation*, 3(3):311–347, 1995.

[BFM97]   In T. Bäck, D. Fogel, and Z. Michalewicz, editors, *Handbook of Evolutionary Computation*. Oxford University Press, New York, NY, USA, 1997.

[BFOS93]   L. Breiman, J. H. Friedman, R. A. Olshen, and C. J. Stone. *Classification and Regression Trees*. Chapman & Hall, New York, NY, USA, 1993.

[BGK+95]   R. S. Barr, B. L. Golden, J. P. Kelly, M. G. C. Resende, and W. R. Stewart. Designing and reporting on computational experiments with heuristic methods. *Journal of Heuristics*, 1(1):9 – 32, 1995.

[Bib93a]   W. Bibel. *Deduction: Automated Logic*. Academic Press, London, UK, 1993.

[Bib93b]   W. Bibel. *Wissensrepräsentation und Inferenz*. Vieweg, Braunschweig, Germany, 1993.

[Bir04]   M. Birattari. *The Problem of Tuning Metaheuristics as Seen from a Machine Learning Perspective*. PhD thesis, Université Libre de Bruxelles, Brussels, Belgium, 2004.

[BKM94]   K. D. Boese, A. B. Kahng, and S. Muddu. A new adaptive multi-start technique for combinatorial global optimization. *Operations Research Letters*, 16(2):101 – 113, 1994.

[BKNS04]   J. A. Bagnell, S. Kakade, A. Y. Ng, and J. Schneider. Policy search by dynamic programming. In S. Thrun, L. Saul, and B. Schölkopf, editors, *Advances in Neural Information Processing Systems 16*. MIT Press, Cambridge, MA, 2004.

[BL98]   B. Berger and T. Leight. Protein folding in the hydrophobic-hydrophilic (HP) model is NP-complete. *Journal of Computational Biology*, 5(1):27 – 40, 1998.

[BM98]   J. A. Boyan and A. W. Moore. Learning evaluation functions for global optimization and Boolean satisfiability. In *Proceedings of the Fifteenth National Conference on Artificial Intelligence*, pages 3 – 10. AAAI Press / The MIT Press, Menlo Park, CA, USA, 1998.

[BM00]   J. Boyan and A. Moore. Learning evaluation functions to improve optimization by local search. *Journal of Machine Learning Research*, 1:77 – 112, 2000.

[BM03]   A. Barto and S. Mahadevan. Recent advances in hierarchical reinforcement learning. *Discrete Event Systems Journal*, 13(4):41 – 77, 2003.

[Boe96]   K. D. Boese. *Models for Iterative Global Optimization*. PhD thesis, University of California, Computer Science Department, Los Angeles, USA, 1996.

[Boy99]   J. A. Boyan. Least-squares temporal difference learning. In I. Bratko and S. Dzeroski, editors, *Proceedings of the Sixteenth International Conference on Machine Learning*, pages 49 – 56. Morgan Kaufmann Publishers, San Francisco, CA, USA, 1999.

[Boy02]   J. A. Boyan. Technical update: Least-squares temporal difference learning. *Machine Learning*, 49(2–3):233 – 246, 2002.

[BP97]   R. Battiti and M. Protasi. Reactive search, a history-based heuristic for MAX-SAT. *ACM Journal of Experimental Algorithmics*, 2:2, 1997.

[BP99]   R. Battiti and M. Protasi. Reactive local search techniques for the maximum k-conjunctive constraint satisfaction problem. *Discrete Applied Mathematics*, 96–97:3 – 27, 1999.

[BP01]   R. Battiti and M. Protasi. Reactive local search for the maximum clique problem. *Algorithmica*, 29(4):610 – 637, 2001.

[BR03]   C. Blum and A. Roli. Metaheuristics in combinatorial optimization: Overview and conceptual comparison. *ACM Computing Surveys*, 35(3):268 – 308, 2003.

[Bri90]   J. S. Bridle. Training stochastic model recognition algorithms as networks can lead to maximum mutual information estimates of parameters. In D. S. Touretzky, editor, *Advances in Neural Information Processing Systems 2*, pages 211 – 217. Morgan Kaufmann Publishers, San Mateo, CA, USA, 1990.

[BRJ97]   G. Booch, J. Rumbaugh, and I. Jacobson. *Unified Modeling Language User Guide*. Addison-Wesley, 1997.

[Bro01]   T. X. Brown. Switch packet arbitration via queue-learning. In T. G. Dietterich, S. Becker, and Z. Ghahramani, editors, *Advances in Neural Information Processing Systems 14*, pages 1337 – 1344. MIT Press, Cambridge, MA, USA, 2001.

[Bru98]   P. Brucker. *Scheduling Algorithms*. Springer Verlag, Heidelberg, Germany, 1998.

[BS03]   D. Bagnell and J. Schneider. Covariant policy search. In G. Gottlob and T. Walsh, editors, *Proceedings of the Eighteenth International Joint Conference on Artificial Intelligence*, pages 1019 – 1024. Morgan Kaufmann, 2003.

[BSPV02] M. Birattari, T. Stützle, L. Paquete, and K. Varrentrapp. A racing algorithm for configuring metaheuristics. In W. B. Langdon, E. Cantú-Paz, K. Mathias, R. Roy, D. Davis, R. Poli, K. Balakrishnan, V. Honavar, G. Rudolph, J. Wegener, L. Bull, M. A. Potter, A. C. Schultz, J. F. Miller, E. Burke, and N. Jonoska, editors, *GECCO 2002: Proceedings of the Genetic and Evolutionary Computation Conference*, pages 11 – 18. Morgan Kaufmann Publishers, San Francisco, CA, USA, 2002.

[BT94]   R. Battiti and G. Tecchiolli. The reactive tabu search. *ORSA Journal on Computing*, 6(2):126 – 140, 1994.

[BT96]   D. P. Bertsekas and J. N. Tsitsiklis. *Neuro-Dynamic Programming*. Athena Scientific, Belmont, MA, USA, 1996.

[BTW97]   D. Bertsekas, J. Tsitsiklis, and C. Wu. Rollout algorithms for combinatorial optimization. *Journal of Heuristics*, 3:245 – 262, 1997.

[CB96]   R. H. Crites and A. G. Barto. Improving elevator performance using reinforcement learning. In D. S. Touretzky, M. C. Mozer, and M. E. Hasselmo, editors, *Advances in Neural Information Processing Systems 8*, pages 1017 – 1023. MIT Press, Cambridge, MA, USA, 1996.

[CB01]   R. Collobert and S. Bengio. Svmtorch: Support vector machines for large-scale regression problems. *Journal of Machine Learning Research*, 1:143 – 160, 2001.

[CB04]   R. Collobert and S. Bengio. Svmtorch: Support vector machines for large-scale regression problems. Available at: http://www.idiap.ch/learning/SVMTorch.html, 2004. Version visited last on 14 February 2005.

[CCPS97]   W. J. Cook, W. H. Cunningham, W. R. Pulleybank, and A. Schrijver. *Combinatorial Optimization*. John Wiley & Sons, New York, NY, USA, 1997.

[CDG99a]   D. Corne, M. Dorigo, and F. Glover. Introduction. In D. Corne, M. Dorigo, and F. Glover, editors, *New Ideas in Optimization*, pages 11 – 32. McGraw Hill, London, UK, 1999.

[CDG99b]   D. Corne, M. Dorigo, and F. Glover, editors. *New Ideas in Optimization*. McGraw Hill, London, UK, 1999.

[CDS03]   M. Chiarandini, I. Dumitrescu, and T. Stützle. Local search for the graph colouring problem — a computational study. Technical Report AIDA-03-01, Fachgebiet Intellektik, Fachbereich Informatik, Technische Universität Darmstadt, Darmstadt, Germany, 2003.

[Çel98]   E. Çela. *The Quadratic Assignment Problem: Theory and Algorithms*. Kluwer Academic Publishers, Dordrecht, The Netherlands, 1998.

[CFG⁺96]   D. Clark, J. Frank, I. P. Gent, E. MacIntyre, N. Tomov, and T. Walsh. Local search and the number of solutions. In E. C. Freuder, editor, *Proceedings of the Second International Conference on Principles and Practice of Constraint Programming*, volume 1118 of *Lecture Notes in Computer Science*, pages 119 – 133. Springer Verlag, Berlin, Germany, 1996.

[Che97]   K. Chen. A simple learning algorithm for the traveling salesman problem. *Physical Review E*, 55:7809 – 7812, 1997.

[CL04]   C.-C. Chang and C.-J. Lin. LIBSVM: a library for support vector machines. Available at: http://www.csie.ntu.edu.tw/ cjlin/libsvm, 2004. Version visited last on 14 February 2005.

[CLL96]   M. Carter, G. Laporte, and S. Lee. Examination timetabling: Algorithmic strategies and applications. *Journal of the Operational Research Society*, 47(3):373 – 383, 1996.

[CLPL03]   S. W. Choia, D. Leeb, J. H. Park, and I. Leea. Nonlinear regression using RBFN with linear submodels. *Chemometrics and Intelligent Laboratory Systems*, 65(2):191 – 208, 2003.

[CW64]   G. Clarke and J. W. Wright. Scheduling of vehicles from a central depot to a number of delivery points. *Operations Research*, 12(4):568 – 581, 1964.

[Day92]   P. Dayan. The convergence of TD(λ) for general λ. *Machine Learning*, 8:341 – 362, 1992.

[dBJV97]   J. S. de Bonet, C. L. Isbell Jr., and P. Viola. Mimic: Finding optima by estimating probability densities. In M. C. Mozer, M. I. Jordan, and T. Petsche, editors, *Advances in Neural Information Processing Systems 10*, pages 424 – 431. MIT Press, Cambridge, MA, USA, 1997.

[DD99]   M. Dorigo and G. Di Caro. The ant colony optimization meta-heuristic. In D. Corne, M. Dorigo, and F. Glover, editors, *New Ideas in Optimization*, pages 11 – 32. McGraw Hill, London, UK, 1999.

[DDG99]   M. Dorigo, G. Di Caro, and L. M. Gambardella. Ant algorithms for discrete optimization. *Artificial Life*, 5(2):137 – 172, 1999.

[DGP97]   D. Du, J. Gu, and P. M. Pardalos, editors. *Satisfiability problem: Theory and Applications*, volume 35 of *DIMACS Series on Discrete Mathematics and Theoretical Computer Science*. American Mathematical Society, Providence, RI, USA, 1997.

[Die00]   T. G. Dietterich. Hierarchical reinforcement learning with the MAXQ value function decomposition. *Journal of Artificial Intelligence Research*, 13:227 – 303, 2000.

[Dig96]   B. L. Digney. Emergent hierarchical control structures: Learning reactive/hierarchical relationships in reinforcement environments. In P. Maes, M. J. Mataric, J.-A. Meyer, J. Pollack, and S. W. Wilson, editors, *Proceedings of the Fourth Conference on the Simulation of Adaptive Behavior*. MIT Press, Cambridge, MA, USA, 1996.

[Dig98]   B. Digney. Learning hierarchical control structure for multiple tasks and changing environments. In R. Pfeifer, B. Blumberg, J.-A. Meyer, and S. W. Wilson, editors, *Proceedings of the Fifth Conference on the Simulation of Adaptive Behavior*. MIT Press, Cambridge, MA, USA, 1998.

[DJW02]   S. Droste, T. Jansen, and I. Wegner. Optimization with randomized search heuristics — the (A)NFL theorem, realistic scenarios, and difficult functions. *Theoretical Computer Science*, 287:131 – 144, 2002.

[Dot05]   A. Dotor. Das Flow Shop Problem im GAILS Framework: Integration und Analyse. Master's thesis, Fachgebiet Intellektik, Fachbereich Informatik, Technische Universität Darmstadt, Darmstadt, Germany, 2005.

[Dow93]   K. A. Dowsland. Simulated annealing. In C. R. Reeves, editor, *Modern Heuristic Techniques for Combinatorial Problems*. Blackwell Scientific Publications, Oxford, UK, 1993.

[DS02]   M. Dorigo and T. Stützle. The ant colony optimization metaheuristic: Algorithms, applications and advances. In F. Glover and G. Kochenberger, editors, *Handbook of Metaheuristics*, pages 251 – 285. Kluwer Academic Publishers, Norwell, MA, USA, 2002.

[DS04]   M. Dorigo and T. Stützle. *Ant Colony Optimization*. MIT Press, Cambridge, MA, USA, 2004.

[DV99]   A. Dean and D. Voss. *Design and Analysis of Experiments*. Springer Verlag, New York, NY, USA, 1999.

[DV04]   R. Dorne and C. Voudouris. HSF: The iOpt's framework to easily design metaheuristic methods. In M. G. C. Resende and J. P. de Sousa, editors, *Metaheuristics: Computer Decision-Making*, pages 237 – 256. Kluwer Academic Publishers, Boston, MA, USA, 2004.

[dW85]   D. de Werra. An introduction to timetabling. *European Journal of Operational Research*, 19(2):151 – 162, 1985.

[EEE01]   J. R. G. Evans, M. J. Edirisinghe, and P. V. C. J. Eames. Combinatorial searches of inorganic materials using the inkjet printer: Science philosophy and technology. *Journal of the European Ceramic Society*, 21:2291 – 2299, 2001.

[Ehr00]   M. Ehrgott. *Multicriteria Optimization*, volume 491 of *Lecture Notes in Economics and Mathematical Systems*. Springer Verlag, Heidelberg, Germany, 2000.

[EMM02] Y. Engel, S. Mannor, and R. Meir. Sparse online greedy support vector regression. In T. Elomaa, H. Mannila, and H. Toivonen, editors, *Proceedings of the 13th European Conference on Machine Learning 2002*, pages 84 – 96. Springer Verlag Heidelberg, Germany, 2002.

[Fay99] M. E. Fayad. *Implementing Application Frameworks: Object-Oriented Frameworks at Work*. John Wiley & Sons, New York, NY, USA, 1999.

[FR95] T. A. Feo and M. G. C. Resende. Greedy randomized adaptive search procedures. *Journal of Global Optimization*, 6:109 – 133, 1995.

[FR01] P. Festa and M. G. C. Resende. GRASP: An annotated bibliography. In P. Hansen and C. C. Ribeiro, editors, *Essays and Surveys on Metaheuristics*, pages 325 – 367. Kluwer Academic Publishers, Boston, MA, USA, 2001.

[Fre82] S. French. *Sequencing and Scheduling: An Introduction to the Mathematics of the Job Shop*. Horwood, Chichester, UK, 1982.

[FS87] M. E. Fayad and D. C. Schmidt. Object-oriented application frameworks. *Communications of the ACM*, 40(10):32 – 38, 1987.

[FSJ99] M. E. Fayad, D. C. Schmidt, and R. E. Johnson, editors. *Building Application Frameworks: Object-Oriented Foundations of Framework Design*. John Wiley & Sons, New York, NY, USA, 1999.

[FTA94] J. L. Ribeiro Filho, P. C. Treleaven, and C. Alippi. Genetic-algorithm programming environments. *IEEE Computer*, 27(6):28 – 43, 1994.

[FV02] A. Fink and S. Voß. Hotframe: A heuristic optimization framework. In S. Voß and D. Woodruff, editors, *Optimization Software Class Libraries*, volume 18 of *Operations Research/Computer Science Interfaces Series*, pages 81 – 154. Kluwer Academic Publishers, Boston, MA, USA, 2002.

[FVW02] A. Fink, S. Voß, and D.L. Woodruff. Metaheuristic class libraries. In F. Glover and G. Kochenberger, editors, *Handbook of Metaheuristics*, pages 515 – 535. Kluwer Academic Publishers, Norwell, MA, USA, 2002.

[GC96] J. Gratch and S. Chien. Adaptive problem-solving for large-scale scheduling problems: A case study. *Journal of Artificial Intelligence Research*, 4:365 – 396, 1996.

[GD95] L. M. Gambardella and M. Dorigo. Ant-Q: A reinforcement learning approach to the traveling salesman problem. In A. Prieditis and S. Russell, editors, *Proceedings of the Twelfth International Conference on Machine Learning*, pages 252 – 260. Morgan Kaufmann Publishers, Palo Alto, CA, USA, 1995.

[GHJV95] E. Gamma, R. Helm, R. Johnson, and J. Vlissides. *Design Patterns*. Addison-Wesley, 1995.

[Gim05] J. Gimmler. Lernen im GAILS Framework mittels Regressiontrees und Featureselection. Master's thesis, Fachgebiet Intellektik, Fachbereich Informatik, Technische Universität Darmstadt, Darmstadt, Germany, 2005.

[GJ79] M. R. Garey and D. S. Johnson. *Computers and Intractability: A Guide to the Theory of NP-Completeness*. Freeman, NY, USA, 1979.

[GJS76]   M. R. Garey, D. S. Johnson, and R. Sethi. The complexity of flowshop and jobshop
             scheduling. *Mathematics of Operations Research*, 1:117 – 129, 1976.

[GK02]   F. Glover and G. Kochenberger, editors. *Handbook of Metaheuristics*. Kluwer Aca-
             demic Publishers, Norwell, MA, USA, 2002.

[GL97]   F. Glover and M. Laguna. *Tabu Search*. Kluwer Academic Publishers, Boston, MA,
             USA, 1997.

[GM03a]   M. Ghavamzadeh and S. Mahadevan. Hierarchical policy gradient algorithms. Morgan
             Kaufmann, 2003.

[GM03b]   M. Ghavamzadeh and S. Mahadevan. Hierarchical policy gradient algorithms. In
             Proceedings of the Twentieth International Conference on Machine Learning, 2003.

[GP96]   C. Glass and C. N. Potts. A comparison of local search methods for flow shop schedul-
             ing. *Annals of Operations Research*, 63:489 – 509, 1996.

[GP02]   G. Gutin and A. P. Punnen, editors. *The Traveling Salesman Problem and Its Varia-
             tions*. Kluwer Academic Publishers, Dordrecht, The Netherlands, 2002.

[GPFW97]   J. Gu, P. Purdom, J. Franco, and B. Wah. Algorithms for the satisfiability (SAT)
             problem: A survey. In D. Du, J. Gu, and P. M. Pardalos, editors, *Satisfiability prob-
             lem: Theory and Applications*, volume 35 of *DIMACS Series on Discrete Mathematics
             and Theoretical Computer Science*, pages 19 – 151. American Mathematical Society,
             Providence, RI, USA, 1997.

[GS01]   L. Di Gaspero and A. Schaerf. A case-study for EasyLocal++: The course timetabling
             problem. Technical Report UDMI/13/2001/RR, University of Udine, Udine, Italy,
             2001.

[GS03]   L. Di Gaspero and A. Schaerf. EasyLocal++: An object-oriented framework for flex-
             ible design of local search algorithms. *Software – Practice and Experience*, 33(8):733
             – 765, 2003.

[GTdW93]   F. Glover, É. D. Taillard, and D. de Werra, editors. *Tabu Search*, volume 43 of *Annals
             of Operations Research*. Baltzer Science Publishers, 1993.

[Har05]   R. Harder. OpenTS. Available at: http://www.coin-or.org/OpenTS/index.html,
             2005. Version visited last on 14 February 2005.

[Hen99]   Pascal Van Hentenryck. *The OPL optimization programming language*. MIT Press,
             1999.

[Hen02]   B. Hengst. Discovering hierarchy in reinforcement learning with HEXQ. In C. Sammut
             and A. G. Hoffmann, editors, *Proceedings of the Nineteenth International Conference
             on Machine Learning*. Morgan Kaufmann, 2002.

[HGM00]   N. Hernandez-Gardiol and S. Mahadevan. Hierarchical memory-based reinforcement
             learning. In S. A. Solla, T. K. Leen, and K.-R. Müller, editors, *Advances in Neural
             Information Processing Systems 12*, pages 1047 – 1053. MIT Press, Cambridge, MA,
             USA, 2000.

[HM99]   P. Hansen and N. Mladenović. An introduction to variable neighborhood search.
             In S. Voss, S. Martello, I. H. Osman, and C. Roucairol, editors, *Meta-Heuristics:
             Advances and Trends in Local Search Paradigms for Optimization*, pages 433 – 458.
             Kluwer Academic Publishers, Boston, MA, USA, 1999.

[HM01]   P. Hansen and N. Mladenović. Variable neighborhood search: Principles and applica-
            tions. *European Journal of Operational Research*, 130(3):449 – 467, 2001.

[HM02]   P. Hansen and N. Mladenović. Variable neighborhood search. In F. Glover and
            G. Kochenberger, editors, *Handbook of Metaheuristics*, pages 145 – 184. Kluwer Aca-
            demic Publishers, Norwell, MA, USA, 2002.

[HMK+98]   M. Hauskrecht, N. Meuleau, L. P. Kaelbling, T. Dean, and C. Boutilier. Hierar-
            chical solution of Markov decision processes using macro-actions. In G. F. Cooper
            and S. Moral, editors, *Proceedings of the Fourteenth Conference on Uncertainty in
            Artificial Intelligence*, pages 220 – 229. Morgan Kaufmann, 1998.

[Hoo98]   H. H. Hoos. *Stochastic Local Search — Methods, Models, Applications*. PhD the-
            sis, Fachgebiet Intellektik, Fachbereich Informatik, Technische Universität Darmstadt,
            Darmstadt, Germany, 1998.

[Hoo99a]   H. H. Hoos. On the run-time behaviour of stochastic local search algorithms for SAT.
            In *Proceedings of the Sixteenth National Conference on Artificial Intelligence*, pages
            661 – 666. AAAI Press / The MIT Press, Menlo Park, CA, USA, 1999.

[Hoo99b]   H.H. Hoos. *Stochastic Local Search – Methods, Models, Applications*. Infix-Verlag,
            Sankt Augustin, Germany, 1999.

[Hoo02]   H. H. Hoos. An adaptive noise mechanism for WalkSAT. In *Proceedings of the Eigh-
            teenth National Conference on Artificial Intelligence*, pages 655 – 660. AAAI Press /
            The MIT Press, Menlo Park, CA, USA, 2002.

[How60]   R. Howard. *Dynamic Programming and Markov Processes*. MIT Press, Cambridge,
            MA, USA, 1960.

[How71]   R. A Howard. *Dynamic Probabilistic Systems: Semi-Markov and Decision Processes*.
            John Wiley & Sons, New York, NY, USA, 1971.

[HR01]   P. Hansen and C. C. Ribeiro, editors. *Essays and Surveys on Metaheuristics*. Kluwer
            Academic Publishers, Boston, MA, USA, 2001.

[HS98]   H. H. Hoos and T. Stützle. Evaluating Las Vegas algorithms — pitfalls and remedies.
            In G. F. Cooper and S. Moral, editors, *Proceedings of the Fourteenth Conference on
            Uncertainty in Artificial Intelligence*, pages 238 – 245. Morgan Kaufmann Publishers,
            San Francisco, CA, USA, 1998.

[HS99]   H. H. Hoos and T. Stützle. Characterising the behaviour of stochastic local search.
            Artificial Intelligence, 112(1–2):213 – 232, 1999.

[HS00a]   H. H. Hoos and T. Stützle. Local search algorithms for SAT: An empirical evaluation.
            Journal of Automated Reasoning, 24(4):421 – 481, 2000.

[HS00b]   H. H. Hoos and T. Stützle. SATLIB: An Online Resource for Research on SAT. In
            H. van Maaren I. P. Gent and T. Walsh, editors, *SAT2000 — Highlights of Satisfiability
            Research in the Year 2000*, pages 283 – 292. IOS Press, Amsterdam, The Netherlands,
            2000.

[HS04]   H. H. Hoos and T. Stützle. *Stochastic Local Search — Foundations and Applications*.
            Morgan Kaufmann Publishers, CA, USA, 2004.

[ILO04]   ILOG S.A., 9, Rue de Verdun, BP 85, 94253 Gentilly Cedex, France. *ILOG Solver 6.0 User's Manual*, 2004.

[JF95]   T. Jones and S. Forrest. Fitness distance correlation as a measure of problem difficulty for genetic algorithms. In L. J. Eshelman, editor, *Proceedings of the Sixth International Conference on Genetic Algorithms*, pages 184 – 192. Morgan Kaufmann Publishers, San Mateo, CA, USA, 1995.

[JM97]   D. S. Johnson and L. A. McGeoch. The traveling salesman problem: A case study in local optimization. In E. H. L. Aarts and J. K. Lenstra, editors, *Local Search in Combinatorial Optimization*, pages 215 – 310. John Wiley & Sons, Chichester, UK, 1997.

[JM02]   D. S. Johnson and L. A. McGeoch. Experimental analysis of heuristics for the STSP. In G. Gutin and A. Punnen, editors, *The Traveling Salesman Problem and its Variations*, pages 369 – 443. Kluwer Academic Publishers, Dordrecht, The Netherlands, 2002.

[JMRS98]   M. Jones, G. McKeown, and V. Rayward-Smith. Templar: An object oriented framework for distributed combinatorial optimization. In *In UNICOM Seminar on Modern Heuristics for Decision Support*, 1998.

[Joa99]   T. Joachims. Making large-scale SVM learning practical. In B. Schölkopf, C. Burges, and A. Smola, editors, *Advances in Kernel Methods: Support Vector Machines*, chapter 11. MIT Press, Cambridge, MA, USA, 1999.

[Joa04]   T. Joachims. Svm^{light}. Available at: `http://svmlight.joachims.org/`, 2004. Version visited last on 14 February 2005.

[Joh74]   D. S. Johnson. Approximation algorithms for combinatorial problems. *Journal of Computer and System Science*, 9(3):256 – 278, 1974.

[Jon00]   M. S. Jones. *An Object-Oriented Framework for the Implementation of Search Techniques*. PhD thesis, University of East Anglia, Norwich, UK, 2000.

[JT94]   T. R. Jensen and B. Toft. *Graph Coloring Problems*. John Wiley & Sons, New York, NY, USA, 1994.

[Jul95]   B. A. Julstrom. What have you done for me lately? Adapting operator probabilities in a steady-state genetic algorithm. In L. J. Eshelman, editor, *Proceedings of the Sixth International Conference on Genetic Algorithms*, pages 81 – 87. Morgan Kaufmann Publishers, San Mateo, CA, USA, 1995.

[Kak02]   S. Kakade. A natural policy gradient. In T. G. Dietterich, S. Becker, and Z. Ghahramani, editors, *Advances in Neural Information Processing Systems 14*. MIT Press, Cambridge, MA, USA, 2002.

[KH05]   J. Knowles and E. J. Hughes. Multiobjective optimization on a budget of 250 evaluations. In *Evolutionary Multi-Criterion Optimization (EMO-2005)*. Springer Verlag, Berlin, Germany, 2005.

[KLM96]   L. P. Kaelbling, M. L. Littman, and A. P. Moore. Reinforcement learning: A survey. *Journal of Artificial Intelligence Research*, 4:237 – 285, 1996.

[Kno04]   J. Knowles. ParEGO: A hybrid algorithm with on-line landscape approximation for expensive multiobjective optimization problems. Technical Report TR-COMPSYSBIO-2004-01, University of Manchester, Manchester, UK, 2004.

[Kor04]   O. Korb. Das Traveling Salesman Problem im GAILS Framework: Integration und Analyse. Master's thesis, Fachgebiet Intellektik, Fachbereich Informatik, Technische Universität Darmstadt, Darmstadt, Germany, 2004.

[KP94]   A. Kolen and E. Pesch. Genetic local search in combinatorial optimization. *Discrete Applied Mathematics*, 48(3):273 – 284, 1994.

[KPL+98]   N. Krasnogor, D. Pelta, P. M. Lopez, P. Mocciola, and E. de la Canal. Genetic algorithms for the protein folding problem: A critical view. In C. F. E. Alpaydin, editor, *Procedings of Engineering of Intelligent Systems*. ICSC Academic Press, 1998.

[Lar82]   H. J. Larson. *Introduction to Probability Theory and Statistical Inference*. John Wiley & Sons, New York, NY, USA, 1982.

[LC98]   F. Laburthe and Y. Caseau. SALSA: A language for search algorithms. In M. Maher and J.-F. Puget, editors, *Proceedings of the 4th International Conference on Principles and Practice of Constraint Programming*, volume 1520 of *Lecture Notes in Computer Science*, pages 310 – 324. Springer Verlag, Berlin, Germany, 1998.

[Leh54]   D. H. Lehmer. Random number generation on the BRL highspeed computing machines. *Math. Rev.*, 15:559, 1954.

[LELP00]   P. Larrañaga, R. Etxeberria, J. A. Lozano, and J. M. Peña. Combinatorial optimization by learning and simulation of bayesian networks. In G. Boutilier and M. Goldszmidt, editors, *Proceedings of the sixteenth Conference on Uncertainty in Artificial Intelligence*, pages 343 – 352. Morgan Kaufmann, 2000.

[LK73]   S. Lin and B.W. Kernighan. An effective heuristic algorithm for the traveling salesman problem. *Operations Research*, 21(2):498 – 516, 1973.

[LMS01]   H. R. Lourenço, O. Martin, and T. Stützle. A beginner's introduction to iterated local search. In *Proceedings of MIC 2001*, pages 1 – 6, Porto, Portugal, 2001.

[LMS02]   H. R. Lourenço, O. Martin, and T. Stützle. Iterated local search. In F. Glover and G. Kochenberger, editors, *Handbook of Metaheuristics*, pages 321 – 353. Kluwer Academic Publishers, Norwell, MA, USA, 2002.

[LO96]   G. Laporte and I.H. Osman, editors. *Meta-Heuristics in Combinatorial Optimization*, volume 63 of *Annals of Operations Research*. Baltzer Science Publishers, 1996.

[Luc59]   D. Luce. *Individual Choice Behavior*. Wiley, NY, USA, 1959.

[LW92]   J. Lehn and H. Wegmann. *Einführung in die Statistik*. B.G. Teubner, Stuttgart, Germany, 1992.

[LWJ03]   H. C. Lau, W. C. Wan, and X. Jia. Generic object-oriented tabu search framework. In *Proceedings of the 5th Metaheuristics International Conference*, 2003.

[LWL04]   H. C. Lau, W. C. Wan, and M. K. Lim. A development framework for rapid metaheuristics hybridization. In S. Halim, editor, *Proceedings of the 28th Annual International Computer Software and Applications Conference*, pages 362 – 367, 2004.

[MA93]   A. W. Moore and C. G. Atkeson. Prioritzed sweeping: Reinforcement learning with less data and less real time. *Machine Learning*, 13:103 – 130, 1993.

[MA04]   M.Milano and A.Roli. MAGMA: A multiagent architecture for metaheuristics. *IEEE Transactions on Systems, Man and Cybernetics – Part B*, 34(2), 2004.

[MB01]   A. McGovern and A. G. Barto. Automatic discovery of subgoals in reinforcement
             learning using diverse density. In C. E. Brodley and A. P. Danyluk, editors, *Proceedings
             of the Eighteenth International Conference on Machine Learning*, 2001.

[MBPS97]   R. Moll, A. Barto, T. Perkins, and R. Sutton. Reinforcement and local search: A case
             study. Technical Report UM-CS-1997-044, University of Massachusetts, Amherst,
             MA, USA, 1997.

[MBPS98]   R. Moll, A. Barto, T. Perkins, and R. Sutton. Learning instance-independent value
             functions to enhance local search. In M. J. Kearns, S. A. Solla, and D. A. Cohn,
             editors, *Advances in Neural Information Processing Systems 11*, pages 1017 – 1023.
             MIT Press, Cambridge, MA, USA, 1998.

[MF99]   P. Merz and B. Freisleben. Fitness landscapes and memetic algorithm design. In
             D. Corne, M. Dorigo, and F. Glover, editors, *New Ideas in Optimization*, pages 244 –
             260. McGraw Hill, London, UK, 1999.

[MGSK88]   H. Mühlenbein, M. Gorges-Schleuter, and O. Krämer. Evolution algorithms in com-
             binatorial optimization. *Parallel Computing*, 7:65 – 85, 1988.

[MH99]   L. Michel and P. Van Hentenryck. Localizer: A modeling language for local search.
             INFORMS Journal on Computing, 11(1):1 – 14, 1999.

[MH00]   L. Michel and P. Van Hentenryck. OPL++: A modeling layer for constraint pro-
             gramming libraries. Technical Report CS-00-07, Brown University, Department of
             Computer Science, Providence, RI, USA, 2000.

[MH01]   L. Michel and P. Van Hentenryck. Localizer++: An open library for local search. Tech-
             nical Report CS-01-02, Brown University, Department of Computer Science, Provi-
             dence, RI, USA, 2001.

[MH02]   L. Michel and P. Van Hentenryck. A constraint-based architecture for local search. In
             *Proceedings of the 17th ACM SIGPLAN conference on Object-oriented programming,
             systems, languages, and applications*, pages 83 – 100. ACM Press, 2002.

[MH05]   L. Michel and P. Van Hentenryck. Localizer++. Available at:
             http://www.cse.uconn.edu/ ldm/projects.html, 2005. Version visited last
             on 14 February 2005.

[Mit96]   M. Mitchell. *An Introduction to Genetic Algorithms*. MIT Press, Cambridge, MA,
             USA, 1996.

[MM94]   O. Maron and A. W. Moore. Hoeffding races: Accelerating model selection search for
             classification and function approximation. In J. D. Cowan, G. Tesauro, and J. Al-
             spector, editors, *Advances in Neural Information Processing Systems 6*, pages 59 – 66.
             Morgan Kaufmann Publishers, San Francisco, CA, USA, 1994.

[MM97]   O. Maron and A. W. Moore. The racing algorithm: Model selection for lazy learners.
             Artificial Intelligence Review, 11(1–5):193 – 225, 1997.

[MM98]   A. McGovern and E. Moss. Scheduling straight-line code using reinforcement learning
             and rollouts. In M. J. Kearns, S. A. Solla, and D. A. Cohn, editors, *Advances in
             Neural Information Processing Systems 11*, pages 903 – 909. MIT Press, Cambridge,
             MA, USA, 1998.

[MMB02]   A. McGovern, E. Moss, and A. G. Barto. Building a basic block instruction scheduler using reinforcement learning and rollouts. *Machine Learning*, 49(2/3):141 – 160, 2002.

[MMG01]   R. Makar, S. Mahadevan, and M. Ghavamzadeh. Hierarchical multi-agent reinforcement learning. In J. P. Müller, E. Andre, S. Sen, and C. Frasson, editors, *Proceedings of the Fifth International Conference on Autonomous Agents*, pages 246 – 253, Montreal, Canada, 2001. ACM Press.

[MMHK04]   S. Mannor, I. Menache, A. Hoze, and U. Klein. Dynamic abstraction in reinforcement learning via clustering. In *Proceedings of the Twenty-First International Conference on Machine Learning*, 2004.

[MN98]   M. Matsumoto and T. Nishimura. Mersenne twister: A 623-dimensionally equidistributed uniform pseudorandom number generator. *ACM Transactions on Modeling and Computer Simulations*, 8(1):3 – 30, 1998.

[MO95]   O. C. Martin and S. W. Otto. Partitoning of unstructured meshes for load balancing. *Concurrency: Practice and Experience*, 7(4):303 – 314, 1995.

[MOF91]   O. C. Martin, S. W. Otto, and E. W. Felten. Large-step Markov chains for the traveling salesman problem. *Complex Systems*, 5(3):299 – 326, 1991.

[Mon00]   D. C. Montgomery. *Design and Analysis of Experiments*. John Wiley & Sons, New York, NY, USA, 5th edition, 2000.

[Mos99]   P. Moscatom. Memetic algorithms: A short introduction. In D. Corne, M. Dorigo, and F. Glover, editors, *New Ideas in Optimization*, pages 219 – 234. McGraw Hill, London, UK, 1999.

[MP96]   H. Mühlenbein and G. Paaß. From recombination of genes to estimations of distributions. In H.-M. Voigt, W. Ebeling, I. Rechenberg, and H.-P. Schwefel, editors, *Proceedings of PPSN-IV, Fourth International Conference on Parallel Problem Solving from Nature*, volume 1141 of *Lecture Notes in Computer Science*, pages 178 – 187. Springer Verlag, Berlin, Germany, 1996.

[MP99a]   V. V. Miagkikh and W. F. Punch III. An approach to solving combinatorial optimization problems using a population of reinforcement learning agents. In W. Banzhaf, J. Daida, A. E. Eiben, M. H. Garzon, V. Honavar, M. Jakiela, and R. E. Smith, editors, *Proceedings of the Genetic and Evolutionary Computation Conference (GECCO-1999)*, volume 1, pages 1358 – 1365. Morgan Kaufmann Publishers, San Francisco, CA, USA, 1999.

[MP99b]   V. V. Miagkikh and W. F. Punch III. Global search in combinatorial optimization using reinforcement learning algorithms. In P. J. Angeline, Z. Michalewicz, M. Schoenauer, X. Yao, and A. Zalzala, editors, *Proceedings of the 1999 Congress on Evolutionary Computation (CEC'99)*, volume 1, pages 623 – 630, 1999.

[MSK97]   D. McAllester, B. Selman, and H. Kautz. Evidence for invariants in local search. In *Proceedings of the Fourteenth National Conference on Artificial Intelligence*, pages 321 – 326. AAAI Press / The MIT Press, Menlo Park, CA, USA, 1997.

[MT00]   P. Mills and E. Tsang. Guided local search for solving SAT and weighted MAX-SAT problems. In I. P. Gent, H. van Maaren, and T. Walsh, editors, *SAT2000 — Highlights of Satisfiability Research in the Year 2000*, pages 89 – 106. IOS Press, Amsterdam, The Netherlands, 2000.

[MTF03] P. Mills, E. Tsang, and J. Ford. Applying an extended guided local search to the quadratic assignment problem. *Annals of Operations Research*, 118:121 – 135, 2003.

[Müh91] H. Mühlenbein. Evolution in time and space – the parallel genetic algorithm. In G. J. Rawlins, editor, *Foundations of Genetic Algorithms*, pages 316 – 337. Morgan Kaufmann Publishers, San Mateo, CA, USA, 1991.

[Mye90] R. H. Myers. *Classical and Modern Regression with Applications*. Duxburry Press, Belmont, CA, USA, 1990.

[Nar01] A. Nareyek, editor. *Local Search for Planning and Scheduling*, volume 2148 of *Lecture Notes in Artificial Intelligence*. Springer Verlag, Berlin, Germany, 2001.

[Nar03] A. Nareyek. Choosing search heuristics by non-stationary reinforcement learning. In M. G. C. Resende and J. P. de Sousa, editors, *Metaheuristics: Computer Decision-Making*, pages 523 – 544. Kluwer Academic Publishers, 2003.

[NJ00] A. Y. Ng and M. I. Jordan. Pegasus: A policy search method for large MDPs and POMDPs. In C. Boutilier and M. Goldszmidt, editors, *Proceedings of the sixteenth Conference on Uncertainty in Artificial Intelligence*, pages 406 – 466. Morgan Kaufmann Publishers, CA, USA, 2000.

[NJH83] M. Nawaz, E. Enscore Jr., and I. Ham. A heuristic algorithm for the m-machine, n-job flow-shop sequencing problem. *OMEGA International Journal of Management Science*, 11(1):91 – 95, 1983.

[NW88] G. L. Nemhauser and L. A. Wolsey. *Integer and Combinatorial Optimization*. John Wiley & Sons, New York, NY, USA, 1988.

[OK96] I. H. Osman and J. P. Kelly, editors. *Meta-Heuristics: Theory & Applications*. Kluwer Academic Publishers, Boston, MA, USA, 1996.

[OL96] I. H. Osman and G. Laporte. Metaheuristics: A bibliography. *Annals of Operations Research*, 63:513 – 628, 1996.

[Par98] R. Parr. *Hierarchical Control and Learning for Markov Decision Processes*. PhD thesis, University of California, Berkeley, California, USA, 1998.

[PBV02a] D. Pelta, A. Blanco, and J. L. Verdegay. Applying a fuzzy sets-based heuristic to the protein structure prediction problem. *International Journal of Intelligent Systems*, 17(7):629 – 643, 2002.

[PBV02b] D. Pelta, A. Blanco, and J. L. Verdegay. A fuzzy valuation-based local search framework for combinatorial problems. *Journal of Fuzzy Optimization and Decision Making*, 1(2):177 – 193, 2002.

[PE02] J. E. Pettinger and R. M. Everson. Controlling genetic algorithms with reinforcement learning. In W. B. Langdon, E. Cantú-Paz, K. E. Mathias, R. Roy, D. Davis, R. Poli, K. Balakrishnan, V. Honavar, G. Rudolph, J. Wegener, L. Bull, M. A. Potter, A. C. Schultz, J. F. Miller, E. Burke, and N. Jonoska, editors, *Proceedings of the Genetic and Evolutionary Computation Conference (GECCO-2002)*, page 692. Morgan Kaufmann Publishers, San Francisco, CA, USA, 2002.

[PE05] J. E. Pettinger and R. M. Everson. Controlling genetic algorithms with reinforcement learning. Available at: http://citeseer.ist.psu.edu/pettinger03controlling.html, 2005. Version visited last on 14 February 2005.

[PGCP99]   M. Pelikan, D. E. Goldberg, and E. Cantú-Paz. BOA: The bayesian optimization algo-
           rithm. In W. Banzhaf, J. Daida, A. E. Eiben, M. H. Garzon, V. Honavar, M. Jakiela,
           and R. E. Smith, editors, *Proceedings of the Genetic and Evolutionary Computation
           Conference (GECCO-1999)*, volume 1, pages 525 – 532. Morgan Kaufmann Publishers,
           San Francisco, CA, USA, 1999.

[PGL99]   M. Pelikan, D. E. Goldberg, and F. Lobo. A survey of optimization by building
           and using probabilistic models. Technical Report IlliGAL Report No. 99018, Illinois
           Genetic Algorithms Laboratory, University of Illinois, IL, USA, 1999.

[Pin95]   M. Pinedo. *Scheduling: Theory, Algorithms, and Systems*. Prentice Hall, Englewood
           Cliffs, NJ, USA, 1995.

[PK01]   D. J. Patterson and H. Kautz. Auto-WalkSAT: A self-tuning implementation of Walk-
           SAT. In H. Kautz and B. Selman, editors, *LICS 2001 Workshop on Theory and Ap-
           plications of Satisfiability Testing (SAT 2001)*, volume 9. Elsevier, Amsterdam, The
           Netherlands, 2001.

[PM99]   Hansen P. and N. Mladenović. First improvement may be better than best improve-
           ment: an empirical study. Technical Report G-99-54, GERAD and École des Hautes
           Études Commerciales, Montréal, Canada, 1999.

[PR97]   R. Parr and S. Russell. Reinforcement learning with hierarchies of machines. In
           M. I. Jordan, M. J. Kearns, and S. A. Solla, editors, *Advances in Neural Information
           Processing Systems 10*. MIT Press, Cambridge, MA, USA, 1997.

[PR00]   M. Prais and C. C. Ribeiro. Reactive GRASP: An application to a matrix decom-
           position problem in TDMA traffic assignment. *INFORMS Journal on Computing*,
           12(3):164 – 176, 2000.

[Pra02]   M. Pranzo. *Algorithms and Applications for Complex Job Shop Scheduling Problems*.
           PhD thesis, Università La Sapienza, Dipartimento di Statistica, Probabilità e Statis-
           tiche Applicate, Roma, Italy, 2002.

[PRB69]   W. H. Payne, J. R. Rabung, and T. P. Bogyo. Coding the lehmer pseudo-random
           number generator. *Communications of the ACM*, 12(2):85 – 86, 1969.

[Pre00]   D. Precup. *Temporal Abstraction in Reinforcement Learning*. PhD thesis, Department
           of Computer Science, University of Massachusetts, Amherst, USA, 2000.

[PRG+03]   J.A. Parejo, J. Racero, F. Guerrero, T. Kwok, and K.A. Smith. FOM: A framework
           for metaheuristic optimization. In P. M. A Sloot, D. Abramson, A. V. Bogdanov, J. J.
           Dongarra, A. Y. Zomaya, and Y. E. Gorbachev, editors, *Proceedings of International
           Conference on Conputational Science*, volume 2660 of *Lecture Notes in Computer
           Science*, pages 886 – 895. Springer Verlag, Berlin, Germany, 2003.

[PS04]   V. Phan and S. Skiena. An improved time-sensitive metaheuristic framework for
           combinatorial optimization. In C. C. Ribeiro and S. L. Martins, editors, *Experimental
           and Efficient Algorithms, Third International Workshop*, volume 2059 of *Lecture Notes
           in Computer Science*, pages 432 – 445. Springer Verlag, Berlin, Germany, 2004.

[PW93]   J. Peng and R. J. Williams. Efficient learning and planning within the Dyna frame-
           work. *Adaptive Behavior*, 1(4):437 – 454, 1993.

[PW94]   P. M. Pardalos and H. Wolkowicz, editors. *Quadratic Assignment and Related Problems*, volume 16 of *DIMACS Series on Discrete Mathematics and Theoretical Computer Science*. American Mathematical Society, Providence, RI, USA, 1994.

[RdS03]   M. G. C. Resende and J. P. de Sousa, editors. *Metaheuristics: Computer Decision-Making*. Kluwer Academic Publishers, Boston, MA, USA, 2003.

[RDSB+03]   O. Rossi-Doria, M. Sampels, M. Birattri, M. Chiarandini, M. Dorigo, L. M. Gambardella, J. Knowles, M. Manfrin, M. Mastrolilli, B. Paechter, L. Paquete, and T. Stützle. A comparison of the performance of different metaheuristics on the timetabling problem. In E. Burke and P. Causmaecker, editors, *Practice and Theory of Automated Timetabling IV*, volume 2740 of *Lecture Notes in Computer Science*, pages 329 – 351. Springer Verlag, Berlin, Germany, 2003.

[Rei91]   G. Reinelt. TSPLIB – A traveling salesman problem library. *ORSA Journal on Computing*, 3:376 – 384, 1991.

[RHHB97]   C. D. Rosin, R. S. Halliday, W. E. Hart, and R. K. Belew. A comparison of global and local search methods in drug docking. In T. Bäck, editor, *Proceedings of the Seventh International Conference on Genetic Algorithms*. Morgan Kaufmann Publishers, San Francisco, CA, USA, 1997.

[RJB98]   J. Rumbaugh, I. Jacobson, and G. Booch. *The Unified Modeling Language Reference Manual*. Addison-Wesley, 1998.

[RN94]   G. A. Rummery and M. Niranjan. On-line Q-learning using connectionist systems. Technical Report CUED/F-INFENG/TR 166, Engineering Department, Cambridge University, Cambridge, UK, 1994.

[RN03]   S. Russell and P. Norvig. *Artificial Intelligence: A Modern Approach*. Prentice-Hall/Pearson Education, Upper Saddle River, NJ, USA, 2nd edition, 2003.

[Roj96]   R. Rojas. *Neural Networks – A Systematic Introduction*. Springer Verlag, Berlin, Germany, 1996.

[RR00]   M. Ryan and M. Reid. Learning to fly: An application of hierarchical reinforcement learning. In P. Langley, editor, *Proceedings of the Seventeenth International Conference on Machine Learning*, pages 807 – 814. Morgan Kaufmann Publishers, CA, USA, 2000.

[RR02]   M. G. C. Resende and C. C. Ribeiro. Greedy randomized adaptive search procedures. In F. Glover and G. Kochenberger, editors, *Handbook of Metaheuristics*, pages 219 – 249. Kluwer Academic Publishers, Norwell, MA, USA, 2002.

[RR03]   C. R. Reeves and J. E. Rowe. *Genetic Algorithms – Principles and Perspectives*. Kluwer Academic Publishers, Dordrecht, The Netherlands, 2003.

[Rüp01]   S. Rüping. Incremental learning with support vector machines. In N. Cercone, T. Y. Lin, and X. Wu, editors, *Proceedings of the 2001 IEEE International Conference on Data Mining*, pages 641 – 642. IEEE, IEEE Computer Society, 2001.

[Rüp02]   S. Rüping. Incremental learning with support vector machines. Technical report, SFB475, Universität Dortmund, Dortmund, Germany, 2002.

[Rüp04] S. Rüping. mySVM – a support vector machine. Available at: `http://www-ai.cs.uni-dortmund.de/SOFTWARE/MYSVM/`, 2004. Version visited last on 14 February 2005.

[SB96] S. Singh and D. Bertsekas. Reinforcement learning for dynamic channel allocation in cellular telephone systems. In M. C. Mozer, M. I. Jordan, and T. Petsche, editors, *Advances in Neural Information Processing Systems 9*, pages 974 – 980. MIT Press, Cambridge, MA, USA, 1996.

[SB98] R. S. Sutton and A. G. Barto. *Reinforcement Learning: An Introduction.* MIT Press, Cambridge, MA, USA, 1998.

[SB04] Ö. Simsek and A. G. Barto. Using relative novelty to identify useful temporal abstractions in reinforcement learning. In *Proceedings of the Twenty-fFirst International Conference on Machine Learning*, pages 751 – 758. ACM Press, 2004.

[SBM98] J. G. Schneider, J. A. Boyan, and A. W. Moore. Value function based production scheduling. In J. Shavlik, editor, *Proceedings of the Fifteenth International Conference on Machine Learning*, pages 522 – 530. Morgan Kaufmann Publishers, San Francisco, CA, USA, 1998.

[SBMRD02] M. Sampels, C. Blum, M. Mastrolilli, and O. Rossi-Doria. Metaheuristics for Group Shop Scheduling. In J.J. Merelo Guervós et al., editor, *Proceedings of PPSN-VII, Seventh International Conference on Parallel Problem Solving from Nature*, number 2439 in Lecture Notes in Computer Science, pages 631 – 640. Springer Verlag, Berlin, Germany, 2002.

[Sch93] A. Schwartz. A reinforcement learning method for maximizing undiscounted rewards. In *Proceedings of the Tenth International Conference on Machine Learning*, pages 298 – 305. Morgan Kaufmann Publishers, San Francisco, CA, USA, 1993.

[Sch99] A. Schaerf. A survey of automated timetabling. *Artificial Intelligence Review*, 13(2):87 – 127, 1999.

[SCL00] A. Schaerf, M. Cadoli, and M. Lenzerini. Local++: A C++ framework for local search algorithms. *Software – Practice and Experience*, 30(3):233 – 257, 2000.

[SD01] T. Stützle and M. Dorigo. Local search and metaheuristics for the quadratic assignment problem. Technical Report AIDA-01-01, Fachgebiet Intellektik, Fachbereich Informatik, Technische Universität Darmstadt, Darmstadt, Germany, 2001.

[SdBS01] M. Samples, M. den Besten, and T. Stützle. Report on milestone 1.2 – definition of the experimental protocols. Available via `http://www.metaheuristics.net/`, 2001. Version visited last on 14 February 2005.

[SF97] J. E. Smith and T. C. Fogarty. Operator and parameter adaptation in genetic algorithms. *Soft Computing*, 1(2):81 – 87, 1997.

[SH00] T. Stützle and H. H. Hoos. $\mathcal{MAX}$–$\mathcal{MIN}$ Ant System. *Future Generation Computer Systems*, 16(8):889 – 914, 2000.

[SH01] T. Stützle and H. H. Hoos. Analysing the run-time behaviour of iterated local search for the travelling salesman problem. In P. Hansen and C. C. Ribeiro, editors, *Essays and Surveys on Metaheuristics*, pages 589 – 611. Kluwer Academic Publishers, Boston, MA, USA, 2001.

[SH03]   A. Shmygelska and H. H. Hoos. An improved ant colony optimisation algorithm for the 2d hp protein folding problem. In *Proceedings of the 16th Canadian Conference on Artificial Intelligence*, volume 2671, pages 400 – 417. Springer Verlag, Berlin, Germany, 2003.

[She00]   D.J. Sheskin. *Handbook of Parametric and Nonparametric Statistical Procedures*. Chaoman & Hall/CRC, Boca Raton, Florida, 2nd edition, 2000.

[SKC94]   B. Selman, H. Kautz, and B. Cohen. Noise strategies for improving local search. In *Proceedings of the 12th National Conference on Artificial Intelligence*, pages 337 – 343. AAAI Press / The MIT Press, Menlo Park, CA, USA, 1994.

[SLS99a]   N. A. Syed, H. Liu, and K. Sung. Incremental learning with support vector machines. In *Proceedings of the Workshop on Support Vector Machines at the International Joint Conference on Articial Intelligence (IJCAI-99), Stockholm, Sweden*, 1999.

[SLS99b]   N. A. Syed, H. Liu, and K. K. Sung. Handling concept drifts in incremental learning with support vector machines. In *Proceedings of the fifth ACM SIGKDD international conference on Knowledge discovery and data mining*, pages 317 – 321. ACM Press, San Diego, California, USA, 1999.

[SM99]   G. R. Schreiber and O. C. Martin. Cut size statistics of graph bisection heuristics. *SIAM Journal on Optimization*, 10(1):231 – 251, 1999.

[SM01]   M. Strens and A. Moore. Direct policy search using paired statistical tests. In C. E. Brodley and A. P. Danyluk, editors, *Proceedings of the Eighteenth International Conference on Machine Learning*, pages 545 – 552. Morgan Kaufmann, San Francisco, CA, 2001.

[SMSM00]   R. S. Sutton, D. A. McAllester, S. P. Singh, and Y. Mansour. Policy gradient methods for reinforcement learning with function approximation. In T. K. Leen, T. G. Dietterich, and V. Tresp, editors, *Advances in Neural Information Processing Systems 12*, pages 1057 – 1063. MIT Press, Cambridge, MA, USA, 2000.

[SPS98]   R. S. Sutton, D. Precup, and S. Singh. Intra-option learning about temporally abstract actions. In J. Shavlik, editor, *Proceedings of the Fifteenth International Conference on Machine Learning*, pages 556 – 564. Morgan Kaufmann Publishers, San Francisco, CA, USA, 1998.

[SPS99]   R. S. Sutton, D. Precup, and S. Singh. Between MDPs and semi-MDPs: A framework for temporal abstraction in reinforcement learning. *Artificial Intelligence*, 112:181 – 211, 1999.

[SR02]   R. Schoknecht and M. Riedmiller. Speeding-up reinforcement learning with multi-step actions. In J. Dorronsoro, editor, *Proceedings of the Twelfth International Conference on Artificial Neural Networks (ICANN)*, volume 2415, pages 813 – 818. Springer Verlag, Berlin, Germany, 2002.

[SS02]   B. Schölkopf and A. Smola. *Learning with Kernels: Support Vector Machines, Regularization, Optimization and Beyond*. MIT Press, Cambridge, MA, USA, 2002.

[SS03]   T. Schiavinotto and T. Stützle. Search space analysis for the linear ordering problem. In G. R. Raidl, J.-A. Meyer, M. Middendorf, S. Cagnoni, J. J. R. Cardalda, D. W. Corne, J. Gottlieb, A. Guillot, E. Hart, C. G. Johnson, and E. Marchiori, editors, *Applications of Evolutionary Computing*, volume 2611 of *Lecture Notes in Computer Science*, pages 322 – 333. Springer Verlag, Berlin, Germany, 2003.

[SS04] A. J. Smola and B. Schölkopf. A tutorial on support vector regression. *Statistics and Computing*, 13(3):199 – 222, 2004.

[Sta95] P. F. Stadler. Towards a theory of landscapes. In R. Lopéz-Peña, R. Capovilla, R. García-Pelayo, H. Waelbroeck, and F. Zertuche, editors, *Complex Systems and Binary Networks*, volume 461 of *Lecture Notes in Physics*, pages 77–163. Springer Verlag, Berlin, Germany, 1995.

[Ste78] D. M. Stein. An asymptotic probabilistic analysis of a routing problem. *Mathematics of Operations Research*, 3:89 – 101, 1978.

[Ste86] R. E. Steuer. *Multiple Criteria Optimization: Theory, Computation and Application.* John Wiley & Sons, New York, NY, USA, 1986.

[Str91] B. Stroustrup. *The C++ Programming Language*. Addison-Wesley, 2nd edition, 1991.

[Stü98] T. Stützle. *Local Search Algorithms for Combinatorial Problems — Analysis, Improvements, and New Applications.* PhD thesis, Fachgebiet Intellektik, Fachbereich Informatik, Technische Universität Darmstadt, Darmstadt, Germany, 1998.

[Stü99] T. Stützle. *Local Search Algorithms for Combinatorial Problems: Analysis, Improvements, and New Applications.* Infix-Verlag, Sankt Augustin, Germany, 1999.

[Sut88] R. S. Sutton. Learning to predict by the method of temporal differences. *Machine Learning*, 3:9 – 44, 1988.

[Sut96] R. S. Sutton. Generalizations in reinforcement learning: Successful examples using sparse coarse coding. In D. S. Touretzky, M. C. Mozer, and M. E. Hasselmo, editors, *Advances in Neural Information Processing Systems 8*, pages 1038 – 1044. MIT Press, Cambridge, MA, USA, 1996.

[Swe85] R. E. Sweet. The Mesa programming environment. *ACM SIGPLAN Notices*, 20(7):216 – 229, 1985.

[Tai05] E. Taillard. Scheduling instances. Available at: `http://ina.eivd.ch/collab-orateurs/etd/problemes.dir/ordonnancement.dir/ordonnancement.html`, 2005. Version visited last on 14 February 2005.

[TFH05] D. Thomas, C. Fowler, and A. Hunt. *Programming Ruby. The Pragmatic Programmer's Guide.* The Pragmatic Programmers, Dallas, TX, USA, 2nd edition, 2005.

[Tho91] E. L. Thorndike. *Animal Intelligence.* Hafner, Darien, CO, USA, 1991.

[Tor99] L. Torgo. *Inductive Learning of Tree-based Regression Models.* PhD thesis, Department of Computer Science, Faculty of Sciences, University of Porto, Porto, Portugal, 1999.

[TPM03] J. Theiler, S. Perkins, and J. Ma. Accurate online support vector regression. *Neural Computation*, 15(11):2683 – 2703, 2003.

[TS01] O. Telelis and P. Stamatopoulos. Combinatorial optimization through statistical instance-based learning. In *Proceedings of the IEEE 13th International Conference on Tools with Artificial Intelligence*, pages 203 – 209, 2001. Online publication: `http://computer.org/proceedings/ictai/1417/1417toc.htm`, version visited last on 14 February 2005.

[TS02]   O. Telelis and P. Stamatopoulos. Guiding constructive search with statistical instance-based learning. *International Journal on Artificial Intelligence Tools*, 11(2):247 – 266, 2002.

[Url05a]   Center for scientific computing home page. http://www.csc.uni-frankfurt.de/, 2005. Version visited last on 14 February 2005.

[Url05b]   GAlib. Available at: http://lancet.mit.edu/ga/, 2005. Version visited last on 14 February 2005.

[Url05c]   ILOGSolver. Available at: http://www.ilog.com/products/solver/, 2005. Version visited last on 14 February 2005.

[Url05d]   Merriam-webster online dictionary. Available at: http://www.m-w.com/, 2005. Version visited last on 14 February 2005.

[Url05e]   TSP home page. http://www.tsp.gatech.edu, 2005. Version visited last on 14 February 2005.

[Url05f]   TSPLIB home page. http://www.iwr.uni-heidelberg.de/groups/comopt/software/TSPLIB95/, 2005. Version visited last on 14 February 2005.

[Url05g]   www.kernel-machines.org. Available at: http://www.kernel-machines.org/, 2005. Version visited last on 14 February 2005.

[VBKG03]   S. Vaidyanathan, D. I. Broadhurst, D. B. Kell, and R. Goodacre. Explanatory optimization of protein mass spectrometry via genetic search. *Analytical Chemistry*, 75(23):6679 – 6686, 2003.

[VD02]   C. Voudouris and R. Dorne. Integrating heuristic search and one-way constraints in the iopt toolkit. In S. Voß and D. Woodruff, editors, *Optimization Software Class Libraries*, volume 18 of *Operations Research/Computer Science Interfaces Series*, pages 177–191. Kluwer Academic Publishers, Boston, MA, USA, 2002.

[VDLL01]   C. Voudouris, R. Dorne, D. Lesaint, and A. Liret. iOpt: A software toolkit for heuristic search methods. In *Proceedings of the 7th International Conference on Principles and Practice of Constraint Programming*, pages 716 – 719. Springer-Verlag, 2001.

[VHE03]   K. Varrentrapp and J. Henge-Ernst. TefoA – Testbed for Algorithms. Technical Report AIDA–03–10, Fachgebiet Intellektik, Fachbereich Informatik, Technische Universität Darmstadt, Darmstadt, Germany, 2003.

[VMOR99]   S. Voß, S. Martello, I. H. Osman, and C. Roucairol, editors. *Meta-Heuristics: Advances and Trends in Local Search Paradigms for Optimization*. Kluwer Academic Publishers, Boston, MA, USA, 1999.

[Vou97]   C. Voudouris. *Guided Local Search for Combinatorial Optimization Problems*. PhD thesis, University of Essex, Department of Computer Science, Colchester, UK, 1997.

[VT95]   C. Voudouris and E. Tsang. Guided local search. Technical Report CSM-247, Department of Computer Science, University of Essex, Colchester, UK, 1995.

[VT02]   C. Voudouris and E. Tsang. Guided local search. In F. Glover and G. Kochenberger, editors, *Handbook of Metaheuristics*, pages 185 – 218. Kluwer Academic Publishers, Norwell, MA, USA, 2002.

[VW02]   S. Voß and D. Woodruff, editors. *Optimization Software Class Libraries*, volume 18
           of *Operations Research/Computer Science Interfaces Series*. Kluwer Academic Pub-
           lishers, Boston, MA, USA, 2002.

[Wal98]   T. Walters. Repair and brood selection in the traveling salesman problem. In A. E.
           Eiben, T. Bäck, M. Schoenauer, and H.-P. Schwefel, editors, *Proceedings of PPSN-
           V, Fifth International Conference on Parallel Problem Solving from Nature*, volume
           1498 of *Lecture Notes in Computer Science*, pages 813 – 822. Springer Verlag, Berlin,
           Germany, 1998.

[Wat89]   C. J. C. H. Watkins. *Learning from Delayed Rewards*. PhD thesis, Cambridge Uni-
           versity, 1989.

[Wat92]   C. J. C. H. Watkins. Q-learning. *Machine Learning*, 8:279 – 292, 1992.

[WD00]   X. Wang and T. G. Dietterich. Efficient value function approximation using regression
           trees. In J. Boyan, W. Buntine, and A. Jagota, editors, *Proceedings of IJCAI-99
           Workshop on Statistical Machine Learning for Large-Scale Optimization*, pages 51 –
           54, 2000.

[Wei90]   E. D. Weinberger. Correlated and uncorrelated fitness landscapes and how to tell the
           difference. *Biological Cybernetics*, 63(5):325 – 336, 1990.

[Wil92]   R. J. Williams. Simple statistical gradient-following algorithms for connectionist re-
           inforcement learning. *Machine Learning*, 8:229 – 256, 1992.

[WM97]   D. Wolpert and W. Macready. No free lunch theorems for optimization. *IEEE Trans-
           actions on Evolutionary Computation*, 1(1):67 – 82, 1997.

[Wu01]   F. Wu. A framework for memetic algorithms. Master's thesis, University of Auckland,
           Department of Computer Science, Auckland, NZ, 2001.

[ZD95]   W. Zhang and T. G. Dietterich. A reinforcement learning approach to job-shop
           scheduling. In C. S. Mellish, editor, *Proceedings of the Fourteenth International Joint
           Conference on Artificial Intelligence*, pages 1114 – 1120. Morgan Kaufmann Publish-
           ers, San Francisco, CA, USA, 1995.

[ZD96]   W. Zhang and T. G. Dietterich. High-performance job-shop scheduling with A time-
           delay TD(λ) network. In D. S. Touretzky, M. C. Mozer, and M. E. Hasselmo, editors,
           Advances in Neural Information Processing Systems 8, pages 1024 – 1030. MIT Press,
           Cambridge, MA, USA, 1996.

[ZD97]   W. Zhang and T. G. Dietterich. Solving combinatorial oprimization tasks by reinforce-
           ment learning: A general methodology applied to rescource-constrained scheduling.
           Technical report, Technical Report, Department of Computer Science, Oregon State
           University, 1997.

[ZDD94]   M. Zweben, B. Daun, and M. Deale. Scheduling and rescheduling with iterative repair.
           In M. Zweben and M. S. Fox, editors, *Intelligent Scheduling*, pages 241 – 255. Morgan
           Kaufmann Publishers, San Francisco, CA, USA, 1994.

[Zha95]   W. Zhang. *Reinforcement Learning for Job-Shop Scheduling*. PhD thesis, Oregon
           State University, Oregon, USA, 1995.

Index

www.ingramcontent.com/pod-product-compliance
Lightning Source LLC
LaVergne TN
LVHW022300060326
832902LV00020B/3184